# Phage Display

# Practical Approach Series

For full details of the Practical Approach titles currently available, please go to www.oup.com/pas.

The following titles may be of particular interest:

**Liposomes (second edition) (No 264)**
*Edited by Vladimir Torchillin and Volkmar Weissig*

"This [Liposome research] is an important and growing area, and the second edition is justified . . . individual scientists working with liposomes or facing a difficult formulation challenge will find much of value" ***Doody's Electronic Journal***
**June 2003   0-19-963655-9 (Hbk) 0-19-963654-0 (Pbk)**

**Epitope Mapping (No 248)**
*Edited by Olwyn Westwood and Frank Hay*

Epitope mapping is a unique compendium of current methodology on epitope mapping that is both comprehensive and authoritative. As such, it is an invaluable laboratory manual and reference guide for all researchers involved in epitope mapping.
**March 2001   0-19-983653-2 (Hbk) 0-19-963652-4 (Pbk)**

**DNA- Protein Interactions (No 231)**
*Edited by Andrew Travers and Malcolm Buckle*

This book covers all the major tools that are required for the study of the large macromolecular enzymatic machines that manipulate DNA, with particular emphasis on biophysical techniques applied to the analysis of transcription and its relation to chromatin structure. Knowledge of basic techniques is assumed, although advances in fundamental fields are covered.
**August 2000   0-19-963692-3 (Hbk) 0-19-963691-5 (Pbk)**

**The Internet for Molecular Biologists (No.269)**
*Edited by Clare Sansom and Robert Horton*

This book is designed to help molecular biologists who are more at home at a laboratory bench than in front of a computer keyboard, to use the internet more effectively. Sansom and Horton provide a broad introduction to using Internet based computing resources including core databases, online resources plus tools and techniques for exploiting and authoring internet-distributed information.
**January 2004   0-19-963887-X (Hbk) 0-19-963888-8 (Pbk)**

# Phage Display

## A Practical Approach

Edited by

## Tim Clackson

ARIAD Pharmaceuticals Inc.,
26 Landsdowne Street, Cambridge,
MA 02139, USA

## Henry B. Lowman

Genentech Inc.,
1 DNA Way,
South San Francisco,
CA 94080, USA

OXFORD
UNIVERSITY PRESS

# OXFORD

UNIVERSITY PRESS

Great Clarendon Street, Oxford OX2 6DP

It furthers the University's objective of excellence in research, scholarship,
and education by publishing worldwide in

Oxford New York

Auckland Bangkok Buenos Aires Cape Town Chennai
Dar es Salaam Delhi Hong Kong Istanbul
Karachi Kolkata Kuala Lumpur Madrid Melbourne Mexico City Mumbai
Nairobi São Paulo Shanghai Taipei Tokyo Toronto

Oxford is a registered trade mark of Oxford University Press
in the UK and in certain other countries

Published in the United States
by Oxford University Press Inc., New York

© Oxford University Press 2004

The moral rights of the author have been asserted

Database right Oxford University Press (maker)

First published 2004

A catalogue record for this title is available from the British Library

Library of Congress Cataloging in Publication Data
(Data available)
ISBN 0 19 963874 8 (hbk)
      0 19 963873 X (pbk)

10 9 8 7 6 5 4 3 2 1

Typeset by Newgen Imaging Systems (P) Ltd., Chennai, India
Printed in Great Britain
on acid-free paper by Antony Rowe Ltd, Chippenham, Wiltshire

# Preface

Display of peptides and proteins on filamentous phage — phage display — is an *in vitro* selection technique that enables polypeptides with desired properties to be extracted from a large collection of variants. Since the original pioneering description of the concept by George Smith in 1985, phage display has evolved into a powerful technology for identifying polypeptides with novel properties, and altering the properties of existing ones. Phage display has diverse uses as a basic research tool in protein engineering and cell biology, allowing dissection of protein function and the identification of novel receptor ligands and antibodies. It has also been embraced by the pharmaceutical and biotechnology industries, and recombinant proteins and antibodies discovered or optimized by phage display are now approved for clinical use.

Despite this broad use, phage display can be complicated and technically challenging to implement, and as a result many of its most interesting applications remain relatively inaccessible to mainstream laboratories. In planning *Phage Display — A Practical Approach*, we sought to generate a comprehensive experimental guide that could bring the full potential of the technology within the reach of most researchers. The aim has been to generate a volume suitable for both the novice and the expert. For those new to the field, key concepts and procedures are described, and chapters include detailed information on experimental design, protocols, and in many cases trouble-shooting guides. For the more seasoned practitioner, the book represents a single point of reference on phage display, providing updates to familiar procedures, alternative protocols, and introductions to new applications.

A particular aim of the volume was to broaden coverage beyond the two most familiar applications of the technology: peptide and antibody libraries. While these are major applications, and are covered by specific chapters in the book, phage display has also been widely used as a more general protein engineering tool for altering or optimizing protein function. In addition, each of the steps in a phage display experiment — creating or obtaining a library, performing selections, and analyzing the results — can be carried out in a wide variety of ways. The process of selection itself (enriching for phage that display peptides or proteins with desired properties) is particularly open to creativity, and essentially limited only by the imagination of the investigator. Our aim was to enable readers of this book to move beyond the specific examples given, to customize the procedures to their own protein or selection system of interest.

To this end, the book begins in a modular fashion with individual chapters on the library creation, selection and analysis steps of phage display. Chapter 1 provides an overview of phage biology and phage display, with the aim of introducing the general principles, experimental strategies and choices to be made when embarking on a phage display project. It is also designed as a navigational tool for the rest of the book, indicating where specific procedures or approaches are to be found and listing alternative approaches at each stage. Chapters 2 and 3 then outline two common approaches to preparing diversified libraries: oligo-nucleotide-directed mutagenesis and DNA shuffling. Chapter 4 provides guidance and procedures for designing a selection process and performing binding selections. A generally applicable method for analyzing the binding properties of selected phage clones (the parameter most commonly of interest), using a simple phage ELISA procedure, is described in Chapter 5.

Individual chapters then follow on specific applications, including selection of ligands from peptide libraries (Chapter 6), selection of peptide enzymatic substrates (Chapter 7), selection of DNA binding proteins (Chapter 9), expression cloning using cDNA display (Chapter 12), and generation of phage antibody libraries and isolation and optimization of antibodies (Chapters 13 and 14). Further chapters describe alternative selection approaches, including selection for stability (Chapter 8), direct selection *in vivo* (Chapter 10), and selection by direct binding to sera (Chapter 11). Together, these chapters provide direct guidance on major established and emerging applications of the technology, and illustrate the range of potential uses to which the approach can be put.

In general we have tried to avoid duplication of protocols that can easily be adapted to different vectors or systems, instead cross-referencing common procedures (such as transformation of *E. coli* by electroporation) between chapters. However, in some cases the alternative protocols used by different groups have been included, to indicate the types of variation that are possible, and to preserve the integrity of established experimental schemes in the authors' laboratories.

We hope that this book proves to be a useful experimental guide for all those interested in exploiting the broad potential of phage display. We are most grateful to the chapter authors for their excellent contributions, and for their time and effort during this project. We also thank the editorial staff of Oxford University Press for their support.

Tim Clackson
Henry B. Lowman

# Contents

## 2 Constructing phage display libraries by oligonucleotide-directed mutagenesis  *27*
*Sachdev S. Sidhu and Gregory A. Weiss*

## 3 *In vitro* DNA recombination  *43*
*Kentaro Miyazaki and Frances H. Arnold*

# Protocol list

# List of abbreviations

| | |
|---|---|
| A | absorbance |
| $A_{xxx}$ | absorbance at xxx nm |
| ABTS | 2,2′-azinobis(3-ethylbenzothiazoline-6-sulfonic acid) |
| $Amp^R$ | ampicillin or carbenicillin resistance |
| ANP | atrial naturetic peptide |
| AP | alkaline phosphatase |
| BLAST | basic local alignment search tool |
| bp | base pair(s) |
| BSA | bovine serum albumin |
| CCC-dsDNA | covalently closed circular double-stranded DNA |
| CD | circular dichroism |
| cDNA | complementary DNA |
| CDR | complementarity determining region |
| cfu | colony forming unit(s) |
| $C_H$ | heavy chain constant region (of an antibody) |
| $C_L$ | light chain constant region (of an antibody) |
| DEPC | diethylpyrocarbonate |
| DMEM | Dulbecco's modified eagle' medium |
| DMSO | dimethyl sulfoxide |
| dNTPs | deoxyribonucleotide triphosphates |
| DS-Ab | disease-specific antibodies |
| dsDNA | double-stranded DNA |
| DTT | dithiothreitol |
| dU-ssDNA | Uracil-containing single-stranded DNA |
| EDTA | ethylenediaminetetraacetic acid |
| EGF | epidermal growth factor |
| ELISA | enzyme-linked immunosorbant assay |
| EPO | erythropoeitin |
| EST | expressed sequence tag |
| Fab | antigen-binding antibody fragment comprising $V_H C_H 1$ and $V_L C_L$ |
| Fc | constant region of an antibody |
| Ff | family of filamentous *E. coli* phage comprising f1, M13 and fd |
| FR | framework region (of an antibody $V_H$ or $V_L$ domain) |
| FVIIa | Factor VIIa |
| Hepes | *N*-[2-hydroxyethyl]piperazine-*N*′-[2-ethanesulfonic acid] |

| | |
|---|---|
| hGH | human growth hormone |
| HRP | horse radish peroxidase |
| IAA | isoamyl alcohol |
| IAS phage | immunity affinity selected phage |
| IG | intergenic regions (of Ff phage genome) |
| IgG | immunoglobulin G |
| IgM | immunoglobulin M |
| IMAC | immobilized metal ion affinity chromatography |
| IPTG | isopropyl β-D-thiogalactoside |
| $J_H$ | heavy chain joining fragment |
| Kan | kanamycin |
| $K_d$ | dissociation constant |
| $k_{off}$ | dissociation rate constant |
| $k_{on}$ | association rate constant |
| LB | Luria-Bertani medium |
| mAb | monoclonal antibody |
| MBP | maltose binding protein |
| MLB | magnesium containing lysis buffer |
| moi | multiplicity of infection |
| MTA | materials transfer agreement |
| MTI | mustard trypsin inhibitor |
| NHS | $N$-hydroxysuccinimide |
| Ni-NTA | nickel-nitrilo-triacetic acid |
| OD | optical density (absorbance) |
| $OD_{xxx}$ | optical density (absorbance) at xxx nm |
| OPD | $o$-phenylenediamine |
| pIII | filamentous phage minor capsid protein |
| pVIII | filamentous phage major capsid protein |
| $P_{phoa}$ | alkaline phosphatase promoter |
| PAGE | polyacrylamide gel electrophoresis |
| PBL | peripheral blood lymphocyte(s) |
| PBS | phosphate buffered saline |
| PCR | polymerase chain reaction |
| PDB | protein data bank |
| PEG | polyethylene glycol |
| pfu | plaque forming unit(s) |
| pNPP | p-nitrophenyl phosphate |
| PS | packaging signal (of Ff phage genome) |
| R | resistance/resistant |
| RACE | rapid amplification of cDNA ends |
| RF | replicative form (of Ff phage genome) |
| RPAS | recombinant phage antibody system |
| RT | room temperature |
| RU | resonance units |
| scFv | single-chain Fv $(V_H + V_L)$ antibody fragment |

| | |
|---|---|
| SDS | sodium dodecyl sulfate |
| SDS-PAGE | SDS polyacrylamide gel electrophoresis |
| SH3 | Src homology 3 |
| SPR | surface plasmon resonance |
| ssDNA | single-stranded DNA |
| StEP | staggered extension process |
| $Tet^R$ | tetracycline resistance |
| TF | tissue factor |
| TMB | $3,3',5,5'$-tetramethyl benzidine |
| TU | transducing unit(s) |
| UV | ultraviolet |
| V-gene | antibody variable region gene |
| $V_H$ | heavy chain variable region (of an antibody) |
| $V_L$ | light chain variable region (of an antibody) |
| wt | wild type |
| WT-UBQ-phage | wild-type ubiquitin-phage |
| X-gal | 5-bromo-4-chloro-3-indolyl-β-D-galactoside |

# List of contributors

**Frances H. Arnold**
Division of Chemistry and Chemical
Engineering, 210–41
California Institute of Technology
1200 East California Boulevard
Pasadena, CA 91125, USA
E-mail: frances@cheme.caltech.edu

**Marcus D. Ballinger**
Sunesis Pharmaceuticals Inc.
341 Oyster Point Boulevard
South San Francisco, CA 94080, USA
E-mail: ballinger@sunesis.com

**Michele Bernasconi**
The Burnham Institute and Program
in Molecular Pathology
Department of Pathology
University of California San Diego
School of Medicine
10901 North Torrey Pines Road
La Jolla, CA 92037, USA
E-mail: mbernasconi@burnham.org

**Andrew R.M. Bradbury**
Biosciences Division
Los Alamos National Laboratory
Los Alamos, New Mexico, USA
*and*
International School for
Advanced Studies (SISSA)
Via Beirut 2–4
Trieste 34012, Italy
E-mail: amb@lanl.gov

**Yen Choo**
MRC Laboratory of Molecular Biology
Hills Road
Cambridge CB2 2QH, UK
*Present address:*
Plasticell Ltd.
10 Sydney Street
London SW3 6pp, UK
E-mail: yen@plasticell.com

**Tim Clackson**
ARIAD Pharmaceuticals, Inc.
26 Landsdowne Street
Cambridge, MA 02139, USA
E-mail: clackson@ariad.com

**Riccardo Cortese**
Istituto di Ricerche di
Biologia Molecolare
P. Angeletti (IRBM)
Via Pontina km 30,6
00040 Pomezia (Rome), Italy
E-mail: riccardo_cortese@merck.com

**Brian C. Cunningham**
Sunesis Pharmaceuticals Inc.
341 Oyster Point Boulevard
South San Francisco, CA 94080, USA
E-mail: bcc@sunesis.com

**Steven E. Cwirla**
XenoPort, Inc.
3410 Central Expressway
Santa Clara, CA 95051, USA
E-mail: scwirla@xenoport.com

**Warren L. DeLano**
Sunesis Pharmaceuticals Inc.
341 Oyster Point Boulevard
South San Francisco, CA 94080, USA
E-mail: warren@sunesis.com

**Mark S. Dennis**
Department of Protein Engineering
Genentech Inc.
1 DNA Way
South San Francisco, CA 94080, USA
E-mail: msd@gene.com

**William J. Dower**
XenoPort, Inc.
3410 Central Expressway
Santa Clara, CA 95051, USA
E-mail: bdower@xenoport.com

**Michael D. Finucane**
Centre for Biomolecular Design and
Drug Development
School of Biological Sciences
Biology Building
University of Sussex
Falmer, Brighton BN1 9QG, UK
*Present address*:
Department of Molecular Pharmacology
Center for Clinical Sciences
Research
Stanford University, CA 94305, USA
E-mail: mikef@stanford.edu

**Marc Fransen**
Katholieke Universiteit Leuven
Faculteit Geneeskunde
Campus Gasthuisberg
Departement Moleculaire
Celbiologie
Afdeling Farmacologie
Herestraat 49, B-3000 Leuven, Belgium
E-mail:
marc.fransen@med.kuleuven.ac.be

**Christian M. Gates**
Amersham Biosciences
800 Centennial Avenue
Piscataway, NJ 08855, USA

**Jason A. Hoffman**
The Burnham Institute and Program
in Molecular Pathology
Department of Pathology
University of California San Diego
School of Medicine
10901 N. Torrey Pines Road
La Jolla, CA 92037, USA
E-mail: jhoffman@burnham.org
*and*
Program in Molecular Pathology
Department of Pathology
University of California
San Diego School of Medicine

**Mark Isalan**
MRC Laboratory of Molecular Biology
Hills Road
Cambridge CB2 2QH, UK
*Present address*:
European Molecular Biology
Laboratory (EMBL)
Meyerhofstrasse 1
Heidelberg 69117
Germany
E-mail: mark.isalan@embl-heidelberg.de

**Laurent Jespers**
MRC Laboratory of Molecular Biology
Hills Road
Cambridge CB2 2QH, UK
E-mail: lsj@mrc-lmb.cam.ac.uk

**Pirjo Laakkonen**
The Burnham Institute and Program
in Molecular Pathology
Department of Pathology
University of California San Diego
School of Medicine
10901 North Torrey Pines Road
La Jolla, CA 92037, USA
E-mail: plaakkonen@burnham.org

**Henry B. Lowman**
Department of Protein Engineering
Genentech, Inc.
1 DNA Way
South San Francisco, CA 94080, USA
E-mail: hbl@gene.com

**James D. Marks**
Departments of Anesthesiology and
Pharmaceutical Chemistry
University of California, San Francisco
Rm 3C-38, San Francisco General
Hospital
1001 Potrero
San Francisco, CA 94110, USA
E-mail: marksj@anesthesia.ucsf.edu

**David J. Matthews**
Exelixis Inc.
170 Harbor Way
South San Francisco, CA 94083, USA
E-mail: matthews@exelixis.com

**Kentaro Miyazaki**
Division of Chemistry and Chemical
Engineering, 210-41
California Institute of Technology
1200 East California Boulevard
Pasadena, CA 91125, USA
*Present address*:
National Institute of Bioscience and
Human
Technology
1-1 Higashi, Tsukuba, Ibaraki
305-8566, Japan
E-mail: miyaken@nibh.go.jp

**Paolo Monaci**
Istituto di Ricerche di Biologia
Molecolare
P. Angeletti (IRBM)
Via Pontina km 30,6
00040 Pomezia (Rome), Italy
E-mail: paolo_monaci@merck.com

**Ulrik Nielsen**
Departments of Anesthesiology and
Pharmaceutical Chemistry
University of California, San Francisco
Rm 3C-38, San Francisco General
Hospital
1001 Potrero
San Francisco, CA 94110, USA
E-mail: nielsenu@anesthesia.ucsf.edu

**Kimmo Porkka**
The Burnham Institute and Program
in Molecular Pathology
Department of Pathology
University of California San Diego
School of Medicine
10901 North Torrey Pines Road
La Jolla, CA 92037, USA
E-mail: kporkka@burnham.org

**Erkki Ruoslahti**
The Burnham Institute and Program in
Molecular Pathology
Department of Pathology
University of California San Diego
School of Medicine
10901 North Torrey Pines Road
La Jolla, CA 92037, USA
E-mail: ruoslahti@burnham.org

**Marjorie Russel**
The Rockefeller University
1230 York Avenue
New York, NY 10021, USA
E-mail:
russelm@rockvax.rockefeller.edu

**Peter J. Schatz**
Affymax Inc.
4001 Miranda Avenue
Palo Alto, CA 94304, USA
E-mail: Peter_Schatz@affymax.com

**Sachdev S. Sidhu**
Department of Protein Engineering
Genentech, Inc.
1 DNA Way
South San Francisco, CA 94080, USA
E-mail: sidhu@gene.com

**Mihriban Tuna**
Centre for Biomolecular Design and
Drug Development
School of Biological Sciences
Biology Building
University of Sussex
Falmer, Brighton BN1 9QG, UK
E-mail: m.tuna@susx.ac.uk

xxiii

**Natalie G. M. Vlachakis**
Centre for Biomolecular Design and
Drug Development
School of Biological Sciences
Biology Building
University of Sussex
Falmer, Brighton BN1 9QG, UK
E-mail: nataliev@biols.susx.ac.uk

**Christopher R. Wagstrom**
Affymax Inc.
4001 Miranda Avenue
Palo Alto,
CA 94304, USA
*Present address*:
Zyomyx, Inc.
26101 Research Road
Hayward,
CA 94545, USA
E-mail: cwagstrom@zyomyx.com

**Gregory A. Weiss**
Department of Protein Engineering
Genentech, Inc.
1 DNA Way
South San Francisco, CA 94080, USA
*Present address*:
516 Rowland Hall
Department of Chemistry
University of California
Irvine, CA 92697–2025, USA
E-mail: gweiss@uci.edu

**Derek N. Woolfson**
Centre for Biomolecular Design and
Drug Development
School of Biological Sciences
Biology Building
University of Sussex
Falmer, Brighton BN1 9QG, UK
E-mail: dek@biols.susx.ac.uk

# Chapter 1

# Introduction to phage biology and phage display

## Marjorie Russel
The Rockefeller University, 1230 York Avenue, New York,
NY 10021, USA.

## Henry B. Lowman
Department of Protein Engineering, Genentech, Inc., 1 DNA Way,
South San Francisco, CA 94080, USA.

## Tim Clackson
ARIAD Pharmaceuticals, Inc., 26 Landsdowne Street,
Cambridge, MA 02139, USA.

## 1 Introduction

Display of peptides and proteins on filamentous phage—phage display—is an *in vitro* selection technique that enables polypeptides with desired properties to be extracted from a large collection of variants. A gene of interest is fused to that of a phage coat protein, resulting in phage particles that display the encoded protein and contain its gene, providing a direct link between phenotype and genotype. This allows phage libraries to be subjected to a selection step (e.g. affinity chromatography), and recovered clones to be identified by sequencing and re-grown for further rounds of selection. Since the initial description of the approach by Smith [1], it has become established as a powerful method for identifying polypeptides with novel properties, and altering the properties of existing ones (for reviews, see 2–5).

Filamentous phages are ideal in many ways for use as cloning vehicles and for display in particular. The genome is small and tolerates insertions into non-essential regions; cloning and library construction are facilitated by the ability to isolate both single- and double-stranded DNA (ssDNA and dsDNA), and by the availability of simple plasmid-based vectors; coat proteins can be modified with retention of infectivity; phage can accumulate to high titers since their production does not kill cells; and phage particles are stable to a broad range of potential selection conditions.

This chapter is intended to provide the background needed to initiate a phage display project. In the first portion (Sections 2 and 3), the life cycle, genetics, and structural biology of filamentous phages are summarized, with a focus on aspects that are relevant to phage display. We then go on to describe general considerations to be made when approaching a new phage display project, including choice of display format, experimental design, and common pitfalls (Sections 4–6). Finally, summaries of commercial sources of phage display vectors, kits, and alternative display systems are provided for those cases in which such reagents can provide a head-start for investigators (Sections 7 and 8). Cross-references are provided to later chapters in the book that provide detailed procedures.

## 2 Biology of filamentous phage

### 2.1 Introduction

Filamentous phages constitute a large family of bacterial viruses that infect many gram-negative bacteria. Their defining characteristic is a circular, ssDNA genome encased in a long, somewhat flexible tube composed of thousands of copies of a single major coat protein, with a few minor proteins at the tips. The genome is small—a dozen or fewer closely packed genes and an intergenic (IG) region that contains sequences necessary for DNA replication and encapsidation. Unlike most bacterial viruses, filamentous phages are produced and secreted from infected bacteria without cell killing or lysis (*Figure 1*). Readers are referred to several excellent reviews on filamentous phage (6–9) for more comprehensive information and citations of the primary literature than are given here.

Most information about filamentous phages derives from those that infect *E. coli*: f1/M13/fd, and to a lesser extent IKe. These phages are characterized by a fivefold rotation axis combined with a twofold screw axis. Phages f1, M13, and fd are those that have been used for display. Their genomes are more than 98% identical and their gene products interchangeable, and the phages are usually referred to collectively, as Ff phages. Unless specified, the properties of filamentous phage described below refer to them together.

### 2.2 Structure of the phage particle

Filamentous phages have a fixed diameter of about 6.5 nm and a length determined by the size of their genome. The 6400-nucleotide ssDNA of Ff is encapsidated in a 930 nm particle, while a 221-nucleotide "microphage" variant is 50 nm long (10). Cloning DNA into a nonessential region of the genome can create longer phage, although the longer they are, the more sensitive the particles are to breakage (e.g. from vortexing).

Phage particles are composed of five coat proteins (*Figure 2*). The hollow tube that surrounds the ssDNA is composed of several thousand copies of the 50-residue major coat protein, pVIII, oriented at a 20° angle from the particle axis and overlapped like fish scales to form a right-handed helix (8). The filament is held together by interactions between the hydrophobic midsections of adjacent subunits. Except for five surface-exposed N-terminal residues, pVIII forms a single, continuous α-helix. The four positively charged residues near

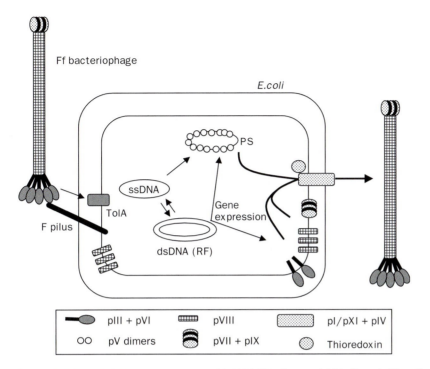

**Figure 1** Life cycle of filamentous phage f1 (M13/fd). Sequential binding of pIII to the tip of the F-pilus and then the host Tol protein complex results in depolymerization of the phage coat proteins, their deposition in the cytoplasmic membrane (where they are available for re-utilization), and entry of the ssDNA into the cytoplasm. The ssDNA is converted by host enzymes to a double-stranded RF, the template for phage gene expression. Progeny ssDNA, coated by pV dimers (except for the packaging sequence hairpin (PS) that protrudes from one end), is the precursor of the virion. A multimeric complex that spans both membranes—composed of pI, pXI, pIV, and the cytoplasmic host protein thioredoxin—mediates conversion of the pV–ssDNA complex to virions and secretion of virions from the cell. This process involves removal of pV dimers and their replacement by the five coat proteins that transiently reside in the cytoplasmic membrane.

the C-terminus of pVIII are at the inner surface of the tube and interact with phosphates of the viral ssDNA.

The ends of the particle are distinguishable in electron micrographs. The blunt end contains several (3–5) copies each of pVII and pIX, two of the smallest ribosomally translated proteins known (33 and 32 residues, respectively). Neither their structure nor disposition in the particle is known. However, immunological evidence indicates that at least some of pIX is exposed (11) and antibody variable regions have been successfully displayed on the amino termini of pVII and pIX (12). Phage assembly begins at the pVII–pIX end, and in the absence of either protein, no particle is formed.

The pointed end of the particle contains about five copies each of pIII and pVI, both of which are needed in order for the phage to detach from the cell membrane; pVI is degraded in cells that lack pIII, which suggests that these proteins

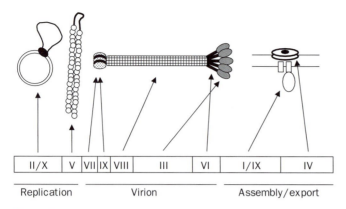

| II/X | V | VII | IX | VIII | III | VI | I/IX | IV |
|------|---|-----|----|----|-----|----|------|----|

Replication         Virion         Assembly/export

**Figure 2** Filamentous phage f1 (M13/fd) genes and gene products. *Gene II* encodes pII, which binds in the IG region (located between genes IV and II/X; not shown) of dsDNA and makes a nick in the + strand, initiating replication by host proteins. pX is required later in infection for the switch to ssDNA accumulation. *Gene V* encodes the ssDNA binding protein pV. *Genes VII* and *IX* encode two small proteins located at the tip of the virus that is first to emerge from the cell during assembly. *Gene VIII* encodes the major coat protein, and *genes III* and *VI* encode pIII and pVI, which are located at the end of the virion and mediate termination of assembly, release of the virion, and infection. *Gene I* encodes two required cytoplasmic membrane proteins, pI and pXI, and *gene IV* encodes pIV, a multimeric outer membrane channel through which the phage exits the bacterium. Note that the genome is in fact circular, but is shown in a linear presentation here for clarity.

assemble in the cell membrane before their incorporation into phage particles (13). They can be isolated from phage as a complex (14). The disposition of pVI in the particle is not known, but pVI with fusions to the C-terminus can be incorporated into phage, suggesting that this portion of the 112-residue pVI may be surface exposed (13).

More is known about the 406-residue pIII, the most commonly used coat protein for display (*Figure 3*). Its N-terminal domain, which is necessary for phage infectivity, is surface exposed and forms the small "knobs" that can often be seen to emanate from the pointed end of the particle in electron micrographs. Three pIII domains have been defined, the two N-terminal of which (N1 and N2) are believed to interact intramolecularly, based on crystallographic analysis (16, 17). The three domains are separated by two long, presumably flexible linkers characterized by repeats of a glycine-rich sequence. The final 132 residues within the C-terminal CT domain are necessary and sufficient for pIII to be incorporated into the phage particle and to mediate termination of assembly and release of phage from the cell; this domain is likely to be buried within the particle (13).

The single-stranded phage genome is oriented within the phage particle. Its orientation is determined by the packaging signal (PS), located in the non-coding IG region of the genome. The PS, an imperfect but extremely stable hairpin, is positioned at the pVII–pIX end of the particle and is necessary and sufficient for efficient encapsidation of circular ssDNA into phage particles. Certain amino acid substitutions in pVII, pIX, and pI (see below) enable single strands that lack a PS to be encapsidated; it is not known whether the DNA is randomly oriented in such particles or if some small duplex region serves as a secondary PS.

## 2.3 Infection

All filamentous phages that have been characterized use pili, which are long and slender cell surface appendages that resemble the phage themselves, as receptors. There are many different kinds of pili. *E. coli* phage use self-transmissible pili that mediate transfer of the plasmid that encodes them to recipient bacteria. Ff phage bind F pili, and IKe uses N and P pili. Phage can infect cells that lack appropriate pili, but the process is extremely inefficient; the efficiency is improved two to four orders of magnitude by agents that concentrate the phage or promote its adherence to the cell surface, such as $CaCl_2$ and polyethylene glycol (18).

Infection normally begins when the N2 domain of pIII (*Figure 3*) binds to the tip of a pilus (19) (see *Figure 4*). As might be expected from the low abundance of F pili (only a few per cell) and the small target size that their ends present, the rate and efficiency of infection of a bacterial culture is improved at high multiplicity of phage per cell and high cell density. However, if high density is achieved by growing cells past log phase (as opposed to concentrating them gently), pilus expression is decreased and infectivity compromised. Furthermore, since F pili do not assemble at low temperatures, efficient infection (and therefore plaque formation) by Ff requires incubation at or above 34 °C.

Pili normally assemble and disassemble continuously, and this, possibly stimulated by phage binding, brings the phage close to the cell surface. Upon pilus binding to N2, the N1 domain is released from its normal interaction with N2, making it available to bind to the host TolA protein, which extends into the periplasm from the cytoplasmic membrane (20, 21). Thus, the infection process appears to conform to a classic model involving a receptor (F-pilus) and a co-receptor (TolA). How the phage penetrates the outer membrane and the

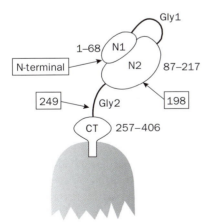

**Figure 3** Domain structure of pIII and fusion points for display. The three domains of pIII are shown: from N- to C-terminus, the N1, N2, and CT (C-terminal) domains are each separated by glycine-rich linker regions (Gly1 and Gly2). CT is buried within the phage particle and is required for virion assembly. Three positions at which display fusions are commonly created are indicated (boxed); numbers refer to residues in the mature protein sequence. Note that fusion at position 198 (corresponding to a convenient *Bam*HI restriction site) leaves an unpaired cysteine residue in the N2 domain (C201) that can interfere with display (15).

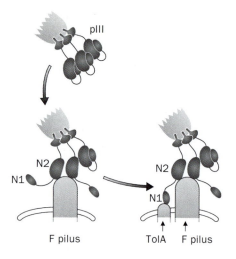

**Figure 4** Infection of *E. coli* by Ff bacteriophage. Infection is initiated by interaction of the N2 domain of pIII with an F pilus projecting from an *E. coli* cell. This interaction releases the N1 domain from its intramolecular interaction with N2, allowing it to bind to a discrete domain (D3) of the bacterial TolA protein. Subsequent steps are not yet characterized. The bacterial outer membrane is omitted from the diagram for clarity—it is not yet clear how infecting phage particles penetrate this membrane.

underlying peptidoglycan layer is not known. Three Tol proteins (Q, R, and A), all integral cytoplasmic membrane proteins, are absolutely required for phage infection (22, 23). They mediate depolymerization of the phage coat proteins into the cytoplasmic membrane and the translocation of the viral ssDNA into the bacterial cytoplasm, although the molecular details of how this is accomplished remain to be determined.

## 2.4 Replication

Upon entry of the viral single-stranded phage DNA (the + strand, which has the same polarity as the mRNA), host RNA and DNA polymerases and topoisomerase convert it to a double-stranded, super-coiled molecule called Replicative Form (RF; see *Figure 1*). The RF serves as template for phage gene expression, and this expression—in particular, synthesis of pII, a site-specific nicking–closing enzyme—is necessary for further replication. Through a rolling circle mechanism, pII nicks the + strand of the RF at a specific site in the non-coding IG region of the phage genome, and the 3′ end of the nick is elongated by host DNA polymerase III using the − strand as template. The original + strand is displaced by Rep helicase as the new + strand is synthesized, and when a round of replication is complete, the displaced + strand is recircularized by the nicking–closing activity of pII and again converted to RF. Synthesis of the − strand requires an RNA primer. The primer is generated by RNA polymerase, which initiates synthesis at an unusual site in the IG region of the + strand consisting of two adjacent hairpins that include promoter-like −35 and −10 motifs separated by a single-stranded region (24).

During the early phase of infection, when the concentration of the phage ssDNA-binding protein (pV) is low, newly synthesized single strands are immediately converted to RF, and both RF and phage proteins increase exponentially. As its concentration increases, pV binds cooperatively to newly generated + strands (*Figure 1*), preventing polymerase access and blocking their conversion to RF. The restart protein pX, which is identical to the carboxy-terminal 111 residues of pII, is required for the stable accumulation of single strands at this stage, but the mechanism by which it acts is not known (25). pV is dimeric, with the interaction surface of the subunits opposite the DNA-binding surface. Thus upon binding, the back-to-back arrangement of the dimers collapses the circular single strand into a rod-like structure. The DNA is oriented in the complex, with the PS hairpin protruding from one end, presumably because pV bind dsDNA only weakly. The pV/ssDNA complex is the substrate for phage assembly.

## 2.5 Genes and gene expression

The ~ 6400 nucleotide Ff genome contains nine closely packed genes (*Table 1*), and one major non-coding region (the IG) which contains the replication origins for + and − strand synthesis and the PS. Two of the phage genes (I and II) have internal translational initiation sites from which in-frame restart proteins are produced. In each case, both the full-length and the restart protein (whose sequence is identical to the carboxy terminal third of the full-length protein) are necessary for successful phage production. Of the 11 phage-encoded proteins, three (pII, pX, pV) are required to generate ssDNA, three (pI, pXI, pIV) are required for phage assembly, and five (pIII, pVI, pVII, pVIII, pIX) are components of the phage particle (*Figure 2*).

All phage proteins are synthesized simultaneously, although diverse mechanisms ensure that each is produced at an appropriate rate. There are differences in promoter and ribosome binding site strength or accessibility. A weak rho-dependent termination signal at the beginning of gene I limits its transcription, and the large number of infrequently used codons reduces its rate of translation. At the other end of the spectrum, overlapping transcripts from multiple

**Table 1** F-specific filamentous phage genes/proteins and properties

| Gene | Protein | Size (aa) | Function | Location | Used for display? |
|------|---------|-----------|----------|----------|-------------------|
| I | I | 348 | Assembly | Inner membrane | |
| | XI | 108 | Assembly | Inner membrane | |
| II | II | 409 | Replication (nickase) | Cytoplasm | |
| | X | 111 | Replication | Cytoplasm | |
| III | III | 406[a] | Virion component | Virion tip (end) | Yes (N-term) |
| IV | IV | 405[a] | Assembly (exit channel) | Outer membrane | |
| V | V | 87 | Replication (ssDNA bp) | Cytoplasm | |
| VI | VI | 112 | Virion component | Virion tip (end) | Yes (C-term) |
| VII | VII | 33 | Virion component | Virion tip (start) | Yes (N-term) |
| VIII | VIII | 50* | Virion component | Virion filament | Yes (N- and C-term) |
| IX | IX | 32 | Virion component | Virion tip (start) | Yes (N-term) |

[a] Mature protein without signal sequence.

promoters (there are only two terminators) and multiple RNA processing events increase the abundance of RNAs for the genes closest to the terminators (26). This results in high levels of pV and pVIII, the proteins that are required in the greatest quantities.

At later times after infection, the rates of phage protein and DNA synthesis taper off. This occurs when the high concentration of pV sequesters the + strands and prevents their conversion to RF, the template for gene expression. In addition, excess pV binds to a tetraplex structure in the gene II and gene X mRNAs (27), repressing their translation, and the reduction in pII levels leads to lower rates of + strand synthesis (25, 28). At these lower synthetic rates, a steady-state level of phage products is maintained by the secretion of progeny phage and by the continued growth and division of the infected cells, and phage production continues at a linear rate.

## 2.6 Physiology of phage assembly

The unique aspect of filamentous phage assembly is that it is a secretory process. Assembly occurs in the cytoplasmic membrane, and nascent phages are secreted from the cell as they assemble. All eight of the phage-encoded proteins that are directly involved in assembly are integral membrane proteins (*Figure 1*). This includes three non-virion proteins—pI and its restart partner, pXI, in the cytoplasmic membrane and pIV in the outer membrane—and the five viral coat proteins, which reside in the cytoplasmic membrane prior to their incorporation into phage. Two (pIII and pVIII) are synthesized as precursors and, after signal sequence cleavage, span the membrane with their C-termini in the cytoplasm. The orientation of the other coat proteins is not known with certainty. pVI is particularly hydrophobic (57% of its residues are hydrophobic) and could span the membrane more than once.

The first progeny phage particles appear in the culture supernatant about 10 min after infection (at 37 °C). Their numbers increase exponentially for about 40 min, after which the rate becomes linear. About 1000 phage per cell are produced in the first hour. Under optimal conditions, the infected cells can continue to grow and divide—and produce phage—indefinitely. Persistent infection is possible because the five capsid proteins and the viral single strands are removed from the cell (by assembly and secretion) at a rate commensurate with their synthesis, and over-accumulation of the non-secreted proteins is prevented by the pV/pII regulatory loop and by dilution resulting from continued cell growth and division. The continued growth of phage-producing cells (at a rate significantly lower than uninfected bacteria) accounts for the turbidity of Ff plaques. When the plaques are clear rather than turbid, the infected cells are being killed. This occurs when there is an imbalance between component synthesis and phage assembly. For example, wild type Ff phage form clear plaques at 39 °C (and very small plaques at 42 °C) because the intrinsic temperature-sensitivity of pV results in reduced accumulation of ssDNA and thus a lower rate of phage production. Even modest perturbations of the phage life cycle can lead to the eventual death of infected cells. Phage into which additional DNA has been cloned often give rise to smaller and/or clearer plaques; this may be due to the imbalance caused by the increased utilization of major compared to minor coat proteins in such particles.

## 2.7 The mechanics of phage assembly

Phage assembly can be divided into five stages: preinitiation, initiation, elongation, pretermination, and termination. Preinitiation is defined as the formation of an assembly site, a region visible by electron microscopy where the cytoplasmic and outer membranes are in close contact (29). Assembly sites are composed of the three morphogenetic proteins, pI, pXI, and pIV, which interact via their periplasmic domains (N-terminal for pIV and C-terminal for pI and, presumably, pXI); the sites form independently of any other phage proteins.

pIV is a cylindrical structure with a central cavity about 8 nm in diameter; it is composed of 12–14 identical subunits with a combined mass of about 600 kDa (30). In the cryo-EM structure of pIV (MR et al., unpublished data), there is some density within the cavity, which presumably explains why the normal state of the pIV channel is closed. Certain mutant forms of pIV, however, open frequently and allow entry of foreign substances into the bacterial periplasm (31, 32). Phage that lack pIII and are thus unable to be released from the cell block such entry, indicating that phage do pass through the pIV channel (33). Nothing is known about the interior of the pIV channel, except that it can accommodate heterologous pVIII from IKe (34), as well as an occasional pVIII subunit carrying a large N-terminal extension or pVIII uniformly substituted with short (6–8 residue) N-terminal extensions, as in phage display (35, 36). The dimensions and/or properties of the channel may be a factor limiting the size of polypeptides that can be displayed on pVIII.

pI and pXI also form a multimeric complex composed of about 5–6 copies of each; its shape and dimensions are not known (J. N. Feng, P. Model and M.R., unpublished data). In the absence of the other phage proteins, pI/pXI causes membrane depolarization, which suggests that the complex may also be a channel. Thus the assembly site may be an extremely large channel that traverses both bacterial membranes. The cytoplasmic N-terminal domain of pI (absent from pXI) contains a conserved nucleotide-binding motif; since this motif is essential for phage assembly (37), and phage assembly requires ATP hydrolysis (38), pI is likely to be an ATPase.

Initiation takes place only if the assembly site, the two minor coat proteins (pVII and pIX) located at the beginning tip of the particle, and the ssDNA substrate are present (37). Genetic analyses suggest that pVII and pIX (in the membrane) interact with the PS, which protrudes from one end of the pV–ssDNA complex (in the cytoplasm), and that the PS associates with the cytoplasmic domain of pI. Host-encoded thioredoxin, a small, cytoplasmic protein known as a potent reductant of protein disulfides, also interacts with pI. Although phage assembly does not utilize this redox activity, thioredoxin appears to be part of the initiation complex, and may confer processivity to the elongation reaction.

Elongation involves the successive replacement of pV dimers that cover the viral DNA by membrane-embedded pVIII and translocation of the DNA across the membrane. The process continues until the end of the viral DNA has been coated by pVIII. If either pIII or pVI is absent, the largely extracellular phage particle remains tethered to the cytoplasmic membrane where it remains competent to resume elongation when another pV–ssDNA complex enters the assembly site; ultimately, tethered phage filaments of more than 10 times unit length accumulate (39). Even in normal infections when pIII and pVI are present, about 5% of progeny phage particles are double length. Such secondary rounds of elongation do not require reinitiation—ssDNA without a PS can be efficiently incorporated (40).

Pretermination (13) is the incorporation of the membrane-embedded pIII–pVI complex at the terminal end of the nascent phage particle. A fragment containing only the C-terminal 83 residues of pIII is sufficient to mediate this step, but cannot effect detachment of the phage from the cell. Termination or release of the phage, which requires a 93 residue C-terminal segment of pIII, has been proposed to consist of a conformational change in the pIII–pVI complex that detaches the complex (and the phage) from the cytoplasmic membrane (13). A still longer portion of pIII (the 132 C-terminal residues) is required for the formation of stable virus particles.

# 3  Coat proteins used for display

All five capsid proteins have been used to display proteins or peptides, to varying degrees (*Table 1*). One report has described the fusion of antibody fragments to the amino termini of both pVII and pIX, which constitute the "start" end of the virion tip (12). The pVI protein, which interacts with pIII at the "end" virion tip, has also been used to display polypeptides through a carboxy-terminal fusion—until recently the only way to directly create fusions in this orientation ((41); see Chapter 12). C-terminal linkage is particularly desirable for display of polypeptides encoded by cDNA fragments, since the inclusion of the stop codon at the end of the cDNA will not prevent display. Hence, pVI display has proved effective for the expression cloning of protein–protein interactors from cDNA libraries, as described in Chapter 12. However, by far the most commonly used virion proteins for phage display are pVIII and pIII.

## 3.1  pVIII

pVIII, the major coat protein, is present in several thousand copies in phage particles. Sequences for display are typically inserted at the N-terminus, between the signal sequence and the beginning of the mature protein coding sequence. However, only short peptides sequences (6–8 residues) can be displayed on every copy of pVIII in a virion—larger sizes prevent packaging of the particles, probably because of the size restrictions of the pIV channel through which phage pass during extrusion (see Section 2.7). Display of larger polypeptides on pVIII requires expression of the fusion protein from a phagemid vector, yielding hybrid virions bearing mainly wild-type pVIII (see Section 4.2).

An emerging trend (for pVIII display but potentially more generally) is the use of protein engineering to modify the protein to broaden applicability: engineered pVIII proteins have been described that permit the display of large polypeptides at high copy number (42), or the display of proteins fused at the C-terminus of the protein (43).

## 3.2 pIII

pIII, present in five copies at the "end" tip of the virion, is the protein of choice for most phage display fusions due to its tolerance for large insertions, its compatibility with monovalent display (see Section 4.3), and the wide availability of suitable vectors. Although pIII is more tolerant than pVIII to substantial insertions, infectivity of the resulting phage can be reduced, sometimes dramatically. As with pVIII, this can be overcome by using phagemid constructs, resulting in the production of hybrid virions that also bear wild-type pIII (see Section 4.2). Since such virions no longer rely on the infectivity of the pIII fusion protein, proteins can instead be fused to truncated pIIIs designed with the structure of the protein in mind (*Figure 3*). These can confer more efficient display, by reducing or eliminating proteolysis of the fusion protein, as well as reducing the size of the phagemid vector. Potential disadvantages include the possibility of sterically hindering access to the displayed protein. Fusions have been reported at pIII residue 198 (44), which deletes the N1 domain and some of N2, or to residue 249 (45), which deletes N1 and N2 and fuses the displayed protein to a short portion of the second glycine-rich linker. C-terminal pIII display through fusion to a linker at the C-terminus of the pIII is also possible (46).

# 4 Starting a phage display project

## 4.1 Feasibility of display

The first step in a phage display project is establishing that display of the polypeptide of interest is feasible. In the early days of phage display, there were worries that the limitations of virion extrusion and infection would restrict display to small peptides and proteins, and furthermore that only proteins that are normally extracellular would be suitable, since display involves secretion of the fusion protein into the oxidizing environment of the bacterial periplasm. Both concerns have proved false, and intracellular and extracellular proteins of a wide range of sizes and structures have been functionally displayed (see *Table 2*). Although some proteins may be recalcitrant to display due to individual properties (such as toxicity to *E. coli*, or interference with phage production), size alone does not appear to be a major factor, and display of most proteins should be feasible.

For proteins that are normally intracellular, precautions can be taken to try to preserve the native structure of the molecule—for example, addition of zinc

**Table 2** Examples of proteins displayed on filamentous phage[a]

| Protein displayed | Molecular weight (kDa) | Format | Reference |
|---|---|---|---|
| *Secreted proteins* | | | |
| Z-domain (Protein A) | 6.5 | pIII | (47) |
| Mustard trypsin inhibitor (MTI-2) | 7 | pIII | (48) |
| C5a | 8 | pIII (Fos–Jun) | (49, 50) |
| Insulin-like growth factor (IGF)-1 | 14.5 | pIII | (51) |
| Lipocalins | 17.5 | pIII | (52) |
| hGH | 22 | pIII, pVIII, pVIII (C-terminal) | (42–44) |
| Trypsin | 24 | pIII, pVIII | (53) |
| Antibody scFv fragments | 25 | pIII, pVIII | [b](54) |
| Subtilisins | 28 | pIII, pVIII | (55, 56) |
| Insulin-like growth factor binding protein (IGFBP)-2 | 31 | pIII | (57) |
| Peptide–β2m–MHC complex | 41 | pIII | (58) |
| Vascular endothelial growth factor (VEGF) | $2 \times 11.5$ | pIII | (59) |
| Antibody Fab fragments | $2 \times 25$ | pIII, pVIII | (60) |
| Alkaline phosphatase | $2 \times 60$ | pIII | (61) |
| Streptavidin | $4 \times 15$ | pVIII | (42) |
| *Intracellular proteins* | | | |
| WW domain | 4 | pIII | (62) |
| Src Homology 3 (SH3) domain | 6.5 | pIII | (63) |
| FRAP/mTOR FRB domain | 9.5 | pIII | [c] |
| Zif268 zinc finger | 10 | pIII | [d]64 |
| Cytochrome $b_{562}$ | 11 | pIII | (65) |
| FK506 binding protein (FKBP)12 | 12 | pIII, T7 display | [c](66) |
| Glutathione S-transferase (GST) | $2 \times 25.5$ | pIII | (67) |

[a] This table is intended to indicate the range of proteins that can be displayed, but is neither complete nor exhaustive. Additional examples can be found in (2).

[b] See Chapter 13 for additional references.

[c] T. Clackson, unpublished data.

[d] See Chapter 9 for additional references.

during preparation of phage displaying zinc finger proteins (see Chapter 9). For proteins with lone cysteine residues, such as FKBP12, selections can be performed in the presence of dithiothreitol (DTT) (to which phage particles are stable—see Section 5.2) to maintain a reducing environment (T.C., unpublished data). If necessary, displayed proteins can even be refolded prior to selection by exposure of the particles to denaturants followed by dialysis (68).

It has also proved feasible to display multi-subunit proteins, by a variety of methods (*Table 2*). For heterodimers, such as antibody Fab fragments, one chain can be displayed and another provided as a soluble protein cosecreted into the periplasm (60). Homo-oligomeric proteins can also be displayed by relying on proteolysis of the displayed fusion protein to release sufficient soluble protein, or by invoking interactions between displayed proteins, either inter- or intra-phage (e.g. see 61).

## 4.2 Phage or phagemid vector?

Proteins can be displayed using vectors based on the natural Ff phage sequence—phage vectors—or using plasmid-based "phagemid" vectors that contain only the fusion protein gene, and no other phage genes (for a review, see 3). The two alternatives are illustrated in *Figure 5* for display on pIII, although similar principles apply for display on other proteins. In phage vectors, the heterologous sequence for display is inserted directly into the coding sequence for pIII or

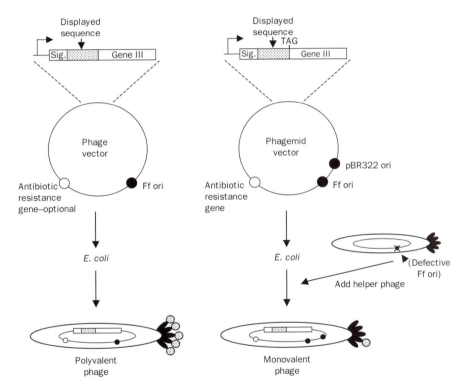

**Figure 5** General scheme for phage display using phage or phagemid vectors. The difference between phage and phagemid vectors is illustrated for pIII display. Sequences for display are inserted between a secretion signal sequence ("sig.") and gene III. Both phage and phagemid vectors carry an Ff origin of replication to permit production of ssDNA and hence virions. Phagemid vectors also have a plasmid origin (here pBR322) and an antibiotic resistance marker to allow propagation as plasmids in *E. coli*. Phage vectors are also often modified with antibiotic resistance markers for convenience, as illustrated here. In many phagemid vectors, an amber stop codon (TAG) is interposed between the displayed sequence and gene III, to allow soluble protein expression by transferring the vector into a non-*supE* suppressor strain.

another coat protein. When introduced into *E. coli*, phage will be produced in which all copies of the coat protein display the heterologous protein (i.e. the protein is displayed polyvalently). Examples of pIII phage display vectors include the fUSE vectors constructed by Smith and coworkers (69), and the M13KE vectors commercially available from New England Biolabs (see Section 8).

In phagemid vectors, the displayed protein fusion gene is cloned into a small plasmid under the control of a weak promoter. In addition to a plasmid origin of replication, the vector also has an Ff origin to allow production of single-stranded vector and subsequent encapsidation into phage particles. To produce such particles, *E. coli* cells harboring the plasmid are infected with helper phage, which is an Ff phage with a compromised origin that leads to its inefficient packaging (see Section 4.4). The infected cells express all the wild-type phage proteins from the helper phage genome, as well as a small amount of the fusion protein encoded by the phagemid, so that phage particles are extruded by the cells that contain both proteins, usually with the wild-type in considerable excess. Because the helper phage genome is poorly packaged, nearly all the phage particles contain the phagemid genome, preserving the linkage between the displayed protein and its gene.

Phagemid vectors have been described for both pIII and pVIII display, although pIII is more common. pIII phagemid vectors are described in more detail in Chapter 4. A major advantage of phagemid vectors is their smaller size and ease of cloning, compared to the difficulties of cloning in phage vectors without disrupting the complex structure of overlapping genes, promoters, and terminators. This generally translates into much higher library sizes for phagemid vectors. In addition, the phagemid approach must be used if monovalent display is desired in order to obtain selection based on true binding affinity (see Section 4.3). For pVIII display, use of phagemids is generally required to achieve display of sequences longer than 6–8 amino acids. Phagemids have also been used for pVI display (see Chapter 12).

## 4.3 Polyvalent or monovalent display?

The choice of polyvalent or monovalent display is related to the choice of vector type. Conventional phage vectors with natural phage promoters will generally produce polyvalent display unless there is extensive proteolysis of the displayed proteins. In addition, phagemid vectors for pVIII display, in which only a fraction of the ~2700 copies are fusion proteins, will typically still display polyvalently. On the other hand, use of phagemid vectors to display protein on pIII under the control of a weak (or uninduced) promoter will typically lead to what is referred to as "monovalent display" (3, 44).

It is important to understand the nature of monovalent display. A typical preparation of phage particles prepared from *E. coli* harboring a pIII phagemid display vector and infected with helper phage will exhibit a Poisson distribution of fusion protein expression: 10% or less of the particles will display one copy of the fusion; a very small percentage will display two copies; and the remaining majority of the particles will display only wild-type pIII. Thus, the major displaying species is monovalent, but most particles do not display at all.

The valency of display is important principally because of its impact on the ability to discriminate binders of differing affinities (see Chapter 4, Section 2.1). Early work (44) showed that polyvalent display prevented the highest-affinity clones in a selection from being identified, because multivalency conferred a high apparent affinity (avidity) on weak-binding clones. Monovalent display allows selection based on pure affinity, and is therefore generally preferred for the many studies where the aim is to identify the tightest binding variant(s) from a library. Conversely, in applications where the initial selectants are of very low affinity—for example, the *de novo* selection of peptides that bind a given target—polyvalency increases the chances of isolating rare and weakly binding clones. A frequent experimental strategy in such projects is to start with polyvalent display, and then move to monovalent display as the affinity of the displayed polypeptide matures (see Chapter 6).

## 4.4 Helper phage

A number of different helper phage for phagemid preparation have been developed and are commercially available, including R408 (70) (Stratagene), M13 K07 (71) (New England Biolabs, Amersham Pharmacia Biotech), and VCSM13 (Stratagene). In general, they can be used interchangeably. All have mutations that reduce packaging efficiency (to ensure phagemid genomes are preferentially packaged), and mutations that allow productive infection of bacteria harboring plasmids with Ff origins of replication ("interference-resistance"). M13 K07 and its derivative VCSM13 carry a kanamycin resistance gene to allow antibiotic selection of helper-infected cells. A procedure for preparing a helper phage stock starting with a commercial sample is provided in Chapter 2, *Protocol 6*.

## 4.5 General protocols for phage preparation and quantitation

Most phage display projects involve the basic procedures of preparing stocks of phage or phagemid particles from infected bacteria (propagation), and titering those stocks to determine the concentration of infectious particles. For phage vectors (and helper phage), determining the concentration of viable particles involves counting phage plaques on a lawn of host bacteria. A general procedure for propagation and titering of phage stocks is provided in *Protocol 1*. It is advisable to follow this procedure for plaque isolation whenever propagating phage vectors or helpers in order to ensure that a pure culture of viable phage is obtained.

For phagemid vectors, titration of phagemids involves infection of bacteria and counting of resulting antibiotic-resistant colonies. A simple procedure for titration of phagemid particles is given in Chapter 4, *Protocol 5*, and a procedure for phagemid propagation is given in Chapter 4, *Protocol 1*.

A convenient alternative to determine the infectivity of phage or phagemid preparations is estimation of concentrations by absorbance. The extinction coefficient depends on the size of phage particles, which in turn depends on the size of the genome (see Section 2.2). In general, for a helper phage such as VCSM13, 1 $OD_{270} = 5 \times 10^{12}$ particles/ml, and for a 5 kb phagemid, 1 $OD_{270} = 1.1 \times 10^{13}$ particles/ml. However, calculations based on absorbance may over-estimate the concentration of viable, infective particles.

## Protocol 1

## Titering and propagating plaque-purified filamentous phage[a]

### Equipment and reagents

- Phage stock solution to be titered (i.e. an appropriate phage vector, or VCSM13 helper phage (Stratagene))
- 2YT media (10 g bacto-yeast extract, 16 g bacto-tryptone, 5 g NaCl; add water to 1 L, pH 7.0 with NaOH; autoclave and cool)
- 2YT/tet media (2YT, 15 µg/ml tetracycline)
- *E. coli* XL1-Blue stock (Stratagene; Tet[R])
- Luria–Bertani (LB) agar plates (5 g bacto-yeast extract, 10 g bacto-tryptone, 10 g NaCl; add water to 1 L, pH 7.0 with NaOH,

- 15 g agar; autoclave and cool to 55 °C; pour into petri plates)
- 2YT top agar (16 g bacto-tryptone; 10 g bacto-yeast extract; 5 g NaCl; 7.5 g agar; add water to 1 L. Heat to dissolve the agar and autoclave.)
- Sterile 10 ml culture tubes
- Sterile Pasteur pipettes (for propagation)
- Microwave
- 45 °C incubator or water bath (optional)

### Method

### A. Titration and plaque purification

1  If the LB agar plates are stored at 4 °C, prewarm them to room temperature. In this case it may also be helpful to allow them to dry overnight with lids in place.

2  Pick a single colony of XL1-Blue cells into 5–25 ml of 2YT/tet media, and incubate the culture at 37 °C with shaking or rotation to a density of about 1 $OD_{550}$.

3  Perform 10-fold serial dilutions of the phage stock in duplicate in 0.1 ml aliquots of 2YT media to yield estimated concentrations of $10^3$–$10^5$ phage/mL.[b]

4  Melt the 2YT top agar (at least 3 ml per phage dilution) by microwaving, and cool to 45 °C, for example in a water bath or incubator.

5  For each phage dilution, mix 0.01 ml of phage with 0.5 ml of fresh XL1-Blue cells in a 10 ml culture tube.

6   To each tube, add 3 ml of top agar, rolling the tube quickly to mix.

7   Immediately pour the top agar mixture onto the surface of an LB agar plate, rotating the plate to spread the top agar quickly and evenly across the surface. If the top agar is too cool, lumps will form, making plaques difficult to count.

8   Allow the top agar to solidify, then incubate the plates at 37 °C overnight.

9   Plaques should appear as relatively clear discs against a background lawn of cells.

10  Calculate the phage concentration in the original stock solution (plaque-forming units/ml; pfu/ml) by multiplying the number of plaques per plate by 100 (because 0.01 ml of diluted phage is used per plate) and by the appropriate serial dilution factor.

## B. Progagation

1   To propagate phage, grow fresh XL1-Blue cells as above, and pick a single isolated plaque using a Pasteur pipette by placing a finger over the open pipette end after inserting it through a plaque into the agar.

2   Using a pipette bulb, force the agar plug into 1 mL of fresh XL1-Blue cells.

3.  Incubate for 1 h at 37° with shaking or rotation.

4.  Dilute the cells into 25–1000 ml of 2YT media containing an appropriate antibiotic (e.g. 10 μg/ml kanamycin for M13K07 or M13VCS helper phage).

5   Incubate the culture for 12–15 h at 37 °C with shaking. Phage may be harvested and purified as described in Chapter 2, *Protocol 6*.

ᵃ This procedure is a modification of that described by Sambrook *et al.* (72) for plating M13 phage.

ᵇ A typical yield from XL1-Blue phage cultures grown in 2YT would be $10^{11}$–$10^{12}$ phage/ml.

## 5   General principles of a phage display project

The structure of a generic phage display project is depicted in *Figure 6*. In general, the procedure can be divided into three stages: creating (or obtaining) a library of (poly)peptide variants; selection; and analysis of the selected clones. Figure 6 and the text below include cross-references to chapters in the book to illustrate the alternative approaches that can be taken at each stage.

### 5.1   Making a library

For proteins that have not previously been displayed, the first stage is to construct a display vector and confirm functional display. Following this, a wide variety of molecular biology techniques can be deployed to introduce diversity into the displayed protein. For targeted introduction of diversity, oligonucleotide-directed mutagenesis is a powerful and versatile technique, and

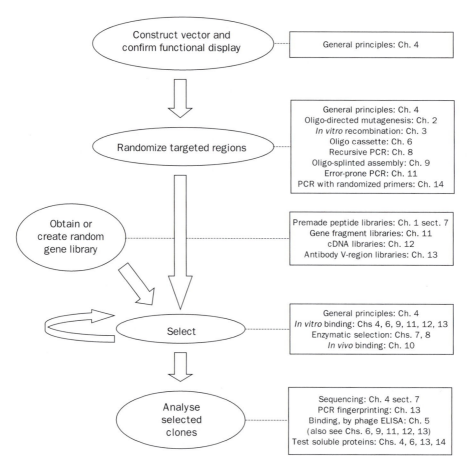

**Figure 6** Flow diagram illustrating a generic phage display project, and examples of techniques that can be used at each stage (boxed).

recent advances have brought larger libraries of $10^{10}$–$10^{11}$ clones within easy reach (see Chapter 2). The diversity introduced can involve complete ("hard") randomization of residues, partial ("soft") randomization in which the wild-type residue is retained in some proportion, or "tailored" randomization, in which only a defined subset of amino acids is specified (see Chapter 2, Section 2, and Chapter 4, Section 3; for an example of tailored randomization, see Chapter 8). In all targeted libraries, it is important to bear in mind the implications of library size limitations on the total amount of diversity that can be surveyed in one library (see Chapter 2, Section 2). Alternatively, introduction of random changes throughout a sequence can be accomplished by *in vitro* recombination (Chapter 3). Numerous alternative approaches for introducing targeted diversity are described in other chapters throughout the book, as indicated in *Figure 6*.

Some specialized phage display applications involve the creation of libraries from collections of genes as opposed to diversifying a single gene. Examples include cDNA libraries (Chapter 12) and antibody V-region libraries (Chapter 13). In these cases, specific protocols have been developed to create these libraries.

The creation of a library can be completely circumvented in some cases by obtaining an appropriate pre-made phage display library. Pre-made peptide display libraries are now commercially available (see Section 8).

## 5.2 Selection

The most common means of selection of desired clones is through an *in vitro* binding incubation, in which the phage library is bound to a target, washed, and then retained phage are eluted. This process is also referred to as "sorting" or "biopanning." The principles and alternatives for selections based on *in vitro* binding are described in detail in Chapter 4, Sections 4 and 5. A wide variety of washing and elution conditions have been used, exploiting the extraordinary stability of the Ff phage virion to extremes of pH, ionic strength, denaturants and even most proteases (with the exception of subtilisin) (see *Table 3*). This stability means that selection approaches are essentially limited only by the imagination of the investigator: for example, selection is even possible based on binding to targets *in vivo* after injection of phage libraries into whole animals (see Chapter 10).

The enrichment for binding clones over nonbinders conferred by a single round of selection can vary widely, from two-fold to more than a 1000-fold, although at least 10-fold is typical for an *in vitro* binding selection. In most cases, the eluted phage are used to reinfect *E. coli* for preparation of new phage ("amplified"), to allow further rounds of selection. Enrichment is typically monitored to help guide the decision to stop selecting and start analyzing clones.

## 5.3 Analysis of clones

In most phage display studies, one is interested in determining the identity of the clones that have been selected, and the properties of the displayed proteins. The most common technique for identifying selected clones is simply to sequence them. The principles of sequence analysis of phage-selected clones,

**Table 3** Examples of conditions used for elution of target-bound phage

| Elution buffer | Reference |
| --- | --- |
| 0.1 N HCl, pH 2.2 with glycine; most common | (1) |
| 20–100 mM HCl | (65, 73) |
| 5–75 mM DTT (with reducable linker) | (55, 74, 75) |
| 100 mM triethylamine | (76) |
| 6 M urea, pH 3 | (77) |
| 50 mM sodium citrate, pH 2–6 | (78) |
| 500 mM KCl, 10 mM HCl, pH 2 | (79) |
| 4 M MgCl$_2$ | (80) |
| Specific ligands | (45, 81) |
| Direct infection of *E. coli* host | (80, 82, 83) |

and considerations for when to stop selecting and start analyzing, are described in Chapter 4, Section 7. An alternative and rapid approach for clone identification is PCR "fingerprinting," in which the insert encoding the displayed protein is amplified by PCR, and then digested with a frequent-cutting restriction enzyme (see Chapter 13, *Protocol 19*). Analysis of the digestion products gives a pattern of bands that may be unique to that clone. PCR fingerprinting techniques are particularly useful in applications such as cDNA library or phage antibody selections, where each clone is likely to have a different restriction pattern.

The most popular and versatile technique for initial characterization of the properties of selected clones, particularly for the majority of studies based on binding selections, is the phage ELISA. In this method, the target of interest is immobilized in wells of a 96-well plate, individual phage supernatants prepared from selected clones are added, and specific binding is detected by use of an anti-phage antibody. In addition to giving "yes–no" information as to whether a given clone binds the target, the assay can be used in a competition format to determine the relative binding affinities of clones, as described in Chapter 5.

Following characterization of the displayed polypeptide "*in situ*" on phage, the next step is often the production of soluble (undisplayed) protein for more in-depth analysis. For most monovalent display vectors this is facilitated by the presence of an amber stop codon between the displayed protein and pIII, allowing soluble non-fusion protein to be expressed alone by transfer to a non-suppressor strain of *E. coli* (45, 60) (see *Figure 5*). Often the expression and subsequent binding analysis can be performed at small scale in 96-well plates (e.g. see Chapter 13). For shorter peptide sequences, expression as fusions to well-characterized affinity proteins such as maltose binding protein (MBP) can allow confirmation of the properties observed for the displayed peptide (see Chapter 6).

With efficient procedures available for library generation and selection, downstream analysis of clones can frequently be the rate-limiting step in a phage display project. This is leading to the increasing use of automation in an effort to improve throughput: for example, the use of Q-pix and Q-bot robots for colony picking is described in Chapter 11.

## 6 Common problems

Many chapters in this book contain troubleshooting hints for specific aspects of phage display, or for particular applications: for example, a guide for phage display binding selections is provided in Chapter 4 (*Table 3*). However, there are several more general potential pitfalls and problems to avoid that can be identified.

### 6.1 Library quality

A key principle of phage display is that the ability to select clones with desired properties is highly dependent on the quality of the initial library—a library that is of insufficient size, poor or inappropriate diversity, or flawed design is unlikely to lead to success. It is worth investing the time to ensure and then confirm the quality and diversity of the starting library—for example, by using a template encoding an inactive protein when creating a library using oligonucleotide-directed mutagenesis (see Chapter 2, Section 3.3).

## 6.2 Expression editing

An assumption made when a diversified library is created for phage display is that all clones will display with similar efficiency. In fact, some sequences will be refractory to display and therefore underrepresented in the displayed library—in the extreme, the optimal clone (e.g. the one with highest affinity) may never be isolated because it fails to display. For example, in peptide libraries, cysteine residues are often rarer than expected based on predictions of randomness. This has been attributed to reduced display due to formation of inappropriate disulfide bonds (see Chapter 7, Section 2.2). Other sequences may be toxic to *E. coli* or interfere with phage assembly, or be sensitive to bacterial proteases. This "expression editing" means that one cannot assume that every clone theoretically contained within a library has indeed been surveyed in a binding selection.

## 6.3 Over-selection

A common observation in monovalent display, but one that understandably often fails to appear in publications, is that apparently successful selections can often yield clones with strange and unexpected structures, or those that upon analysis do not have the property selected for (e.g. higher affinity). In many cases these outcomes can be attributed to over-selection of the library beyond the rounds at which the desired clones are dominant. Under these conditions, the imposed selective pressure for binding affinity (for example) becomes ineffective, since nearly all clones at that stage will be of equivalent affinity, and factors such as expression level and valency start to drive selection. The result is often that bizarre clones are selected—for example those with internal duplications that lead to bivalency, or those of weak affinity but which are displayed at very high levels. Such observations are a testimony to the power of selection techniques, but emphasize that selections should be carefully monitored, and samples from each round of selection preserved for future analysis.

## 7 Alternative display systems

A number of variations on classical filamentous phage display have been described. These are not generally covered in this book, although many of the principles and protocols may be applicable. Fos–Jun fusion display was developed as an early and clever solution to the problem of how to display proteins C-terminally (84). The Jun leucine zipper sequence is displayed conventionally on pIII; a second protein is co-secreted that comprises the Fos leucine zipper with the displayed protein fused at its C-terminus. Fos–Jun interaction and subsequent disulfide formation provides a C-terminally displayed protein system that has been useful for cDNA interaction cloning. A second technique, termed selectively-infective phage (SIP), exploits the modularity of pIII (see *Figure 3*) to establish a "two-hybrid"-related system in which pIII is split into two pieces, rendering the phage noninfectious. A high-affinity interaction between proteins fused to the pIII domains restore infectivity, allowing identification of binding partners without the need for affinity selection (85).

In addition, bacteriophages other than the Ff family have been used for display of polypeptides, including the lambda (86), T4 (87), and T7 phages (e.g. see (88)). Display as C-terminal fusions to the gene 10 capsid protein of phage T7, in particular, is being increasingly used due to the availability of commercial kits and libraries (see Section 8). Because these phage are lytic, in many cases display does not involve the secretion of the fusion proteins, potentially conferring an advantage for display of some normally intracellular proteins (although see Section 4.1). An application of T7 display phage is described in Chapter 10. Lytic phage are also useful for cDNA display (see Chapter 12, Section 2.2).

Finally, alternative platforms for *in vitro* selection outside of phage display include display on ribosomes following arrest of translation, and display on DNA binding proteins such as the lac repressor (for a review, see (89)). In some cases these techniques have been used together with phage display in a single project aimed at discovering and then improving new binding molecules (see Chapter 6).

## 8 Commercial sources of phage display libraries and kits

A number of vectors, libraries, and complete systems for phage display are now available commercially, as indicated in *Table 4*. New England Biolabs supplies a series of pre-made polyvalent pIII libraries displaying linear or disulfide-constrained peptides, and these are a sensible and efficient starting point for investigators searching for simple peptide ligands. Libraries are provided in the form of kits that include appropriate *E. coli* strains, control reagents, sequencing primers, and protocols. A starting vector is also available for construction of custom polyvalent display libraries.

Amersham Biosciences provides the RPAS system for creation of libraries of single-chain Fv antibody fragments displayed monovalently on pIII (using similar principles and protocols to those outlined in Chapter 13). Premade libraries are not supplied, but a series of modules is available for constructing libraries, performing selections, and expressing the selected antibodies. The RPAS system also represents a commercial source for a pIII phagemid display vector for the construction of custom monovalent display libraries.

Novagen commercializes a series of kits for the creation, packaging, and selection of libraries for display on bacteriophage T7. Several pre-made cDNA display libraries are also available, as well as starting vectors for the creation of custom libraries displayed at a range of valencies.

**Table 4** Commercial suppliers of phage display vectors and libraries

| Manufacturer | Tradename | Vector | Reagents available | Website |
|---|---|---|---|---|
| New England Biolabs | PhD™ phage display system | M13KE pIII phage vector | Pre-made peptide libraries with control target and eluant; starting phage vector | www.neb.com |
| Amersham Biosciences | Recombinant Phage Antibody System (RPAS) | pCANTAB-5E pIII phagemid vector | Kits for library construction, selections, phage enzyme-linked immunosorbant assays (ELISAs), and expression of soluble proteins; starting phagemid vector | www.amershambiosciences.com |
| Novagen | T7Select® System | T7Select vectors: for C-terminal display on gene 10 capsid protein of T7 phage | Kits for library construction, phage in vitro packaging, and selections; pre-made cDNA display libraries | www.novagen.com |

# References

1. Smith, G. P. (1985). *Science*, **228**, 1315.
2. Clackson, T. and Wells, J. A. (1994). *Trends Biotechnol.*, **12**, 173.
3. Lowman, H. B. (1997). *Annu. Rev. Biophys. Biomol. Struct.*, **26**, 401.
4. Smith, G. P. and Petrenko, V. A. (1997). *Chem. Rev.*, **97**, 391.
5. Sidhu, S. S. (2000). *Curr. Opin. Biotechnol.*, **11**, 610.
6. Webster, R. (1996). In *Phage display of peptides and proteins* (ed. B. Kay, J. Winter, and J. McCafferty), p. 1, Academic Press, New York.
7. Makowski, L. and Russel, M. (1997). In *Structural biology of viruses* (ed. W. Chiu, R. M. Burnett, and R. L. Garcea), p. 352, Oxford University Press, New York.
8. Marvin, D. A. (1998). *Curr. Opin. Struct. Biol.*, **8**, 150.
9. Model, P. and Russel, M. (1988). In *The bacteriophages* (ed. R. Calendar), p. 375, Plenum Publishing, New York.
10. Specthrie, L., Bullitt, E., Horiuchi, K., Model, P., Russel, M., and Makowski, L. (1992). *J. Mol. Biol.*, **228**, 720.
11. Endemann, H. and Model, P. (1995). *J. Mol. Biol.*, **250**, 496.
12. Gao, C., Mao, S., Lo, C. H., Wirsching, P., Lerner, R. A., and Janda, K. D. (1999). *Proc. Natl. Acad. Sci. USA*, **96**, 6025.
13. Rakonjac, J., Feng, J. N., and Model, P. (1999). *J. Mol. Biol.*, **289**, 1253.
14. Gailus, V. and Rasched, I. (1994). *Eur. J. Biochem.*, **222**, 927.
15. Cunningham, B. C., Lowe, D. G., Li, B., Bennett, B. D., and Wells, J. A. (1994). *EMBO J.*, **13**, 2508.
16. Holliger, P. and Riechmann, L. (1997). *Structure*, **5**, 265.
17. Lubkowski, J., Hennecke, F., Pluckthun, A., and Wlodawer, A. (1998). *Nature Struct. Biol.*, **5**, 140.
18. Russel, M., Whirlow, H., Sun, T. P., and Webster, R. E. (1988). *J. Bacteriol.*, **170**, 5312.
19. Deng, L. W., Malik, P., and Perham, R. N. (1999). *Virology*, **253**, 271.
20. Riechmann, L. and Holliger, P. (1997). *Cell*, **90**, 351.
21. Lubkowski, J., Hennecke, F., Pluckthun, A., and Wlodawer, A. (1999). *Structure*, **7**, 711.
22. Click, E. M. and Webster, R. E. (1997). *J. Bacteriol.*, **179**, 6464.
23. Click, E. M. and Webster, R. E. (1998). *J. Bacteriol.*, **180**, 1723.
24. Horiuchi, K. (1997). *Genes Cells*, **2**, 425.
25. Fulford, W. and Model, P. (1988). *J. Mol. Biol.*, **203**, 49.
26. Goodrich, A. F. and Steege, D. A. (1999). *RNA*, **5**, 972.
27. Oliver, A. W., Bogdarina, I., Schroeder, E., Taylor, I. A., and Kneale, G. G. (2000). *J. Mol. Biol.*, **301**, 575.
28. Fulford, W. and Model, P. (1988). *J. Mol. Biol.*, **203**, 39.
29. Lopez, J. and Webster, R. E. (1985). *J. Bacteriol.*, **163**, 1270.
30. Linderoth, N. A., Simon, M. N., and Russel, M. (1997). *Science*, **278**, 1635.
31. Russel, M., Linderoth, N. A., and Sali, A. (1997). *Gene*, **192**, 23.
32. Marciano, D. K., Russel, M., and Simon, S. M. (1999). *Science*, **284**, 1516.
33. Marciano, D. K., Russel, M., and Simon, S. M. (2001). *Proc. Natl. Acad. Sci. USA*, **98**, 9359.
34. Russel, M. (1992). *J. Mol. Biol.*, **227**, 453.
35. Greenwood, J., Willis, A. E., and Perham, R. N. (1991). *JMB*, **220**, 821.
36. Iannolo, G., Minenkova, O., Petruzzelli, R., and Cesareni, G. (1995). *J. Mol. Biol.*, **248**, 835.
37. Feng, J. N., Model, P., and Russel, M. (1999). *Mol. Microbiol.*, **34**, 745.
38. Feng, J. N., Russel, M., and Model, P. (1997). *Proc. Natl. Acad. Sci. USA*, **94**, 4068.
39. Rakonjac, J. and Model, P. (1998). *J. Mol. Biol.*, **282**, 25.
40. Russel, M. and Model, P. (1989). *J. Virol.*, **63**, 3284.

41. Jespers, L. S., Messens, J. H., De Keyser, A., Eeckhout, D., Van den Brande, I., Gansemans, Y. G. *et al.* (1995). *Biotechnology*, **13**, 378.
42. Sidhu, S. S., Weiss, G. A., and Wells, J. A. (2000). *J. Mol. Biol.*, **296**, 487.
43. Weiss, G. A. and Sidhu, S. S. (2000). *J. Mol. Biol.*, **300**, 213.
44. Bass, S., Greene, R., and Wells, J. A. (1990). *Proteins*, **8**, 309.
45. Lowman, H. B., Bass, S. H., Simpson, N., and Wells, J. A. (1991). *Biochemistry*, **30**, 10832.
46. Fuh, G. and Sidhu, S. S. (2000). *FEBS Lett.*, **480**, 231.
47. Nord, K., Nilsson, J., Nilsson, B., Uhlen, M., and Nygren, P. A. (1995). *Protein Eng.*, **8**, 601.
48. Volpicella, M., Ceci, L. R., Gallerani, R., Jongsma, M. A., and Beekwilder, J. (2001). *Biochem. Biophys. Res. Commun.*, **280**, 813.
49. Hennecke, M., Otto, A., Baensch, M., Kola, A., Bautsch, W., Klos, A. *et al.* (1998). *Eur. J. Biochem.*, **252**, 36.
50. Cain, S. A., Williams, D. M., Harris, V., and Monk, P. N. (2001). *Protein Eng.*, **14**, 189.
51. Dubaquie, Y. and Lowman, H. B. (1999). *Biochemistry*, **38**, 6386.
52. Beste, G., Schmidt, F. S., Stibora, T., and Skerra, A. (1999). *Proc. Natl. Acad. Sci. USA*, **96**, 1898.
53. Corey, D. R., Shiau, A. K., Yang, Q., Janowski, B. A., and Craik, C. S. (1993). *Gene*, **128**, 129.
54. McCafferty, J., Griffiths, A. D., Winter, G., and Chiswell, D. J. (1990). *Nature*, **348**, 552.
55. Ruan, B., Hoskins, J., Wang, L., and Bryan, P. N. (1998). *Protein Sci.*, **7**, 2345.
56. Atwell, S. and Wells, J. A. (1999). *Proc. Natl. Acad. Sci. USA*, **96**, 9497.
57. Lucic, M. R., Forbes, B. E., Grosvenor, S. E., Carr, J. M., Wallace, J. C., and Forsberg, G. (1998). *J. Biotechnol.*, **61**, 95.
58. Vest Hansen, N., Ostergaard Pedersen, L., Stryhn, A., and Buus, S. (2001). *Eur. J. Immunol.*, **31**, 32.
59. Muller, Y. A., Li, B., Christinger, H. W., Wells, J. A., Cunningham, B. C., and de Vos, A. M. (1997). *Proc. Natl. Acad. Sci. USA*, **94**, 7192.
60. Hoogenboom, H. R., Griffiths, A. D., Johnson, K. S., Chiswell, D. J., Hudson, P., and Winter, G. (1991). *Nucl. Acids Res*, **19**, 4133.
61. McCafferty, J., Jackson, R. H., and Chiswell, D. J. (1991). *Protein Eng*, **4**, 955.
62. Dalby, P. A., Hoess, R. H., and DeGrado, W. F. (2000). *Protein Sci.*, **9**, 2366.
63. Hiipakka, M., Poikonen, K., and Saksela, K. (1999). *J. Mol. Biol.*, **293**, 1097.
64. Rebar, E. J. and Pabo, C. O. (1994). *Science*, **263**, 671.
65. Ku, J. and Schultz, P. G. (1995). *Proc. Natl. Acad. Sci. USA*, **92**, 6552.
66. Sche, P. P., McKenzie, K. M., White, J. D., and Austin, D. J. (1999). *Chem. Biol.*, **6**, 707.
67. Widersten, M. and Mannervik, B. (1995). *J. Mol. Biol.*, **250**, 115.
68. Figini, M., Marks, J. D., Winter, G., and Griffiths, A. D. (1994). *J. Mol. Biol.*, **239**, 68.
69. Parmley, S. F. and Smith, G. P. (1988). *Gene*, **73**, 305.
70. Russel, M., Kidd, S., and Kelley, M. R. (1986). *Gene*, **45**, 333.
71. Vieira, J. and Messing, J. (ed.) (1987). In *Methods in enzymology*, Vol. 153, (ed. ), p. 3, Academic Press, London.
72. Sambrook, I., Fritsch, E., and Maniatis, T. (ed.) (1989). *Molecular cloning: a laboratory manual*, 2nd edn., p. 4.22, Cold Spring Harbor Press, New York.
73. Chen, Y., Wiesmann, C., Fuh, G., Li, B., Christinger, H. W., McKay, P. *et al.* (1999). *J. Mol. Biol.*, **293**, 865.
74. Fairbrother, W. J., Christinger, H. W., Cochran, A. G., Fuh, G., Keenan, C. J., Quan, C. *et al.* (1998). *Biochemistry*, **37**, 17754.
75. Lowman, H. B., Chen, Y. M., Skelton, N. J., Mortensen, D. L., Tomlinson, E. E., Sadick, M. D. *et al.* (1998). *Biochemistry*, **37**, 8870.
76. Hawkins, R. E., Russell, S. J., and Winter, G. (1992). *J. Mol. Biol.*, **226**, 889.

77. Goodson, R. J., Doyle, M. V., Kaufman, S. E., and Rosenberg, S. (1994). *Proc. Natl. Acad. Sci. USA*, **91**, 7129.

78. Roberts, B. L., Markland, W., Ley, A. C., Kent, R. B., White, D. W., Guterman, S. K. *et al.* (1992). *Proc. Natl. Acad. Sci. USA*, **89**, 2429.

79. Dennis, M. S., Eigenbrot, C., Skelton, N. J., Ultsch, M. H., Santell, L., Dwyer, M. A. *et al.* (2000). *Nature*, **404**, 465.

80. Krebs, B., Rauchenberger, R., Reiffert, S., Rothe, C., Tesar, M., Thomassen, E. *et al.* (2001). *J. Immunol. Methods*, **254**, 67.

81. Oldenburg, K. R., Loganathan, D., Goldstein, I. J., Schultz, P. G., and Gallop, M. A. (1992). *Proc. Natl. Acad. Sci. USA*, **89**, 5393.

82. Wojnar, P., Lechner, M., Merschak, P., and Redl, B. (2001). *J. Biol. Chem.*, **276**, 20206.

83. Wind, T., Stausbol-Gron, B., Kjaer, S., Kahns, L., Jensen, K. H., and Clark, B. F. (1997). *J. Immunol. Methods*, **209**, 75.

84. Crameri, R. and Suter, M. (1993). *Gene*, **137**, 69.

85. Krebber, C., Spada, S., Desplancq, D., and Pluckthun, A. (1995). *FEBS Lett.*, **377**, 227.

86. Hoess, R. H. (2002). *Curr. Pharm. Biotechnol.*, **3**, 23.

87. Ren, Z. J., Lewis, G. K., Wingfield, P. T., Locke, E. G., Steven, A. C., and Black, L. W. (1996). *Protein Sci.*, **5**, 1833.

88. Danner, S. and Belasco, J. G. (2001). *Proc. Natl. Acad. Sci. USA*, **98**, 12954.

89. Dower, W. J. and Mattheakis, L. C. (2002). *Curr. Opin. Chem. Biol.*, **6**, 390.

# Chapter 2

# Constructing phage display libraries by oligonucleotide-directed mutagenesis

Sachdev S. Sidhu and Gregory A. Weiss
Department of Protein Engineering, Genentech, Inc., 1 DNA Way,
South San Francisco, CA 94080.

## 1 Introduction

As with any selection technology, the success of phage display depends on library diversity and quality. This chapter details robust methods that can be used to rapidly construct highly diverse phage-displayed libraries. Our protocols optimize previously described methods, allowing for the facile construction of libraries almost 100-fold greater than previously thought practical (1). Libraries with diversities greater than $10^{10}$ can be readily constructed with a single reaction, and multiple reactions can be combined to generate libraries with diversities on the order of $10^{12}$.

## 2 Considerations for library design

### 2.1 Site-directed mutagenesis

In principle, any random or site-directed mutagenesis method can be used to construct a library. We prefer the site-directed approach because the sites of mutation can be precisely controlled. A site-directed library is constructed by replacing unique codons with degenerate codons, that is, codons which encode more than one amino acid. All site-directed mutagenesis methods involve the replacement of a gene segment with synthetic DNA. By using appropriately designed synthetic oligonucleotides, the amount of diversity introduced at each position can be varied in a controlled manner.

### 2.2 Degenerate codon design

The simplest and most complete method of introducing diversity is "hard randomization": the introduction of degenerate codons that encode all twenty natural amino acids. Due to the nature of the genetic code, conventional DNA

**Table 1** Useful degenerate codons[a]

| Codon | Description | Amino acids | Stop codons | No. of codons[b] |
|-------|-------------|-------------|-------------|------------------|
| NNS | All 20 amino acids | All 20 | TAG | 32 |
| NNC | 15 amino acids | A, C, D, F, G, H, I, L, N, P, R, S, T, V, Y | None | 16 |
| NWW | Charged, hydrophobic | D, E, F, H, I, K, L, N, Q, V, Y | TAA | 16 |
| RVK | Charged, hydrophilic | A,D, E, G, H, K, N, R, S, T | None | 12 |
| DVT | Hydrophilic | A, C, D, G, N, S, T, Y | None | 9 |
| NVT | Charged, hydrophilic | C, D, G, H, N, P, R, S, T, Y | None | 12 |
| NNT | Mixed | A, D, G, H, I, L, N, P, R, S, T, V | None | 16 |
| VVC | Hydrophilic | A, D, G, H, N, P, R, S, T, | None | 9 |
| NTT | Hydrophobic | F, I, L, V | None | 4 |
| RST | Small side chains | A, G, S, T | None | 4 |
| TDK | Hydrophobic | C, F, L, W, Y | TAG | 6 |

[a] While the NNS codon provides "hard randomization" or substitution with all 20 amino acids, it is sometimes desirable to substitute with only a limited set of amino acids. Some useful codons for such substitutions are shown here.

[b] The number of unique codons contained in the degenerate codon. Due to the nature of the genetic code, many degenerate codons are redundant. That is, the number of unique codons exceeds the number of unique amino acids encoded.

synthesis methods necessitate that a degenerate codon encoding all twenty natural amino acids must contain at least 32 unique codons (*Table 1*). Thus, hard randomization results in some redundancy; some amino acids are represented by only one codon while others are represented by two or three codons.

Methods also exist for introducing incomplete or biased diversity. "Tailored randomization" refers to the use of degenerate codons which encode only a subset of the natural amino acids. Often, these subsets are chosen to represent a shared characteristic. For example, degenerate codons can be chosen to allow only amino acids with small side chains or, alternatively, only those with hydrophobic side chains (*Table 1*). An even more restrictive method is "soft randomization" or "doping," which strongly biases a degenerate codon towards a single sequence (usually the wild-type) but also allows a low, defined level of variation. This is readily achieved by using synthetic DNA in which each degenerate base is predominantly wild-type, with small, equimolar amounts of the remaining three nucleotides added. The mutation frequency can be readily adjusted during DNA synthesis by simply altering the amount of the non-wild-type nucleotides relative to the wild-type nucleotide.

## 2.3 Theoretical versus actual diversity

Library design requires two decisions. One must select positions to randomize, and, second, one must choose a mutagenesis strategy. A major factor governing both decisions is the resultant, actual diversity of the constructed library. While it is tempting to completely randomize a large number of codons, one must bear in mind that theoretical diversities can easily exceed actual diversities. For example, a library containing seven hard randomized codons contains $3.4 \times 10^{10}$ codon combinations ($32^7$) representing $1.3 \times 10^9$ ($20^7$) unique amino acid combinations. Even a library containing $10^{10}$ unique members is unlikely to provide complete coverage in such a case. The simultaneous randomization of a large number of codons may be better accomplished with tailored or soft randomization strategies that limit the chemical diversity at each site. Of course, different randomization strategies can be employed at different sites within a single library.

## 3 Oligonucleotide-directed mutagenesis

### 3.1 Oligonucleotide-directed mutagenesis versus cassette mutagenesis

As described above, the construction of a site-directed library involves the replacement of a gene segment with a synthetic DNA fragment containing one or more degenerate codons. This can be accomplished with either double-stranded cassette mutagenesis (2) or with single-stranded oligonucleotide-directed mutagenesis. We prefer oligonucleotide-directed mutagenesis (3), because unlike cassette mutagenesis, there is no requirement for unique restriction sites near the targeted region and only one oligonucleotide is required to construct a library. As a result, the method is extremely simple and robust.

### 3.2 The chemistry and biology of oligonucleotide-directed mutagenesis

The method described in this chapter is a highly optimized version of a procedure that was originally developed by Zoller and Smith (3) and significantly improved by Kunkel et al. (4). The phagemid to be mutagenized is first purified in a single-stranded form from an Escherichia coli dut⁻/ung⁻ strain such as CJ236 (4). Such strains produce DNA with a significant amount of uracil incorporated in place of thymine. The uracil-containing single-stranded DNA (dU-ssDNA) serves as a template to which an appropriately designed mutagenic oligonucleotide is annealed (Figure 1(a)). The oligonucleotide is designed with the first 15 bases complementary to the sequence immediately preceding the region to be mutated, and the final 15 bases complementary to the sequence immediately following this region. The oligonucleotide sequence between these two flanking sequences (the variable region) can be of any desired sequence, and thus, this is the region that introduces mutations. Because oligonucleotides greater than 100 bases in length can be readily synthesized, variable regions spanning more than 20 codons can be used to mutagenize large stretches of primary sequence with a single oligonucleotide. The mutation of regions farther apart in the primary sequence can be accomplished in a single library by simultaneously annealing two or more oligonucleotides to a single template. The only

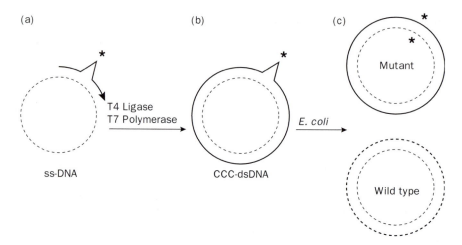

**Figure 1** Oligonucleotide-directed mutagenesis. (a) A synthetic oligonucleotide (solid line) is annealed to the dU-ssDNA template (dashed line). The oligonucleotide is designed to encode mutations (*) in the mismatched variable region which is flanked by perfectly complementary sequences. (b) Heteroduplex CCC-dsDNA is enzymatically synthesized by T7 DNA polymerase and T4 DNA ligase. (c) Heteroduplex CCC-dsDNA is introduced into an *E. coli* host where the mismatched region is repaired to either the wild-type sequence or the mutant sequence.

limit to this process is that the sequences of the different oligonucleotides must not overlap, because an overlap in two different oligonucleotides makes their annealing mutually exclusive.

The annealed oligonucleotide is next used as a primer for T7 DNA polymerase which synthesizes a complete DNA strand complementary to the template strand. T4 DNA ligase then ligates the newly synthesized strand to form covalently closed circular, double-stranded DNA (CCC-dsDNA) (*Figure 1(b)*). Because the variable region of the mutagenic oligonucleotide is not perfectly complementary to the template DNA, the result is a CCC-dsDNA heteroduplex with mismatches in the region targeted for mutagenesis. It should be noted that this mismatch need not involve DNA regions of equivalent length. Deletions or insertions can be introduced by using oligonucleotides with variable regions that are shorter or longer than the corresponding template sequence. This is especially useful for the construction of phage-displayed peptide libraries, because a single template can be used to construct many different libraries, simply by varying the length and sequence of the mutagenic oligonucleotide (5).

The CCC-dsDNA heteroduplex is resolved by introducing the DNA into an *E. coli* host where the mismatch is repaired to either the wild-type or mutant sequence (*Figure 1(c)*). In an *ung*⁺ strain, the uracil-containing template strand is preferentially inactivated and the newly synthesized mutagenic strand is replicated. If highly pure template DNA is used, the mutagenesis is very efficient (>80%). Also, the simultaneous annealing of multiple mutagenic oligonucleotides does not significantly reduce the mutation frequency. We regularly obtain efficient mutagenesis (>50%) with the simultaneous incorporation of three or four oligonucleotides.

## 3.3 Construction of an inactive template

Because oligonucleotide-directed mutagenesis is not 100% efficient, the constructed library will contain significant amounts of the wild-type template. This is the major drawback of the method, especially in cases where the wild-type construct is functional (e.g. affinity maturation of a displayed protein that already binds to the target ligand). In such cases, the wild-type will be selected along with variants. In fact, the wild-type is likely to dominate the selection because it will be the single most prevalent sequence in the naïve library (even in a library constructed with 80% mutation frequency, 20% of the sequences will be wild-type). Fortunately, the effect of wild-type background can be negated with an inactive template that permits the construction of phage-displayed libraries but does not permit the display of proteins translated from wild-type template sequence. This is readily accomplished by using a template with stop codons incorporated within the region targeted for randomization in the library (6). Translation of such a template sequence produces truncated proteins that are not phage-displayed. Successful incorporation of a mutagenic oligonucleotide introduces appropriately designed degenerate codons and simultaneously replaces the stop codons with coding sequences. Only library members with completely restored open reading frames will produce proteins that can be phage-displayed. The wild-type background is still present as DNA, but it is absent at the protein level since the stop codons result in premature translation termination. Thus, the inactive template does not interfere in subsequent phenotypic selections.

The TAA stop codon is preferred for the construction of inactive templates, because it is the most efficient termination codon in *E. coli*. In a library construction involving a single mutagenic oligonucleotide, a single stop codon is probably sufficient, and it can be located anywhere within the region that will be replaced by the library-encoding mutagenic oligonucleotide. However, we generally use three consecutive stop codons to ensure complete termination. For constructions involving multiple mutagenic oligonucleotides, at least one stop codon should be present within each region that is mutagenized. This ensures that library members will only be displayed if they incorporate all of the mutagenic oligonucleotides.

## 4 Library construction and storage

The following methods are optimized to produce highly diverse phage-displayed libraries with phagemid-based systems. With a few minor modifications, the methods are also suitable for phage-based display systems. In particular, while phagemids require the addition of helper phage to package DNA into phage particles, phage vectors are usually self-packaging.

31

## 4.1 Preparation of single-stranded DNA template

Oligonucleotide-directed mutagenesis uses ssDNA as the template. Thus, the method is ideally suited for phage display, because ssDNA can be purified directly from phage particles. Mutagenesis efficiency depends on template purity, and thus, the use of highly pure dU-ssDNA is critical for successful library construction. We use the Qiagen QIAprep Spin M13 Kit as an economical, affinity column-based method for obtaining highly pure ssDNA. *Protocol 1* is a modified version of the Qiagen protocol; it yields at least 20 μg of ssDNA for a medium copy number phagemid (e.g. pBR322 based). This is sufficient for one library construction. The protocol can be scaled up by inoculating a larger overnight culture and purifying the ssDNA with multiple spin columns.

---

## Protocol 1

## Isolation of dU-ssDNA template

### Equipment and reagents

- TAE buffer (40 mM Tris-acetate, 1 mM ethylenediamine tetracetic acid (EDTA); adjust pH to 8.0; autoclave)
- TAE/agarose gel (TAE buffer, 1% w/v agarose, 1:5000 v/v 10% ethidium bromide)
- 2YT media (10 g bacto-yeast extract, 16 g bacto-tryptone, 5 g NaCl; add water to 1 L and adjust pH to 7.0 with NaOH; autoclave)
- 2YT/carb/cmp media (2YT, 50 μg/ml carbenicillin, 5 μg/ml chloramphenicol)
- 2YT/carb/uridine media (2YT, 50 μg/ml carbenicillin, 0.25 μg/ml uridine)

- Phosphate buffered saline (PBS) (137 mM NaCl, 3 mM KCl, 8 mM $Na_2HPO_4$, 1.5 mM $KH_2PO_4$; adjust pH to 7.2 with HCl; autoclave)
- Polyethylene glycol (PEG) NaCl (20% PEG-8000 w/v, 2.5 M NaCl)
- QIAprep Spin M13 Kit (Qiagen, QIAprep spin columns, Buffer MP, Magnesium containing lysis buffer (MLB), Wash Buffer PE, Buffer EB)
- M13K07 helper phage (see *Protocol 6*)

### Method

1  From a fresh plate, pick a single colony of *E. coli* CJ236 (or another *dut⁻/ung⁻* strain) harboring the appropriate phagemid into 1 ml of 2YT media supplemented with appropriate antibiotics to maintain the host F′ episome and the phagemid. For example, 2YT/carb/cmp media contains carbenicillin to select for phagemids that carry the β-lactamase gene and chloramphenicol to select for the CJ236 F′ episome. Shake at 200 rpm and 37 °C for 6–8 h. Add M13K07 helper phage to a final concentration of $10^{10}$ phage/ml (for a final multiplicity of infection (moi) of approximately 10). Shake at 200 rpm and 37 °C for 15 min. Transfer the culture to 30 ml of 2YT/carb/uridine media. Shake overnight at 200 rpm and 37 °C.

2  Centrifuge for 10 min at 15 krpm and 4 °C in a Sorvall SS-34 rotor (27,000 g). Transfer the supernatant to a new tube containing 1/5 volume of PEG/NaCl and incubate for 5 min at room temperature. Centrifuge 10 min at 10 krpm and 4 °C in

an SS-34 rotor (12,000 g). Decant the supernatant. Centrifuge briefly at 4 krpm (2000 g) and aspirate the remaining supernatant.

3  Resuspend the phage pellet in 0.5 ml of PBS. Centrifuge for 5 min at 15 krpm and 4°C in an SS-34 rotor to pellet insoluble matter. Transfer the supernatant to a 1.5 ml microcentrifuge tube.

4  Add 7.0 µl of Buffer MP and mix. Incubate at room temperature for at least 2 min.

5  Apply the sample to a QIAprep spin column in a 2-ml microcentrifuge tube. Centrifuge for 15 s at 8 krpm in a microcentrifuge. Discard the flow-through. The phage particles remain bound to the column matrix.

6  Add 0.7 ml Buffer MLB to the column. Centrifuge for 15 s at 8 krpm. Discard the flow-through.

7  Add another 0.7 ml Buffer MLB. Incubate at room temperature for at least 1 min. Centrifuge at 8 krpm for 15 s. Discard the flow-through. The DNA is separated from the protein coat and remains adsorbed to the matrix.

8  Add 0.7 ml Wash Buffer PE. Centrifuge at 8 krpm for 15 s. Discard the flow-through.

9  Repeat step 8. Residual proteins and salt are removed.

10  Centrifuge at 8 krpm for 30 s. Transfer the column to a fresh 1.5 ml micro-centrifuge tube.

11  Add 100 µl of Buffer EB (10 mM Tris-Cl, pH 8.5) to the center of the column membrane. Incubate at room temperature for 10 min and centrifuge for 30 s at 8 krpm. Save the eluant, as it contains the purified dU-ssDNA.

12  Analyze the DNA by electrophoresing 1.0 µl on a TAE/agarose gel. The DNA should appear as a predominant single band, but faint bands with lower electrophoretic mobility are often visible (lane 2 in *Figure 2*). These are likely caused by secondary structure in the ssDNA. Determine the DNA concentration by measuring absorbance at 260 nm ($A_{260} = 1.0$ for 33 ng/µl of ssDNA). Typical DNA concentrations range from 200–500 ng/µl.

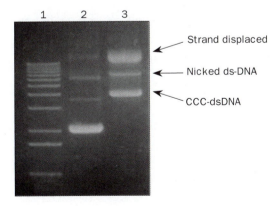

1    2    3

— Strand displaced

← Nicked ds-DNA

← CCC-dsDNA

**Figure 2** *In vitro* synthesis of heteroduplex CCC-dsDNA. The reaction products were electrophoresed on a 1.0% TAE/agarose gel containing ethidium bromide for DNA visualization. Lane 1. 1 kb DNA marker (Gibco BRL). Lane 2. The dU-ssDNA template. Lane 3. Reaction product from *Protocol 2*. The lower band is correctly extended and ligated CCC-dsDNA. The middle band is nicked dsDNA. The upper band is strand displaced DNA.

## 4.2 *In vitro* synthesis of heteroduplex CCC-dsDNA

A three step procedure (*Protocol 2*) is used to incorporate the mutagenic oligo-nucleotide into heteroduplex CCC-dsDNA, using dU-ssDNA as the template. The protocol described here is an optimized, large-scale version of a previously described method (4). The oligonucleotide is first 5'-phosphorylated and then annealed to a dU-ssDNA template. The oligonucleotide is enzymatically extended and ligated to form heteroduplex CCC-dsDNA (lane 3 in *Figure 2*). Finally, the CCC-dsDNA is purified and desalted.

The protocol below produces 20–40 µg of highly pure, low conductance CCC-dsDNA. This is sufficient for the construction of a library containing more than $10^{10}$ unique members. All steps below can be scaled-up considerably, with the possible exception of the annealing step. The annealing protocol described here works well with volumes of 250 µl or less. It may also be possible to scale this for larger volumes.

# Protocol 2

# Phagemid library construction

## Equipment and reagents

- $10 \times$ TM buffer (0.5 M Tris, pH 7.5, 0.1 M MgCl$_2$)
- 10 mM ATP (Amersham Biosciences)
- 100 mM dithiothreitol (DTT) (Sigma)
- T4 polynucleotide kinase (New England Biolabs)
- 25 mM dioxyribonucleotide triphosphate (dNTPs): solution with 25 mM each of dATP, dCTP, dGTP, dTTP (Amersham Biosciences)

- TAE/agarose gel (see *Protocol 1*)
- QIAquick Gel Extraction Kit (Qiagen, QIAquick spin columns, Buffer QG, Buffer PE)
- T4 DNA ligase (Invitrogen)
- T7 DNA polymerase (New England Biolabs)

## A. Phosphorylation of the mutagenic oligonucleotide[a]

1  In a 1.5 ml microcentrifuge tube, combine 0.6 µg oligonucleotide, 2.0 µl $10\times$ TM buffer, 2.0 µL 10 mM ATP, 1.0 µl 100 mM DTT. Add water to a total volume of 20 µl.

2  Add 20 units of T4 polynucleotide kinase. Incubate for 1.0 h at 37 °C.

## B. Annealing the oligonucleotide to the template[b]

1  To the 20 µl phosphorylation reaction mix from *Protocol 2A*, add 20 µg of dU-ssDNA template (from *Protocol 1*), 25 µl $10\times$ TM buffer, and water to a final volume of 250 µl. These DNA quantities provide an oligonucleotide : template molar ratio of 3 : 1, assuming that the oligonucleotide : template length ratio is 1 : 100.

2  Incubate at 90 °C for 2 min, 50 °C for 3 min, and 20 °C for 5 min.

## C. Enzymatic synthesis of CCC-dsDNA[c]

**1** To the annealed oligonucleotide/template mixture from *Protocol 2B*, add 10 µl 10 mM ATP, 10 µl 25 mM dNTPs, 15 µl 100 mM DTT, 6 Weiss units T4 DNA ligase, 30 units T7 DNA polymerase.

**2** Incubate at 20 °C for at least 3 h. Overnight incubation is preferred.

**3** Affinity purify and desalt the DNA using the Qiagen QIAquick DNA purification kit. Add 1.0 ml of buffer QG and mix.

**4** Apply the sample to two QIAquick spin columns placed in 2 ml microcentrifuge tubes. Centrifuge at 13 krpm for 1 min in a microcentrifuge. Discard the flow-through.

**5** Add 750 µl buffer PE to each column. Centrifuge at 13 krpm for 1 min. Discard the flow-through and centrifuge at 13 krpm for 1 min. Place the column in a new 1.5 ml microcentrifuge tube.

**6** Add 35 µl of ultrapure water to the center of the membrane. Incubate at room temperature for 1 min.

**7** Centrifuge at 13 krpm for 1 min to elute the DNA. Combine the eluants. The DNA can be used immediately for *E. coli* electroporation (*Protocol 3*), or it can be stored frozen for later use.

**8** Electrophorese 1.0 µl of the eluted reaction product alongside the dU-ssDNA template. Use a TAE/agarose gel with ethidium bromide for DNA visualization (*Figure 2*). The electrophoretic mobility of circular DNA depends on salt concentrations, pH, and the presence of ethidium bromide. To observe the relative mobilities shown in *Figure 2*, the DNA *must* be electrophoresed on a TAE/agarose gel with ethidium bromide added directly to the molten gel rather than the running buffer. A successful reaction results in the complete conversion of dU-ssDNA to dsDNA, which has a lower electrophoretic mobility. Usually, at least two product bands are visible (*Figure 2*). The faster band is the desired product: correctly extended and ligated CCC-dsDNA which transforms *E. coli* efficiently and provides a high mutation frequency (>80%). The slower band is a strand displaced product resulting from an intrinsic, unwanted activity of T7 DNA polymerase (7). Although the strand displaced product provides a low mutation frequency ( < 20%), it also transforms *E. coli* at least 30-fold less efficiently than CCC-dsDNA. If a significant proportion of the single-stranded template is converted into CCC-dsDNA, a highly diverse library with high mutation frequency will result. Sometimes a third band is visible, with an electrophoretic mobility between the two bands described above (*Figure 2*). This intermediate band is correctly extended but unligated dsDNA (nicked dsDNA) which results from either insufficient T4 DNA ligase activity or from incomplete oligonucleotide phosphorylation.

[a] The mutagenic oligonucleotide must be 5′-phosphorylated to enable ligation by T4 DNA ligase (*Protocol 2C*). While oligonucleotides can be chemically phosphorylated during synthesis, the synthetic reaction is not always 100% efficient. Enzymatic phosphorylation

with T4 polynucleotide kinase is highly reliable and efficient. The phosphorylated oligonucleotide should be used immediately in the subsequent steps.

[b] The protocol given here minimizes the amount of time reactants are incubated at the DNA melting temperature (90 °C) to minimize dephosphorylation of the oligonucleotide. To scale up this reaction, run multiple annealing reactions of 250 μl each.

[c] T7 DNA polymerase uses the annealed mutagenic oligonucleotide as a primer for the template-directed synthesis of a complementary DNA strand. Following complete synthesis of the second strand, T4 DNA ligase seals the nick to form heteroduplex CCC-dsDNA.

## 4.3 *E. coli* electroporation and production of library phage

To complete the library construction, the heteroduplex CCC-dsDNA produced in *Protocol 2* must be introduced into an *E. coli* host that contains an F′ episome to enable M13 bacteriophage infection and propagation. In this way, the heteroduplex can be resolved through DNA repair and replication in the host (*Figure 1(c)*), and the resulting library can be packaged into phage particles. Phage-displayed library diversities are limited by methods for introducing DNA into *E. coli*, with the most efficient method being high-voltage electroporation.

Electroporation is based on the observation that *E. coli* (and many other cells) will take up DNA from surrounding solution when subjected to a strong electric field (8). However, DNA uptake only occurs under conditions of high field strength and low current flow. In practical terms, this means that the *E. coli*/DNA mixture must be extremely low in conductance so that a strong electric field can be established without a high current flow through the solution.

The yield of *E. coli* transformants from an electroporation reaction can be increased by increasing the concentrations of DNA and/or *E. coli*. However, both components must be extremely pure and free from conducting species. The standard method for desalting DNA has been ethanol precipitation, but DNA purified in this way contains significant amounts of charged impurities, and thus it can only be electroporated at concentrations around 10 μg/ml. Affinity purification produces highly pure DNA solutions with extremely low conductance (see *Protocol 2C*), allowing electroporations to be performed with DNA concentrations exceeding 100 μg/ml. Under such saturating conditions, a large percentage of the electrocompetent cells are transformed.

The second component of the electroporation reaction, the *E. coli* cells, must also be highly concentrated for maximum efficiency. While all *E. coli* strains can be made electrocompetent, different strains vary greatly in their tolerance to the washing steps required to produce electrocompetent cells (*Protocol 5*). In many cases, a significant proportion of cells do not survive this step, and thus, the concentration of viable, electrocompetent cells is reduced. High efficiency electroporation strains are tolerant to these washing procedures. The best strains, such as *E. coli* MC1061 (9), remain 100% viable in the electrocompetent state, and therefore, they can be electroporated at extremely high concentrations ($>3 \times 10^{11}$ colony forming units (cfu)/ml).

We have constructed a new strain, *E. coli* SS320, that is ideal for both high efficiency electroporation and phage production (5). Using a standard bacterial mating protocol (10), we transferred the F′ episome from *E. coli* XL1-Blue (11) (Stratagene) to *E. coli* MC1061 (Bio-Rad). The progeny strain was readily selected for double resistance to streptomycin and tetracycline, because *E. coli* MC1061 carries a chromosomal marker for streptomycin resistance and the F′ episome from *E. coli* XL1-blue confers tetracycline resistance. *E. coli* SS320 retains the high electroporation efficiency of *E. coli* MC1061, while the presence of an F′ episome enables infection by M13 phage.

## Protocol 3

## *E. coli* electroporation and phage propagation[a]

### Equipment and reagents

- Electrocompetent SS320 *E. coli* (see *Protocol 5*)
- Electroporation cuvet (0.2 cm gap, BTX or Bio-Rad)
- Electroporator (BTX ECM-600 or Bio-Rad Gene Pulser)
- PBS (see *Protocol 1*)
- PEG/NaCl (see *Protocol 1*)

- SOC media (5 g bacto-yeast extract, 20 g bacto-tryptone, 0.5 g NaCl, 0.2 g KCl; add water to 1.0 L and adjust pH to 7.0 with NaOH; autoclave; add 5.0 ml of autoclaved 2.0 M $MgCl_2$ and 20 ml of filter sterilized 1.0 M glucose)
- 2YT/carb media (see *Protocol 1*)
- M13K07 helper phage (see *Protocol 6*)

### Method

1  Chill the purified DNA (20 μg in a minimum volume, from *Protocol 2*) and a 0.2 cm gap electroporation cuvet on ice. Thaw a 350 μl aliquot of electrocompetent *E. coli* SS320 on ice. Add the cells to the DNA and mix by pipetting several times (avoid introducing bubbles).

2  Transfer the mixture to the cuvet and electroporate. For electroporation, follow the manufacturer's instructions, preferably using a BTX ECM-600 electroporation system with the following settings: 2.5 kV field strength, 129 ohms resistance, and 50 μF capacitance. Alternatively, a Bio-rad Gene Pulser can be used with the following settings: 2.5 kV field strength, 200 ohms resistance, and 25 μF capacitance.

3  Immediately, rescue the electroporated cells by adding 1 ml SOC media and transferring to a 250 ml baffled flask. Rinse the cuvette twice with 1 ml SOC media. Add SOC media to a final volume of 25 ml and incubate for 20 min at 37 °C with shaking at 200 rpm.

4  To determine the library diversity, plate serial dilutions on 2YT/agar plates supplemented with an appropriate antibiotic to select for the library phagemid (e.g. carbenicillin for β-lactamase encoding phagemids). Transfer the culture to a 2 L baffled flask containing 500 ml 2YT media, supplemented with antibiotic for phagemid selection (e.g. 2YT/carb media) and M13K07 helper phage ($10^{10}$ phage/ml final concentration). Incubate overnight at 37 °C with shaking at 200 rpm.

**5** Centrifuge the culture for 10 min at 10 krpm and 4 °C in a Sorvall GSA rotor (16,000$g$). Transfer the supernatant to a fresh tube and add 1/5 volume of PEG/NaCl solution to precipitate the phage. Incubate for 5 min at room temperature.

**6** Centrifuge for 10 min at 10 krpm and 4 °C in a GSA rotor. Decant the supernatant. Respin briefly and remove the remaining supernatant with a pipet. Resuspend the phage pellet in 1/20 volume of PBS. Pellet insoluble matter by centrifuging for 5 min at 15 krpm and 4 °C in an SS-34 rotor (27,000$g$). Transfer the supernatant to a clean tube.

**7** Estimate the phage concentration spectrophotometrically ($OD_{268} = 1.0$ for a solution of $5 \times 10^{12}$ phage/ml).

**8** Use the library immediately, or store it at −70 °C.

[a] The purified CCC-dsDNA from *Protocol 2* is electroporated into *E. coli* SS320. Subsequent infection with VCSM13 helper phage packages the library into phage particles that can be readily purified from the culture supernatant. The quality of the library can be assessed by DNA sequence analysis of individual clones.

### 4.4 Library storage and reinfection

Phage particles can be stored in an infectious state for years at 4 °C or indefinitely at −70 °C. However, the same is not necessarily true for many proteins, and whether frozen phage stocks can be used directly in selection experiments depends on the stability of the displayed polypeptide. Phage-displayed libraries of small peptides are usually very stable, and we regularly use frozen peptide libraries directly for selections. In contrast, many large proteins are prone to denaturation or proteolysis, and the amount of displayed ligand may decrease rapidly with storage. The stability of a particular phage-displayed protein must be determined empirically. The direct use of frozen libraries is very convenient, but it may not be possible for unstable proteins. In such cases, the primary library must be repropagated in *E. coli* to produce fresh phage particles that can be used immediately in selection experiments. While *E. coli* SS320 is a superior electroporation strain (Section 4.3), *E. coli* XL1-Blue is preferred for subsequent phage propagations, because it provides a somewhat greater yield of phage particles. *Protocol 4* results in the repropagation of $5 \times 10^{10}$ unique phage. Ideally, the number of repropagated phage should exceed the original library diversity, and thus, this protocol is sufficient for libraries with diversities less than $5 \times 10^{10}$.

## Protocol 4

# Repropagation of phage-displayed libraries

### Equipment and reagents

- 2YT/carb media (see *Protocol 1*)
- 2YT/tet media: 2YT, 5 µg/ml tetracycline
- *E. coli* XL1-Blue (Stratagene)
- M13K07 helper phage (see *Protocol 6*)
- Phage library stock (see *Protocol 3*)

## Method

1  Determine the infectivity titer of the phage library stock by infecting serial dilutions into *E. coli* XL1-Blue ($OD_{600} = 0.5$–$1.0$) and plating on appropriate media for phagemid selection. Depending on the age and state of the phage stock, the infectivity titre will be less than or equal to the phage concentration determined by measuring the $OD_{268}$ (see *Protocol 3*).

2  Inoculate 100 ml of 2YT/tet media in a 500 ml baffled flask with a large quantity of *E. coli* XL1-Blue from a fresh LB/tet plate (less than two weeks old; the inoculum need not be a single colony). Grow the culture at 37 °C with shaking at 200 rpm to an $OD_{600} = 1.0$ ($5 \times 10^8$ cfu/ml, $5 \times 10^{10}$ cfu total). Add $5 \times 10^{10}$ infective phage particles and incubate at 37 °C with shaking at 200 rpm for 20 min. Remove a small aliquot and titer on selective media to determine the efficiency of infection. Transfer the culture to 1 L of 2YT media supplemented with an appropriate antibiotic forphage-mid selection (e.g. 2YT/carb media) and M13K07 helper phage ($10^{10}$ phage/ml) in a 4 L baffled flask. Grow overnight at 37 °C with shaking at 200 rpm.

3  Precipitate and purify the phage particles as described in *Protocol 3*. Use the phage immediately.

## 5. Biological reagents

In addition to the library itself, phage display selections require two main biological reagents: electrocompetent *E. coli* for library construction and helper phage for library propagation. Although these reagents are available from commercial sources, the cost can be prohibitive for large scale applications. Furthermore, our preferred electroporation strain, *E. coli* SS320, is not yet commercially available. Therefore, we have provided large-scale protocols for the production of electrocompetent *E. coli* (*Protocol 5*) and helper phage (*Protocol 6*).

## Protocol 5

# Preparation of electrocompetent *E. coli* SS320[a]

### Equipment and reagents

- 2YT/tet media (see *Protocol 4*)
- Superbroth/tet (24 g bacto-yeast extract, 12 g bacto-tryptone, 5 ml glycerol; add water to 900 ml; autoclave; add 100 ml of autoclaved 0.17 M $KH_2PO_4$, 0.72 M $K_2HPO_4$ and 5 µg/ml tetracycline)
- Filter sterilized 10% ultrapure glycerol (Gibco BRL: Invitrogen) in ultrapure water
- Magnetic stirbars (2 inch)
- *E. coli* SS320 stock (see Section 4.3)
- Sterile 1.0 mM Hepes (*N*-[2-hydroxyethyl]-piperazine-*N'*-[2-ethanesulfonic acid]) pH 7.4 in ultrapure water

**Protocol 5** continued

## Method

1  Inoculate 1 ml 2YT/tet media with a single colony of *E. coli* SS320 from a fresh 2YT/tet plate (less than one week old). Incubate at 37 °C with shaking at 200 rpm for 6–8 h. Transfer the culture to 50 ml of 2YT/tet media in a 500 ml baffled flask. Incubate overnight at 37 °C with shaking at 200 rpm.

2  Inoculate six 2 L baffled flasks, each containing 900 ml Superbroth/tet, with 5 ml of the overnight culture, and grow the cells at 37 °C with shaking at 200 rpm to an $OD_{600} = 0.6$–0.8 (approximately 3–4 h).

3  Chill three of the flasks for 5 min on ice, with occasional swirling.

*NB: The following steps should be done in a 4 °C cold room, on ice, with prechilled solutions and equipment.*

4  Centrifuge the cells at 5.5 krpm and 4 °C for 5 min in a GS3 rotor (5000 g), and preferably, in a centrifuge with a fast brake (e.g. Sorvall RC-5B Plus). Decant the supernatant. Add culture from the remaining three flasks (these should be chilled while the first set is centrifuging) to the same tubes. Repeat the centrifugation and decant the supernatant.

5  Fill each centrifuge tube with 1.0 mM Hepes, pH 7.4. Add a sterile magnetic stirbar to each tube to facilitate pellet resuspension. Swirl briefly to dislodge the pellet from the tube wall and then stir at a moderate rate until the pellet is completely resuspended.

6  Centrifuge for 10 min at 5.5 krpm and 4 °C in a GS3 rotor. Decant the supernatant. While decanting, leave the stir bar in the centrifuge tube; it will remain on the side opposite the pellet. Be careful to maintain the angle of the centrifuge tube when removing the tube from the rotor to avoid disturbing the pellet.

7  Resuspend the cells in an equal volume of 1.0 mM Hepes (as in step 5). Centrifuge at 5.5 krpm and 4 °C for 10 min in a GS3 rotor. Decant the supernatant. Resuspend each pellet in 150 ml of 10% ultrapure glycerol. Do not combine the pellets.

8  Centrifuge the cells at 5.5 krpm and 4 °C for 15 min in a GS3 rotor. Decant the supernatant and remove the stirbar. Remove remaining traces of the supernatant with a pipet. Add 3.0 ml of 10% ultrapure glycerol to one tube and resuspend the pellet by pipetting. Transfer the suspension to the next tube and repeat until all the pellets are resuspended.

9  Flash freeze 350 μl aliquots in 1.5 ml microcentrifuge tubes. The protocol yields approximately 12 ml of cells at a concentration of $3 \times 10^{11}$ cfu/ml.

[a] This method is a large scale version of a previously described method (8). The protocol yields approximately 12 ml of highly concentrated, electrocompetent *E. coli* SS320 ($3 \times 10^{11}$ cfu/ml), a quantity sufficient for approximately 35 library constructions. Electrocompetent cells can stored indefinitely at −70 °C.

## Protocol 6

## Preparation of helper phage stock

### Equipment and reagents

- 2YT/tet media (see *Protocol 4*)
- 2YT/kan media (2YT, 10 μg/ml kanamycin)
- *E. coli* XL1-Blue (Stratagene)
- PBS (see *Protocol 1*)
- M13K07 helper phage starter stock (New England Biolabs)

### Method

1  Pick a single *E. coli* XL1-blue colony from a fresh 2YT/tet plate (less than 1 week old) into 2 ml of 2YT/tet media. Grow for 6–8 hours at 37 °C with shaking at 200 rpm.

2  Add M13K07 helper phage to a final concentration of $10^{10}$ phage/ml. Incubate for 30 min at 37 °C with shaking at 200 rpm.

3  Transfer the culture to 1 L of 2YT/kan media in a 4 L baffled flask (M13K07 carries a kanamycin resistance marker). Grow overnight at 37 °C with shaking at 200 rpm.

4  Precipitate and purify the phage as described in *Protocol 3*. Resuspend the phage in PBS to a final concentration of $10^{13}$ phage/ml ($OD_{268} = 1.0$ for a solution containing $5 \times 10^{12}$ phage/ml). The protocol yields at least 100 ml of M13K07 helper phage stock.

5  For long term storage, store at −70 °C. For immediate use, store at 4 °C; the stock is stable for at least 6 months at this temperature.

## References

1. Dower, W. J. and Cwirla, S. E. (1992). In *Guide to electroporation and electrofusion* (ed. D. C. Chang, B. M. Chassy, J. A. Saunder, and A. E. Sowers), p. 291, Academic Press, San Diego.

2. Lowman, H. B. and Wells, J. A. (1991). *Methods: A Companion to Methods in Enzymology*, Vol. 3, p. 205, Academic Press, London.

3. Zoller, M. J. and Smith, M. (1987). In *Methods in enzymology*, Vol. 154 (ed. R. Wu and L. Grossman), p. 329, Academic Press, London.

4. Kunkel, T. A., Roberts, J. D., and Zakour, R. A. (1987). In *Methods in enzymology*, Vol. 154 (ed. R. Wu and L. Grossman), p. 367, Academic Press, London.

5. Sidhu, S. S., Lowman, H. B., Cunningham, B. C., and Wells, J. A. (2000). In *Methods in enzymology*, Vol. 328 (ed. J. Thorner, S. D. Emr, and J. N. Abelson), p. 333, Academic Press, London.

6. Lowman, H. B. (1998). In *Methods in molecular biology* (ed. S. Cabilly), p. 249, Humana Press Inc., Totowa, NJ.

7. Lechner, R. L., Engler, M. J., and Richardson, C. C. (1983). *J. Biol. Chem.*, **258**, 11174.

8. Dower, W. J., Miller, J. F., and Ragsdale, C. W. (1988). *Nucl. Acids Res.*, **16**, 6127.

9. Casadaban, M. J. and Cohen, S. (1980). *J. Mol. Biol.*, **138**, 179.

10. Miller, J. H. (1972). *Experiments in molecular biology*, 1st edn., p. 190, Cold Spring Harbor Laboratory Press, NY.

11. Bullock, W. O., Fernandez, J. M., and Short, J. M. (1987). *Biotechniques*, **5**, 376.

# Chapter 3
# *In vitro* DNA recombination

## Kentaro Miyazaki and Frances H. Arnold

Division of Chemistry and Chemical Engineering,
Mail Code 210-41, California Institute of Technology,
1200 East California Boulevard, Pasadena, CA 91125, USA.

## 1 Introduction

This chapter illustrates *in vitro* DNA recombination methods frequently used in directed evolution. Four different methods are presented with full experimental detail: DNA shuffling (Stemmer method), random priming recombination, Staggered extension process (StEP), and *in vitro* heteroduplex formation *in vivo* DNA repair (heteroduplex recombination). One problem in recombination experiments is unwanted nonrecombinant parent sequences and mis-recombined sequences. Protocols to reduce these backgrounds are included.

## 2 Background to *in vitro* DNA recombination
### 2.1 Use of *in vitro* DNA recombination in directed evolution

Directed evolution mimics Darwinian evolution by using iterative mutation and selection (or screening). This simple algorithm enables us to engineer molecules with desired properties (1). Sequence diversity is usually created by point mutagenesis and recombination. Although sequential rounds of random point mutagenesis can be used, this evolutionary strategy suffers two weaknesses. First, only one *best* sequence can parent the next generation even when there are multiple *good* sequences. Discarded *good* mutations have to be rediscovered in future generations to be incorporated. Depending on the screening strategy, the cost to rediscover these mutations can be large compared to collecting them by recombination. Second, the fitness that can be reached by point mutagenesis is possibly limited because deleterious mutations are often accumulated along with beneficial ones. Recombination can overcome these drawbacks by allowing mutations to become associated in different combinations. Multiple sequences are recombined, and, as a result of screening, beneficial mutations accumulate while deleterious ones are eliminated.

DNA recombination *in vitro* as an evolutionary strategy was first reported by Stemmer (2, 3). Stemmer's method, which he called "DNA shuffling," combines homologous recombination of related sequences and a low (and controllable) rate of random point mutagenesis in one experimental operation. The term DNA shuffling has now become synonymous with *in vitro* recombination and its associated point mutation capability. Stemmer's group also demonstrated evolution by recombination of naturally occurring homologous genes ("family shuffling" or "molecular breeding") (4). Gene families, which diverged from a common ancestor and thus share the same structural framework, are considered to be pre-selected for similar folds and function(s). This property—diversity in sequence yet similarity in structure and function—is advantageously used in molecular breeding experiments to minimize the disruption of structure while expanding the sequence space explored for evolution of specific, desirable features. Since Stemmer's shuffling method was first published in 1994, several new methods have been described. These methods are presented and compared in Section 3.

## 2.2 Applications of *in vitro* DNA recombination

The first demonstration of DNA shuffling was to improve drug resistance (2, 3). DNA shuffling in its various forms has since been applied to altering a wide array of protein properties. Folding and solubility were improved for the green fluorescent protein (5) and for single-chain antibody fragments produced in *Escherichia coli* (6). Thermostable enzymes have been generated (7–10). Enzyme substrate specificity was changed (11), and enzyme activity in organic solvent was increased (12). DNA shuffling has also been applied to evolving whole operons (13).

Family shuffling was first employed to recombine members of the cephalosporinase gene family. The activity of the best mutant towards a new antibiotic was far above the level of any parent and also outperformed single-gene shuffling products (4). Similarly, the substrate range of a biphenyl dioxygenase was extended by shuffling a pair of related dioxygenase genes (14, 15). A library prepared by shuffling twenty-six subtilisin sequences contained sequences highly adapted for various properties tested, including activity in acid and alkaline media, stability in dimethylformamide and thermostability (16). For more examples of evolution by *in vitro* recombination, see (17).

## 2.3 Recombination statistics

When recombining multiple sequences to accumulate beneficial mutations, it is useful to consider the recombined library size (12). As the number of sequences and mutations increases, the probability of identifying the rarest recombination species decreases rapidly. Because the least frequent sequences are the ones that contain most mutations from the parent population, those sequences are often the most desired. The experimenter should thus consider the screenable library size when deciding the number of sequences to recombine and the functional improvements that can be obtained.

For the recombination of $N$ sequences with $M$ total mutations, the probability of generating progeny sequences containing $\mu$ mutations ($P_\mu$) is equal to the number of ways a $\mu$-mutation sequence can be generated ($C_\mu^M$) multiplied by the probability of generating any single $\mu$-mutation sequence

$$P_\mu = C_\mu^M \left(\frac{1}{N}\right)^\mu \left(\frac{N-1}{N}\right)^{M-\mu}$$

$$= \frac{M!}{(M-\mu)!\mu!} \left(\frac{1}{N}\right)^\mu \left(\frac{N-1}{N}\right)^{M-\mu}.$$

The probability of generating any specific sequence decreases precipitously with increasing numbers of parents. The rarest sequence will be the one containing all the mutations ($\mu = M$), its probability ($P_M$) is $1/N^M$. In practice, some degree of oversampling is required in order to maximize the chance of discovering a given variant. The sampling S required to achieve a given level of confidence of having sampled the rarest variant in a library is given by $(1 - P_M)^S < 1 -$ [confidence limit]. Generally, for 95% certainty that a specific clone has been sampled, the oversampling is between 2.6 and 3.0.

Given these numbers, it is highly unlikely, for example, to find the best recombination of ten mutations from ten separate parents each having a single mutation in a single round of shuffling and screening ($P_{10} = 0.0000000001$). However, it is possible to find a good triple mutant ($P_3 = 0.0574$).

## 3 Methods for *in vitro* DNA recombination

### 3.1 Stemmer method

Stemmer's published DNA shuffling method (2) uses enzymatic digestion of the parent genes to generate a pool of random DNA fragments. These fragments can be assembled by iterative cycles of denaturation, annealing and extension with thermostable DNA polymerase. This reaction generates a mixture of products in length and combination, from which full-length genes are amplified in a polymerase chain reaction (PCR) reaction with flanking primers. A low (and controllable) rate of random point mutation accompanies recombination. High-fidelity shuffling is advantageously used to recombine beneficial mutations and minimize the generation of potential deleterious mutations (18, 19).

Fragmentation of the DNA is usually performed using a nonspecific endonuclease such as DNase I (an alternative method for random DNA fragmentation is described in section 3.2). In the reaction, magnesium (2, 3) or manganese (18–20) ion is included. These ions affect the digestion differently: magnesium promotes random nicks, while manganese stimulates cleavage of both strands at approximate the same site (21). The products of reaction with magnesium-DNase I may therefore remain annealed in non-denaturation conditions, and the apparent fragment size will depend on the electrophoresis conditions (e.g. buffers and temperatures). To stop the DNase I digestion, it is highly recommended to add an excess of ethylene diamine tetraacetic acid (EDTA) over the divalent cation rather than use thermal inactivation. Divalent cations are known to enhance the thermostability of DNase I, and the reaction continues with increasing rate (22). EDTA inactivation is quicker, technically simpler and more reproducible.

## Protocol 1

# DNA shuffling [a]

### Equipment and reagents

- DNase I (Boehringer)
- 0.5 M EDTA, pH 8.0
- *Taq* DNA polymerase (Boehringer, Promega, or Stratagene)
- 10 × *Taq* buffer (Boehringer, Promega, or Stratagene)
- Deoxyribonucleotide triphosphate (dNTP) mix: 2 mM of each dNTP (Amersham Biosciences)

- QIAEX II gel extraction kit (Qiagen) or Wizard prep (Promega)
- DNA parent samples (2–5 μg/each)
- Forward and reverse PCR primers (50 pmol/each)

### A. Fragmentation

1  Prepare 10 μl (2–5 μg) of parent DNA samples in water, mixed in equal proportions.[b] Equilibrate at 15 °C.

2  Prepare DNase I solution: 50 mM Tris-HCl, pH 7.4, 10 mM $MnCl_2$, 0.2 U DNase I in 100 μl. Equilibrate at 15 °C.

3  Add 40 μl of DNase I solution (from step 2) to 10 μl of parent DNA. Perform the digestion at 15 °C.

4  Take 10 μl aliquots after 1, 2, 3, 5, and 10 min of incubation and immediately mix each of them with 1 μl of 0.5 M EDTA.[c]

5  Separate the DNA fragments by electrophoresis in a 2% (w/v) agarose gel[d] containing ethidium bromide (0.5 μg/μl). Visualize the DNA with a preparative UV illuminator (typically at 366 nm)[e] and excise the DNA in the desired size range.[f]

6  Purify the fragments (e.g. using QIAEXII or Wizard prep kit) and elute in 10 μl water.

### B. Assembly

1  Combine 5 μl of 10 × *Taq* buffer, 5 μl of dNTP mix, 10 μl of the purified fragments (from *Protocol 1A*, step 6), and 2.5 U *Taq* polymerase in a total volume of 50 μl.

2  Run the assembly reaction using the following thermocycler program: 3 min at 94 °C followed by 40 cycles of 30 s at 94 °C, 1 min at 55 °C, and 1 min + 5 s/cycle at 72 °C.[g]

3  Check the extent of reaction by analyzing a small proportion on an agarose gel.[h]

**Protocol 1** continued

### C. Amplification

**1** Combine 10 µl of 10× *Taq* buffer, 10 µl of dNTP mix, 0.5 µM (final) forward and reverse primers, 1 µl of assembly reaction, and 5 U *Taq* polymerase in a total volume of 100 µl.

**2** Run the PCR reaction at 94 °C for 30 s, 55 °C for 30 s, and 72 °C for 30 s. Perform a total of 30 cycles.

**3** Check the amplification by analyzing a small proportion on an agarose gel.[i]

[a] Modified from the method given in (18).

[b] Parent DNA may be plasmids carrying target sequences, sequences excised by restriction endonucleases or amplified by PCR.

[c] DNase I concentration and/or incubation time/temperature can be adjusted if necessary.

[d] For selection of gel concentration, see Section 3.8.3.

[e] Tracking dyes in standard loading buffer (10 × loading buffer: 50% (v/v) glycerol, 0.1% (w/v) bromophenol blue, 0.1% (w/v) xylene cyanol) can mask fluorescence from DNA fragments. For optimal gel transparency, dilute the dyes with glycerol until they are slightly visible. Alternatively, use dye-free loading buffer for the samples and load dyes in empty lanes to monitor the migration.

[f] Contamination with full-length parent sequence sometimes happens, especially when a PCR product is used. Even after digestion for extended periods of time, a trace of full-length parent may remain. Using the unfractionated mixture for the assembly reaction would generate an unacceptably large fraction of parent molecules. Visual control by gel electrophoresis is recommended for reliable separation of fragments in the desired size range.

[g] The PCR program depends on the fragment size. Assembly from small fragments may require more cycles and longer extension time than from large fragments. Small fragments may also require lower annealing temperature, at least during the initial cycles.

[h] You should see a smear extending through the size of your full-length product. Use more cycles, or lower the annealing temperature, if a smear does not appear.

[i] If the amplification is dominated by a smear rather than full-length sequence, repeat the amplification with a lower concentration of assembly products. Another solution to the smear problem is the use of nested primers; see Section 3.7.3.

## 3.2 Random DNA fragmentation with endonuclease V from *E. coli*

DNase I has most frequently been used for random DNA fragmentation. However, the length of fragments varies greatly with minor changes in conditions, including the amount of nuclease, the source (supplier) or lot of nuclease, the reaction temperature, and the purity of DNA substrates. The experiment requires very careful control of the partial digestion reaction and, hence, is labor-intensive and time-consuming. To overcome these drawbacks, an alternative approach was developed, in which random fragmentation is achieved in complete digestion (23). In this method, uracil-containing recombination templates are prepared by PCR in the presence of dUTP and dNTP. The full-length DNA is then digested with endonuclease V from *E. coli*, which is known to cleave uracil-containing DNA at the second or third phosphodiester bond 3' to uracil sites. By adjusting the concentrations of dUTP and dTTP, one can obtain the desired fragment length.

---

## Protocol 2
# Random DNA fragmentation

### Equipment and reagents

- *Taq* DNA polymerase (Boehringer, Promega, or Stratagene)
- 10 × *Taq* buffer (Boehringer, Promega, or Stratagene)
- 2 mM each of dATP, dGTP, dCTP (Amersham Biosciences)
- dTTP/dUTP mix: 2 mM/0 mM, 1.5 mM/ 0.5 mM, 1 mM/1 mM, 0.5 mM/1.5 mM, 0 mM/2 mM (Amersham Biosciences)

- Forward and reverse primers (25 pmol each)
- Template DNA samples (1–50 ng each)[a]
- QIAquick spin column (Qiagen) or QIAEX II gel extraction kit (Qiagen)
- Endonuclease V from *E. coli* (Trevigen)
- 10 × endonuclease V buffer (Trevigen)

### Method

### A. Preparation of uracil-containing recombination templates

1. Combine 5μl of 10 × *Taq* buffer, 0.2 mM each of dATP, dGTP and dCTP, and 0.2 mM of a dTTP/dUTP mixture, 1–50 ng of template DNA, 25 pmol of each forward and reverse primers, and 1.25 U *Taq* DNA polymerase in a total of 50 μl.

2. Heat the PCR mixture at 95°C for 1 min, followed by 25-30 cycles of incubation at 94°C for 30 s, 55°C for 30 s, and 72°C for 30 s, and a final incubation at 72°C for 7 min.

3. Separate the DNA fragments by electrophoresis in a 1% (w/v) agarose gel containing ethidium bromide (0.5 μg/ml). Visualize the DNA with a preparative UV illuminator and excise the DNA in the desired size range.

4. Purify the fragments (e.g. using a QIAspin column or QIAEX II kit) and elute in 27 μl water.

### B. Endonuclease V digestion of the recombination templates

1. Add 3 μl of 10 x endonuclease buffer and 1 U endonuclease V.

2. Incubate the solution at 37°C for 12 h, followed by heating at 95°C for 10 min.

3. Separate the DNA fragments by electrophoresis in a 2% (w/v) agarose gel containing ethidium bromide (0.5 μg/ml). Visualize the DNA with a preparative UV illuminator and excise the DNA in the desired size range.[b]

4.  Purify the fragments (e.g. using a QIAspin column or QIAEX II kit) and elute in 30 μl water. Use 3–10 μl of the purified fragments to continue with the assembly reaction (section B of *Protocol 1*).

[a] Plasmid DNA is preferred.

[b] You should see not only smear DNA fragments but also full-length recombination templates. This is because some recombination templates lack uracil in their sequence.

### 3.3 Random priming recombination

In this method, conventional random priming DNA synthesis (24) is used to generate fragments for assembly (25). DNA synthesis is carried out using Klenow fragment of *E. coli* DNA polymerase I which lacks 3'->5' exonuclease activity (26). The random priming reaction can be carried out using a "wild type" Klenow fragment under slightly acidic conditions (pH 6.6) where problematic 3'->5' exonuclease activity is reduced (25, 27). However, we recommend using a "variant" Klenow fragment, whose 3' ->5' exonuclease activity is abolished by genetic engineering (D355A, E357A mutations). Random hexanucleotides are used for priming; they are long enough to form stable duplexes with template and short enough to ensure random annealing. While longer primers can also be used, annealing may not be random for short genes. Unlike DNase I fragmentation, this method does not require double-stranded DNA and can be used on both double-stranded and single-stranded DNA templates. Moreover, random primers can also be used with RNA templates.

## Protocol 3

# Random priming recombination[a]

### Equipment and reagents

- Klenow fragment (3'->5' exo⁻) (New England Biolabs)
- Klenow fragment buffer (3'->5' exo⁻) (New England Biolabs)
- Template DNA samples (1–50 ng each)[b]
- Random hexanucleotides (Amersham Biosciences)
- 5 mM of each dNTP (Amersham Biosciences)
- QIAquick spin column (Qiagen) or QIAEX II gel extraction kit (Qiagen)

### Method

1.  Combine 0.2–0.5 pmol of each template DNA and 7 nmol of random hexanucleotide primers in a total volume of 78 μl.
2.  Incubate 5 min at 100°C and immediately transfer to an ice water bath.

3. Add 10 μl of 10 x exo⁻ Klenow fragment buffer, 10 μl of dNTP mix, and 10 U Klenow fragment to bring to 100 μl.

4. Incubate 3–6 h at 22°C.

5. Stop the reaction by placing the tube on ice.

6. Separate the DNA fragments by electrophoresis in 2% (w/v) agarose gel containing ethidium bromide (0.5 μg/ml). Visualize the DNA with a preparative UV illuminator and excise the DNA of the desired size.[c]

7. Purify the fragments (e.g. using a QIAspin column or QIAEX II kit) and elute in 30 μl water. Use 3–10 μl of the products to continue with the assembly reaction (section B of *Protocol 1*).

[a] Modified from (25).

[b] Templates can be plasmids carrying target sequences, sequences excised by restriction endonucleases, or amplified by PCR. However, because the products are separated from template by size, it is *not* recommended to use very short DNA templates.

[c] Tracking dyes in standard loading buffer (10 x gel loading buffer: 50% (v/v) glycerol, 0.1% (w/v) bromophenol blue, 0.1% (w/v) xylene cyanol) can mask fluorescence from DNA fragments. For optimal gel transparency, the dyes can be diluted with glycerol until they are slightly visible. Alternatively, use dye-free loading buffer for the samples and load dyes in empty lanes to monitor the migration.

## 3.4 Staggered extension process (StEP)

StEP recombination is based on template switching during polymerase-catalyzed primer extension (28). The denaturation and short annealing–extension cycles limit the primer extension in a single cycle. Extension interrupted by heat denaturation resumes during the next annealing–extension step, where the partially extended primers can anneal to different parent sequences. Multiple cycles of partial extension then create a library of chimeric sequences. As it involves no template digestion or fragment assembly, this protocol is very simple.

The faster the full-length product appears in the extension reaction, the fewer the template switches that have occurred and the lower the crossover frequency. Everything possible should be done to minimize time spent in each cycle: selecting a faster thermocycler, using smaller test tubes with thinner walls, and, if necessary, reducing the reaction volume. DNA polymerases currently used in DNA amplification are very fast. Even very brief cycles of denaturation and annealing provide time for these enzymes to extend primers for hundreds of nucleotides: 70°C, >60 nt/s; 55°C, ∼ 24 nt/s; 37°C, ∼ 1.5 nt/s; 22°C, ∼ 0.25 nt/s (29). Therefore, it is not unusual for the full-length product to appear after only 10–15 cycles. Polymerases are not all equally fast, however. The proofreading activity of *Pfu* and Vent polymerases slows them down, offering another way to increase recombination frequency. Processivity of the polymerase may also affect the crossover frequency. Because of the short extension time, the selection of thermal cycler and cycling program can significantly affect the crossover efficiency. It is thus highly recommended to maintain a fixed time between the set temperatures.

## Protocol 4

## StEP DNA recombination[a]

### Reagents

- *Taq* DNA polymerase (Boehringer, Promega, or Stratagene)
- 10 × *Taq* buffer (Boehringer, Promega, or Stratagene)
- dNTP mix: 2 mM each dNTP (Amersham Biosciences)
- forward and reverse PCR primers (7.5 pmol/each)
- template DNA samples (1–20 ng/each)[b]

### Method

1  Combine 5 μl of 10 × *Taq* buffer, 5 μl of dNTP mix, 1–20 ng of template DNA, 0.15 μM (final) forward and reverse primers, and 2.5 U *Taq* polymerase in a total volume of 50 μl.

2  Run 80–100 polymerase extension cycles: 94 °C for 30 s and 55 °C for 5–15 s.

3  Check a small amount of the reaction on an agarose gel.[c]

4  (*optional*) To remove parent DNA prepared from a *dam*[+] *E. coli* strain (methylation positive strain, e.g. DH5α, XL1-Blue) or *ung*[−] strain (e.g. CJ236), see Section 3.9.

5  (*optional*) Amplify the target sequence in a standard PCR reaction.

6  Check a small amount of the reaction on an agarose gel.[d]

[a] Modified from the method given in (28).

[b] Templates may be plasmids carrying target sequences, sequences excised by restriction endonuclease or amplified by PCR. When the size of parent DNA is close to those of products, enzymatic degradation of the parent is necessary. See Section 3.6 for details.

[c] Possible reaction products are full-length amplified sequence, a smear, or a combination of both. If a discrete band is obtained after the StEP reaction, the products may be cloned into a vector without amplification.

[d] If the amplification reaction is not successful and a smear rather than a discrete band occurs, repeat the amplification with a lower concentration of the StEP reaction mixture. Run the reaction with several dilutions of the StEP reaction: $\frac{1}{10}$, $\frac{1}{20}$, and $\frac{1}{50}$, and select the most successful one for cloning. Another solution to the smear problem is to use nested primers (see Section 3.7.3).

## 3.5 *In vitro* heteroduplex formation and *in vivo* repair (heteroduplex recombination)

This method comprises two simple processes: *in vitro* heteroduplex formation and subsequent mismatch repair *in vivo* (30). In contrast to other methods, recombination occurs in *non*-homologous (mismatched) sequences rather than in regions of homology. Either whole plasmids or target fragments can be used. For creating the plasmid heteroduplex, the plasmids have to be linearized with restriction enzymes (one plasmid with restriction enzyme A and the other with B), whose restriction sites are located outside the target region. This enables one to differentiate a heteroduplex from the homoduplex in a bacterial cell. The linear, homoduplex form is immediately digested by the host and will not contribute to transformation. On the other hand, the circular, heteroduplex form is modified by the mismatch repair system. Sealing nicks by DNA ligase treatment increases the efficiency of recombination several-fold (*Table 1*, (30)).

If short target sequences are used for parents, the discrimination system *in vivo* cannot be used. Further, if double-stranded fragments are used, the probability of forming heteroduplex is at most 50%, and it is hard to isolate it from homoduplex by any physical method (e.g. agarose gel, membrane). To maximize the formation of heteroduplex, asymmetric PCR synthesis of one of the two strands of the target genes is advantageous. The resulting single-stranded sequences are used for annealing. The heteroduplex products are cloned into a vector, and the plasmids are used to transform bacteria.

## Protocol 5

## Heteroduplex recombination: Plasmid heteroduplex[a]

### Reagents

- 20 × SSPE: 0.2 M $NaH_2PO_4$, pH 7.4, 3.6 M NaCl, 20 mM EDTA
- DNA ligase (Stratagene)
- Two plasmids for recombination (2–4 μg/ each)

### Method

1  Combine equal amounts (~2–4 μg) of two plasmids, 5 μl of 20 × SSPE to bring to 100 μl.

2  Incubate the mixture at 100 °C for 10 min, immediately transfer to an ice water bath, and continue incubation at 68 °C for 2 h.

3  Repair the nicks with DNA ligase.

4  Transform bacteria with the products.

[a] Method taken from (30).

## Protocol 6

## Heteroduplex recombination: insert heteroduplex[a]

### Equipment and reagents

- 20 × SSPE: 0.2 M NaH$_2$PO$_4$, pH 7.4, 3.6 M NaCl, 20 mM EDTA
- DNA ligase (Stratagene)
- QIAspin PCR purification kit (Qiagen)
- DNA template samples (5 ng/each)
- Forward and reverse PCR primers (15–25 pmol/each)

### Method

1  Combine 5 µl of 10 × *Taq* buffer, 5 µl of the dNTP mix, 0.3–0.5 µM of one primer, 5 ng of one DNA template and 5 U *Taq* polymerase in a total volume of 50 µl. One DNA template is combined with the forward primer, another with the reverse primer.

2  Synthesize the ssDNA templates in a PCR-like reaction for 100 cycles: 94 °C for 30 s; 56 °C for 30 s; and 72 °C for 1 min.

3  Combine the annealing products together and purify them using a QIAspin PCR purification kit or an equivalent kit. Elute the DNA in water.

4  Combine 47.5 µl of purified products and 2.5 µl of 20 × SSPE.

5  Incubate the mixture at 100 °C for 10 min, immediately transfer to an ice water bath, and continue the incubation at 68 °C for 2 h.

6  Purify the annealed products, digest them with the appropriate restriction enzymes and ligate them into a cloning vector.

7  Transform bacteria with the products.

[a] Method taken from (30).

## 3.6  Choice of recombination method

The recombination methods (DNA shuffling, random priming recombination, StEP, and heteroduplex recombination) each have their own advantages and disadvantages. Experimenters choose a method taking several factors into account, including type of nucleic acid (DNA or RNA), number of parents, length of genes, operational simplicity, secondary structure, and sequence similarity. Relative performance will probably differ from sequence to sequence and optimal conditions have to be determined experimentally.

### 3.6.1 Type of nucleic acids

Double-stranded DNA is usable in all DNA recombination methods. DNA shuffling requires dsDNA, which can be prepared either *in vivo* (e.g. from plasmid) or *in vitro* (e.g. by PCR). Random priming recombination and StEP can use ssDNA and RNA. Heteroduplex formation becomes highly efficient if ssDNA is used (30).

### 3.6.2 Number of parents

None of polymerase-based recombination methods (DNA shuffling, random priming recombination, StEP) are restricted as to the number of parents. By contrast, heteroduplex recombination is limited to recombining two sequences per reaction.

### 3.6.3 Gene length

In general, in polymerase-based reactions such as PCR, shorter sequences serve as a better template for the reproducibility of the reaction. However, in recombination experiments, contamination of full-length parent(s) in the shorter fragments has to be avoided. If one would like to recombine short sequences, the full-length contaminant has to be removed other than by a size separation. See Section 3.6 for details.

The length of the parent DNA can affect recombination efficiency. The StEP method requires repetitive short extensions. Recombination frequency depends on extension length per cycle. In random priming DNA synthesis, it is independent of the length of the DNA template. DNA fragments as small as 200 bases can be primed equally well as large molecules such as λ phage DNA (24). The average size of synthesized DNA is an inverse function of primer concentration to template (31). Based on this guideline, proper conditions can be set for a given gene. However, the optimal conditions may vary from gene to gene.

Although PCR (and related polymerase reactions) in theory can be used to generate very long sequences, in practice the efficiency of amplification decreases significantly for very long sequences. Another drawback of PCR amplification is the introduction of unwanted mutations, which is particularly problematic for long sequences. By contrast, heteroduplex recombination neither suffers from the limitations of PCR-based approaches nor requires transformation with multiple gene fragments. In heteroduplex recombination, a whole plasmid is typically used for operational simplicity. However, because of the kinetics of annealing, long sequences adversely affect the recombination efficiency (30). In this case, short insert fragments can be used. However, because unwanted homoduplex and desired heteroduplex are non-separable, a pair of single strand sequences are used.

### 3.6.4 Operational simplicity

Stemmer's DNA shuffling method is the most widely used, but it requires multiple steps: fragmentation, assembly, and PCR. Each reaction step has to be carefully monitored, and the method is accordingly labor-intensive. This is also the case for the random priming method. Random digestion of uracil-containing DNA with *E. coli* endonuclease V (23) is less laborious because fragments can be prepared in complete digestion. StEP requires fewer steps, and once optimal conditions are found for a particular gene, one can use identical conditions without major modifications. Heteroduplex recombination is quite simple and quick. However, for maximal recombination efficiency, monitoring the formation of heteroduplex may be necessary.

## 3.6.5 Secondary structure

Secondary structure formation in ssDNA may adversely affect the performance of all methods, especially at low temperatures. In DNA shuffling, the randomness of gene fragmentation with DNase I may be reduced. Random priming recombination can also be affected at the most crucial step of the reaction: annealing and extension of the random primers. Small primers will be more sensitive to secondary structure than long primers at this annealing step. The low temperature used in the extension of the random primers also stabilizes secondary structure. Extension at elevated temperature requires longer primers, which would probably lead to less efficient recombination, at least for short genes. Termination of elongation of both short and long primers is sensitive to secondary structure. Formation of stable stem-and-loop structures may be the most important source of nonrandom distribution of extended fragments in random priming recombination. StEP recombination, with its very short annealing period, may be sensitive to secondary structure. Randomness created by *in vitro* heteroduplex formation can also be limited especially at low temperature.

## 3.6.6 Sequence similarity

Although all the techniques are in theory applicable for recombining genes with many mutations (sequence identity above ~50%) genes, high sequence identity seems to be critical for recombination success. Family shuffling has only been reported for the Stemmer method. The original Stemmer protocol has been devised to increase the recombination efficiency and to avoid contamination of parent sequences; the modified protocol uses various frequent-cutter restriction enzymes for fragmentation instead of DNase I (23).

## 3.6.7 Comparison of recombination methods

DNA recombination methods so far reported were compared using the green fluorescent protein system developed by Volkov *et al.* (30). The green fluorescent protein variants containing a stop codon produce nonfluorescent proteins. Generation of fluorescence requires recombination between the sites to restore the wild type sequence. The percentage of fluorescent host *E. coli* colonies obtained by recombining two GFP templates *via* various recombination methods are thereby compared (30, 32). The results are summarized in *Table 1*. In general, two mutations separated by short nucleotides are less efficiently recombined than are distant mutations. The insert heteroduplex recombination, where ssDNA is used for parent, gives the highest recombination efficiency for mutations separated by >99 base pairs (bp). The Stemmer method and StEP are equally efficient. With DNase I fragmentation, using smaller fragments (<100 bp) yield a slightly higher efficiency than larger fragments (100–200 bp). DNA fragments prepared by endonuclease V digestion of uracil-containing DNA gave nearly the same recombination efficiency (10% when two stop codons were separated by 180 bp; data not included in *Table 1*).

**Table 1** Comparison of recombination methods[a]

| Recombination methods | Distance between mutations (bp) | | | | |
|---|---|---|---|---|---|
| | 423 | 315 | 207 | 99 | 24 |
| DNA shuffling (<100 bp fragments)[b] | 20.5 | 14.5 | 11.5 | 9.6 | 5.8 |
| DNA shuffling (100–200 bp fragments)[b] | 19.2 | 9.7 | 8.3 | 8.4 | 5.1 |
| StEP[b] | 18.5 | 13.1 | 9.8 | 8.2 | 4.8 |
| Random priming recombination[b] | 5.0 | 4.8 | 3.1 | 1.2 | 0.9 |
| Heteroduplex recombination (plasmid, nick)[c] | 3 | 3 | 1.3 | 1.1 | 0 |
| Heteroduplex recombination (plasmid, ligase)[c] | 10 | 9 | 7 | 8 | <1 |
| Heteroduplex recombination (insert, dsDNA)[c] | 15–18 | 12–14 | 8–10 | 10–11 | 0.7–1.0 |
| Heteroduplex recombination (insert, ssDNA)[c] | 29 | 25 | 18 | 16 | 1.2 |

[a] Percentage of fluorescent *E. coli* colonies obtained by recombining two GFP templates containing stop codon mutations.

[b] Data taken from (30).

[c] Data taken from (32).

## 3.7 Removal of background

In Stemmer DNA shuffling, undigested parent fragment after DNase I digestion (occurs most frequently when PCR-amplified fragments are used) is the major contaminant. To avoid this, visual monitoring the digestion reaction is effective; gel separation is employed if necessary.

Many polymerase reactions become more reproducible when short DNA is used as a template. Random priming recombination and StEP method are the particular case in recombination experiments. However, if templates are too short and indistinguishable from the products in size, the nonrecombinant full-length genes will be a problem. Because conventional physical separation, such as agarose gel or membrane separation, cannot be employed, chemical and/or enzymatic treatments are applied to reduce the background. *Dpn*I and uracil DNA glycosylase break the phosphodiester backbone of parent DNA in such a way that digested parents will not be amplified in the final PCR.

Another problematic contaminant leading to high concentration of inactive clones are the products of mis-recombination. Reaction conditions that promote recombination usually also generate unwanted recombination. Mis-recombination is common to all polymerase-based recombination methods such as Stemmer's method, random priming recombination, and StEP. Background of this kind can be eliminated by nested PCR.

### 3.7.1 *Dpn*I digestion

The restriction enzyme *Dpn*I cleaves only when its recognition site (GATC) is fully-methylated ($N^6$ methylation of adenine) for both strands, while leaving hemi-methylated DNA intact (33). It thereby specifically removes parent DNA isolated from a *dam*$^+$ *E. coli* strain such as DH5α or XL-1 Blue. This decontamination method can be used in StEP by treating the full-length product.

### 3.7.2 Uracil DNA glycosylase

Another way to remove parent sequences is to incorporate dUTP in parent DNA but not in product DNA. This can be achieved by PCR in the presence of dUTP instead of dTTP. *Taq* polymerase recognizes dUTP equally efficiently and gives an equivalent yield of products. Alternatively, one can isolate plasmid from an *ung*$^-$ *E. coli* strain (e.g. CJ236) and excise the insert fragment by restriction digestion. Uracil N-glycosylase recognizes uracil residues in DNA and removes the base to generate abasic sites (34). Incubating the damaged DNA in alkaline solution at high temperature breaks phosphodiester bonds at the 3'-end of the abasic sites while keeping intact normal DNA (dUTP-free), thereby eliminating parent background. It is important to note that, if the alkaline treatment is insufficient, the parent serves as a highly mutagenic template in subsequent amplification. The resulting library will be full of inactive clones.

---

## Protocol 7

## Decontamination of uracil-containing DNA with uracil DNA glycosylase

### Equipment and reagents

- *Taq* DNA polymerase (Stratagene)
- 10 × *Taq* DNA polymerase (Stratagene)
- dATP, dGTP, dCTP, dUTP (Amersham Biosciences)
- Uracil DNA glycosylase (MBI Fermentas)
- QIAspin column (Qiagen)
- Template DNA (0.2–0.5 pmol)
- Forward and reverse PCR primers (50 pmol/each)

### A. Preparation of parent DNA containing dUTP

1  Combine 10 µl of 10 × *Taq* buffer, 0.2 mM each of dATP, dGTP, dCTP, 0.6 mM dUTP,[a] template DNA, 0.5 µM each primers, and 5 U *Taq* polymerase in a total volume of 100 µl.

2  Amplify the target sequence under normal PCR conditions.

3  Purify the fragment and dissolve it in water.

## B. StEP recombination

1 Run the StEP reaction as in *Protocol 3* using the uracil-containing DNA parents as templates.

## C. Decontamination of parent DNA with uracil DNA glycosylase

1 Add 5 U of uracil DNA glycosylase to the StEP reaction mixtures. Incubate the reaction mixtures for 2 h at 37 °C.

2 Add 50 mM NaOH. Incubate the mixture at 95 °C for 30 min.

3 Recover the mixtures using QIAspin column. Use these mixtures for the subsequent process.

[a] The dUTP concentration is threefold higher than the other dNTP concentrations to ensure optimal PCR efficiency.

### 3.7.3 Nested PCR

The reaction conditions which promote DNA recombination also promotes mis-annealing. Those unwanted products derived from *incorrect* recombination have to be eliminated. Nested PCR increases the yield and specificity of a PCR product. An aliquot of the first-round recombination product, for example 1 µl of a 1/10–1/1000 dilution, is subjected to a PCR amplification. The PCR is performed with two new primers which hybridize to sequences internal to the first-round sequences.

### 3.8 Technical tips

Optimizing DNA recombination conditions is labor-intensive and time-consuming. The reaction is frequently controlled kinetically and thus extremely sensitive to temperature, concentration of reagents, and any other seemingly *minor* conditions. For maximal reproducibility, it is strongly recommended to use the same thermal cycler and thin-wall PCR tubes once good conditions have been identified.

### 3.8.1 Oil overlay and reaction volume

If a thermal cycler does not heat the lids of the reaction tubes, you need to overlay the reaction mixture with light mineral oil to prevent evaporation. In thermal cyclers where the lid of the reaction tubes is heated continuously >96 °C, a PCR can be run without oil overlay. Insufficient oil overlay may result in evaporation, which would lead to higher reagent concentration and to a decrease in temperature. For most applications, 70 µl oil for 100 µl reactions, 40 µl oil for 50 µl reactions, and 30 µl oil for 25 µl reactions is probably sufficient.

Too much oil can interfere with heat transfer. Always use the appropriate amount of oil.

### 3.8.2 Quality control of reagents

DNA polymerases are physically unstable (e.g. to foaming and high temperature) and always need to be kept at $-20\,^{\circ}\text{C}$ for storage and on ice when in use. Deoxynucleotide triphosphates are also thermolabile, and the stock solution must be stored in aliquots at $-20\,^{\circ}\text{C}$. Avoid repeated freezing and thawing. Manganese chloride can form precipitates after long storage. The stock solution must be stored in aliquots at $-20\,^{\circ}\text{C}$. Avoid repeated freezing and thawing.

### 3.8.3 Agarose gel electrophoresis

Agarose gel electrophoresis is frequently used for the analysis and size fractionation of DNA fragments. Efficient range of separation of linear DNA molecules is 0.1–3 kbp in 2% (w/v), 0.2–4 kbp in 1.5% (w/v), 0.4–6 kbp in 1.2% (w/v), and 0.5–7 kbp in 0.9% (w/v) agarose. Commonly used concentration (0.8–1.2%) may not work well for separation of short DNA fragments (50–300 bp) which are typically used in DNA shuffling.

Short DNA fragments are often hard to see on an agarose gel. This is especially problematic in preparative experiments where long UV transillumination (e.g. 366 nm) is used. Moreover, tracking dyes in loading buffer ($10 \times$ loading buffer: 50% (v/v) glycerol, 0.1% bromophenol blue (w/v), 0.1% (w/v) xylene cyanol) can mask the fluorescence from DNA (migration of bromophenol blue is ~300 bp and that of xylene cyanol is ~4000 bp in 0.5–1.4% agarose). For optimal transparency, the loading buffer can be diluted in glycerol until they are slightly visible. Alternatively, use dye-free loading buffer for the samples and load dyes in empty lanes to monitor the migration.

## References

1. Arnold, F. H. (1998). *Acc. Chem. Res.*, **31**, 125.
2. Stemmer, W. P. (1994). *Nature*, **370**, 389.
3. Stemmer, W. P. (1994). *Proc. Natl. Acad. Sci. USA*, **91**, 10747.
4. Crameri, A., Raillard, S. A., Bermudez, E., and Stemmer, W. P. (1998). *Nature*, **391**, 288.
5. Crameri, A., Whitehorn, E. A., Tate, E., and Stemmer, W. P. (1996). *Nature Biotechnol.*, **14**, 315.
6. Crameri, A., Cwirla, S., and Stemmer, W. P. (1996). *Nature Med.*, **2**, 100.
7. Giver, L., Gershenson, A., Freskgard, P.-O., and Arnold, F. H. (1998). *Proc. Natl. Acad. Sci. USA*, **95**, 12809.
8. Hoseki, J., Yano, T., Koyama, Y., Kuramitsu, S., and Kagamiyama, H. (1999). *J. Biochem.*, **126**, 951.
9. Zhao, H. and Arnold, F. H. (1999). *Protein Eng.*, **12**, 47.
10. Miyazaki, K., Wintrode, P. L., Grayling, R. A., Rubingh, D. N., and Arnold, F. H. (2000). *J. Mol. Biol.*, **297**, 1015.
11. Zhang, J. H., Dawes, G., and Stemmer, W. P. (1997). *Proc. Natl. Acad. Sci., USA*, **94**, 4504.
12. Moore, J. C., Jin, H. M., Kuchner, O., and Arnold, F. H. (1997). *J. Mol. Biol.*, **272**, 336.

13. Crameri, A., Dawes, G., Rodriguez, E. Jr., Silver, S., and Stemmer, W. P. (1997). *Nature Biotechnol.*, **15**, 436.
14. Kumamaru, T., Suenaga, H., Mitsuoka, M., Watanabe, T., and Furukawa, K. (1998). *Nature Biotechnol.*, **16**, 663.
15. Bruhlmann, F. and Chen, W. (1999). *Biotechnol. Bioeng.*, **635**, 44.
16. Ness, J. E., Welch, M., Giver, L., Bueno, M., Cherry, J. R., Borchert, T. V. *et al.* (1999). *Nature Biotechnol.*, **17**, 893.
17. Petrounia, I. P. and Arnold, F. H. (2000). *Curr. Opin. Biotech.*, **11**, 325.
18. Zhao, H. and Arnold, F. H. (1997). *Proc. Natl. Acad. Sci. USA*, **94**, 7997.
19. Zhao, H. and Arnold, F. H. (1997). *Nucl. Acids Res.*, **25**, 1307.
20. Lorimer, I. A. and Pastan, I. (1995). *Nucl. Acids Res.*, **23**, 3067.
21. Melgar, E. and Goldthwait, D. A. (1968). *J. Biol. Chem.*, **243**, 4409.
22. Bickler, S. W., Heinrich, M. C., and Bagby, G. C. (1992). *Biotechniques*, **13**, 64.
23. Miyazaki, K. (2002). *Nucl. Acids Res.*, **30**, e139.
24. Feinberg, A. P. and Vogelstein, B. (1983). *Anal. Biochem.*, **132**, 6.
25. Shao, Z., Zhao, H., Giver, L., and Arnold, F. H. (1998). *Nucl. Acids Res.*, **26**, 681.
26. Klenow, H. and Henningsen, I. (1970). *Proc. Natl. Acad. Sci. USA*, **65**, 168.
27. Lehman, I. R. and Richardson, C. C. (1964). *J. Biol. Chem.*, **239**, 233.
28. Zhao, H., Giver, L., Shao, Z., Affholter, J. A., and Arnold, F. H. (1998). *Nature Biotechnol.*, **14**, 258.
29. Innis, M. A., Myambo, K. B., Gelfand, D. H., and Brow, M. A. (1988). *Proc. Natl. Acad. Sci. USA*, **85**, 9436.
30. Volkov, A. A., Shao, Z., and Arnold, F. H. (1999). *Nucl. Acids Res.*, **27**, e18.
31. Hodgson, C. P. and Fisk, R. Z. (1987). *Nucl. Acids Res.*, **15**, 6296.
32. Volkov, A. A. and Arnold, F. H. (2000). In *Methods in enzymology*, Vol. 328 (ed. J. Thorner, S. D. Emr, and J. Abelson), pp. 447–56, Academic Press, London.
33. Nelson, M. and McClelland, M. (1992). In *Methods in enzymology*, Vol. 216 (ed. Wu, R.), p. 279, Academic Press, London.
34. Tomilin, N. V. and Aprelikova, O. N. (1989). *Int. Rev. Cytol.*, **114**, 125.

# Chapter 4

# Phage selection strategies for improved affinity and specificity of proteins and peptides

Mark S. Dennis and Henry B. Lowman
Department of Protein Engineering, Genentech, Inc.,
1 DNA Way, South San Francisco, CA 94080.

## 1  Introduction

Filamentous phage display of peptide or protein variants has been widely used for rapid selection of protein variants that bind with improved affinity and improved specificity to a target molecule (for reviews, see 1, 2). The key feature of such selection schemes is that the genotype of a particular variant is linked through packaging inside a virion particle to the phenotype of a displayed protein or peptide that has been fused to one of the coat proteins, such as gene III protein (pIII), of the phage. Phage particles can be selected by binding to an affinity matrix, propagated in *Escherichia coli*, and identified by DNA sequencing (*Figure 1*).

The optimization of many peptide : protein, protein : protein, and protein : DNA interactions has been described. These have included, for example, hormone : receptor (3, 4), antibody : antigen (5, 6; see Chapter 14), inhibitor : protease (7–11), zinc-finger : DNA (12, 13), and hormone-mimic : receptor (14; see Chapter 6) interactions. The goals of these efforts are usually to improve the non-covalent binding affinity or equilibrium (eq) dissociation constant (i.e. $K_d = [\text{displayed protein}]_{eq}[\text{target}]_{eq}/[\text{complex}]_{eq}$) or the specificity (i.e. the ratio $K_d^{\text{target B}}/K_d^{\text{target A}}$, where binding to target A is preferred over binding to target B) of a molecular interaction. The magnitude of binding affinity can be a limiting factor in the potency and efficacy of therapeutic antibodies (15, 16). For example, an antibody : antigen affinity corresponding to $K_d \leq 1$ nM is often desirable. Additionally, specificity of molecular interactions can be critically important for hormone agonists (4), protease inhibitors (7–11), or DNA recognition (12, 13), because interactions with molecules closely related to the target can have unintended side effects.

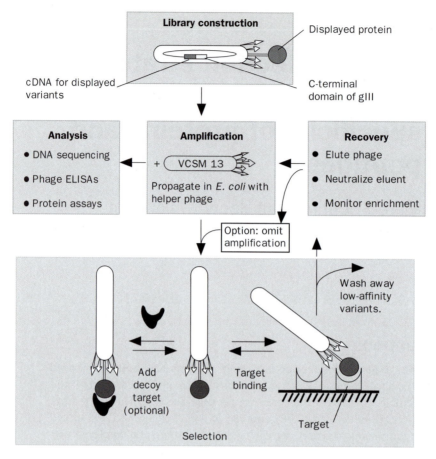

**Figure 1** Schematic diagram of steps in monovalent phage display for optimizing displayed-protein : target interaction. A phagemid library, propagated in *E. coli* with a helper phage, is packaged into particles (phage virions) that display a protein fused to a coat protein of bacteriophage M13. Particles presenting many variants are allowed to bind to an affinity matrix containing the target molecule, optionally in the presence of a competing "decoy" target to select for specificity. After washing away of nonbinding and low-affinity variants, the remaining particles are dissociated from the target affinity matrix, and used for subsequent cycles of binding enrichment (optionally without further amplification), propagation, sequencing, and/or biochemical characterization.

In this chapter, we discuss strategies and protocols for affinity and specificity selections using displayed proteins or peptides ("displayed protein") having some initial affinity for a receptor, enzyme, or other target molecule ("target"). We briefly discuss features of a phage-plasmid vector (phagemid), and assume that the reader will make use of protocols elsewhere (see Chapters 2 and 3) for the generation of large collections ("libraries") of peptide or protein variants displayed on M13 phage particles.

## 2 Vector considerations

Both phage and phagemid vectors have been described for the display of proteins and peptides fused to coat proteins, usually pVIII (the major coat protein) or pIII (a minor coat protein), of filamentous phage. The essential features of a phagemid display vector are a single-strand DNA (ss DNA) origin (phage *ori* region) and a double-strand DNA (ds DNA) origin of replication; a selectable marker (e.g. the *bla* gene, encoding ampicillin resistance (Amp$^R$)); a promoter/operator region with a ribosome binding site; a secretion signal sequence; and the complementary DNA (cDNA) for the protein to be displayed, fused through a linker peptide to a phage coat protein gene (or fragment) to be used for display. In this chapter, we will assume that a similar vector, having effectively "monovalent display" and conferring ampicillin resistance to the *E. coli* host, is used. However, many variations of the essential phagemid components are possible.

### 2.1 Monovalent and polyvalent phage display

Libraries of polyvalent phage, that is, those displaying more than one copy of the foreign protein per phage virion, have been constructed in several forms. Typically, the displayed protein is fused to all copies of pIII or pVIII within a phage vector, or fused to pVIII in a phagemid vector (see 1). A complication arises, however, in attempts to select the highest binding-affinity variants from such libraries (see e.g. 17,18). Multiple copies of displayed protein lead to multiple binding events, giving rise to a chelate or avidity effect (2, 19). This results in over-representation of lower affinity variants during the affinity selection process. A similar problem was observed in early attempts to sort human growth hormone (hGH) phage versus weaker-binding variants (20, 21).

A solution to avidity-limited selection was found by using monovalent phage display (20). In this system, a phagemid vector contains the displayed-protein cDNA fused to a portion of gIII. An appropriate helper phage supplies the remaining copies of wild-type pIII. In such a system, hGH is displayed at limited levels, with 10% or fewer of the phagemid particles displaying one copy of hGH (1, 20). This allowed the selection of hGH variants with as little as three-fold higher receptor-binding affinity than wild-type (3), ultimately leading to the construction of a 400-fold affinity-improved variant with high specificity (22). Monovalently displaying phagemid particles have been used in selecting high-affinity ($K_d < 100$ pM) antibody fragments (Fab's) as well as other peptides and proteins (reviewed in 1, 2, 16).

In some cases, high-affinity protein variants have been successfully recovered from phage libraries that were designed to display polyvalently (see e.g. 8). When the initial (wild-type) protein's binding affinity is weak (e.g. $K_d > 1$ μM), it may even be advantageous to begin with a polyvalently displaying construct, then construct monovalent libraries for final affinity optimization. Polyvalent display results in a far greater representation of rare library members simply because every particle displays at least one copy of protein, compared to < 10% of particles displaying protein in monovalent display. In addition, their recovery from binding selections is enhanced by the avidity effect (see below).

## 2.2 Confirming display

The first goal of vector construction will be to demonstrate that the intended protein is displayed on phage particles. Although one may proceed with the expectation that the displayed protein will bind to the intended target, this control is recommended. controls are usually needed. The fusion of some proteins via their C-termini to phage coat proteins can create structural or steric barriers to folding, secretion, or function.

To establish whether the protein is functionally displayed, phage binding to a panel of monoclonal or polyclonal antibodies directed to the target can be measured using a phage enzyme-linked immunosorbant assay (ELISA) (see below). Alternatively, if these reagents are unavailable, a specific epitope tag can be included for ELISA detection. For example, a $His_6$ peptide, fused to the N-terminus of the displayed protein, can be detected using metal chelating matrices or anti-His-tag antibodies (Qiagen).

If display can not be detected, modification or randomization of the linker region (see below) may be a solution. Proteins can sometimes be displayed via fusion to the C-terminus of displayed proteins or of phage-coat proteins (see, e.g. 23, 24).

## 2.3 Protein expression from phagemid vectors

A phagemid display vector can also be used directly for the expression of soluble (non-phage-linked) protein for biochemical analysis. This becomes especially useful in the later stages of optimization, when variants differing in expression levels, or by small increments in affinity or specificity, may need to be characterized.

Inclusion of an amber (TAG) stop codon between the displayed-protein gene and the phage gene to which it is fused has been used to provide a vector that doubles for both phage display and protein expression (3). In certain strains of *E. coli* which are amber-suppressors (e.g. XL1-Blue (Stratagene), 25), protein translation continues through the amber codon, although at low efficiency, with substitution of Gln in *supE* strains, and production of a fusion protein for phage display. The same construct, however, may be transformed into a "normal" (non-suppressor) strain of *E. coli* for expression and secretion of soluble protein into the periplasmic space.

The feature of partial suppression in amber-suppressor strains also allows for presentation of homodimers or homotrimers in phage display systems involving self-associating proteins (see e.g. 26).

## 2.4 Vector construction and phagemid preparation

A typical vector is the phagemid phGH–M13gIII (20). This construct was used to provide a fused hGH gene (containing no amber codon) with the

C-terminal domain of M13 gene III; phGHam-gIII adds an amber codon (3). For monovalent phage display, helper phage K07 (27) or VCSM13 (Stratagene) have been used to produce particles containing wild-type pIII (see *Figure 1*). Generally only 1–10% of particles contain a copy of fusion protein resulting in phage that are monovalent (Figure 1; 18). This vector contains the $P_{phoA}$ (alkaline phosphatase promoter), *st*II (heat stable enterotoxin) signal sequence, cDNA for human growth hormone, and the C-terminal fragment (beginning at residue 249 of the mature pIII sequence) of bacteriophage M13 gene III. A naturally occurring "linker region" within pIII at this site consists of Gly-Gly-Gly-Ser, or Glu-Gly-Gly-Gly-Ser repeats. The structural flexibility of this region is probably advantageous for phage display. The linker can also be modified by design or by creating a randomized-linker library and selecting for display of protein.

We routinely use the *E. coli* strain XL1-Blue (Stratagene; 22) as the host, although other F' ("male") strains such as SS320 (a non-suppressor strain; see Chapter 2) of *E. coli* can also be used. Competent cells for transformation or electroporation are commercially available (Stratagene). XL1-Blue has the convenient feature of being tetracycline resistant; the Tet[R] gene is on the F', which permits facile selection for infectable cultures of *E. coli*. After subculturing, however, we have occasionally observed the emergence of Tet[R] cells which are not infectable by M13 phage. Therefore, cultures used for phage display should always be grown from fresh single colonies streaked from reliable frozen stocks.

*Protocol 1* describes a typical preparation of phagemid particles using XL1-Blue cells and VCSM13 helper phage, which supplies the necessary packing and nonstructural genes for phage assembly.

## Protocol 1

## Phage preparation from XL1-Blue *E. coli*

### Equipment and reagents

- PBS (137 mM NaCl, 3 mM KCl, 8 mM $Na_2HPO_4$, 1.5 mM $KH_2PO_4$, pH 7.2; alternatively, 140 mM NaCl, 10 mM $Na_2HPO_4$, pH 7.2)
- PEG/NaCl buffer (200 g/L PEG-8000, 2.5 M NaCl)
- Phagemid construct (Amp[R])
- M13K07 helper phage (New England Biolabs) or VCSM13 helper phage (Stratagene; see Chapter 2, *Protocol 6*, for preparation)
- 2YT media (10 g bacto-yeast extract, 16 g bacto-tryptone, 5 g NaCl; add water to 1 L, pH 7.0 with NaOH; autoclave and cool)

- 2YT/carb/tet media (2YT, 50 µg/ml carbenicillin (a functional equivalent of ampicillin), 15 µg/ml tetracycline)
- *E. coli* XL1-Blue stock (Stratagene; Tet[R])
- LB/carb agar plates (5 g bacto-yeast extract, 10 g bacto-tryptone, 10 g NaCl; add water to 1 L, pH 7.0 with NaOH, 15 g agar; autoclave and cool to 55 °C; add carbenicillin (for Amp[R] phagemid constructs only) to 50 µg/ml and pour into Petri plates)

## Method

1 Transform the phagemid construct into *E. coli* XL1-Blue cells using electroporation (or other method) and plate onto LB/carb plates. Incubate overnight at 37 °C.

2 Pick a single colony and inoculate 5 ml of 2YT/carb/tet media containing $10^9$ plaque forming units (pfu)/ml of VCSM13 helper phage. This protocol can be scaled up to 25 ml or larger cultures for library propagation. Allow the culture to grow at 37 °C overnight with shaking or rotation.

3 Remove cells by spinning 10 min at 7500 rpm.

4 Precipitate phage by adding 20% volume of PEG/NaCl to the supernatant and vortexing vigorously. Incubate at room temperature for 1–10 min. The solution usually appears turbid.

5 Spin the precipitated phage at 7500 rpm for 10 min and discard the supernatant. Spin the tube again and remove the excess PEG/NaCl buffer, by inverting the tube for 5 min on a paper towel.

6 Suspend the phage pellet in 0.2 ml of PBS buffer (per 5 ml culture). Transfer the phage-containing supernatant to a clean 1.5 ml microfuge tube.

7 Remove debris by centrifuging at 13000 rpm for 5 min. Transfer the phage-containing supernatant to a clean 1.5 ml microfuge tube.

8 Repeat steps 4–7 to obtain a cleaner phage pellet (optional).

9 The concentration of phage can be estimated by measuring absorbance at 270 nm. The extinction coefficient of the phage depends on the size of the phage (or phagemid) genome (28). For a 5 kb genome, 1 $OD_{270}$ corresponds to approximately $1.1 \times 10^{13}$ phage/ml. The number of infectious phage particles can be estimated by colony-counting as described below (see *Protocol 5*), and whether the protein is displayed can be assessed by ELISA (see *Protocols 2 and 3*).

## 3 Library design

The size and solubility of phage particles as well as the efficiency of transformation and requirements for *E. coli* growth place limits on the size of phage libraries that can be produced and effectively sorted under typical laboratory conditions. Libraries of $>10^{10}$ can now be routinely produced (see Chapter 2). However, the number of libraries to be produced and the initial affinity of the displayed protein for the target should be considered in deciding how much diversity to introduce and by what method. Library diversity can be tailored according to random or structure-function based approaches.

Non-site-specific randomization techniques such as DNA shuffling (29; see Chapter 3) can be used to introduce diversity throughout the wild-type molecule. This approach offers the advantage of sampling substitutions and combinations of substitutions at many positions throughout a protein sequence. Such random substitutions may modulate binding affinity through direct or indirect effects on the binding interface.

Site-specific randomization is often used in cases where information about the target-binding site on the displayed protein is already available from structural or mutagenesis studies (7–10), or where there is general localization of binding determinants (e.g. in the complementarity determining region (CDR) loops of antibodies).

Two site-directed approaches are described here. Complete or "hard" randomization seeks to fully mutate each position while partial or "soft" randomization maintains a bias for the wild-type sequence. It may be useful, particularly in cases where no other structure–function information is available, to begin with soft randomization at many sites and proceed to hard randomization at focused sites.

## 3.1 Hard randomization

Hard randomization libraries can be obtained by synthesizing oligonucleotides containing NNS (N = bases A, G, C, or T; S = G or C) in order to fully randomize portions of a displayed protein while keeping other portions of the sequence constant. NNS degeneracy allows for substitution of all 20 natural amino acids, but other degenerate codons can also be used (see Chapter 2) based upon sequence or structure-function considerations. To construct the template for random mutagenesis, a stop codon (TAA) should be introduced at each position to be mutated in the starting phagemid. This serves to eliminate the display of background wild-type protein because translation of the fusion protein is terminated. The construct can also be used for assessing nonspecific phage binding to the target (see Section 5.6).

Binding affinity and the contribution of individual side chains to binding should be considered in choosing sites for mutation. For example, starting with a displayed protein of high affinity ($K_d \leq 10$ nM), excellent phage recovery can be expected (see below), and six residues may be fully randomized. NNS encodes 20 amino acids with 32 codons (versus 64 codons for NNN). Therefore, the theoretical diversity of the library is $32^6 = 10^9$, and a library size (number of transformants) of $10^{11}$ should adequately represent the library. If all randomized positions are extremely critical to binding, very few mutated variants can be expected to bind the target. In these cases, or with more modest starting affinities ($K_d = 10$ nM–1 μM), recovery will be reduced, and only 3–4 residues should be mutated per library (theoretical diversity of $3 \times 10^4$–$1 \times 10^6$). In this case, multiple libraries of smaller diversity can be used to ensure complete sequence coverage.

Results from hard-randomized libraries can provide convincing evidence that the highest affinity clones possible have been obtained from selection; however, the limitations of library size prevent simultaneous randomization of many positions. The effects of substitution at one position may be altered by substitutions at another position, perhaps distant in the linear sequence. Such "nonadditive" effects can be missed in hard-randomized libraries unless specific residues are covaried.

## 3.2 Soft randomization

In contrast, soft randomization libraries are designed to maintain a bias towards the wild-type protein by introducing a partial rate of mutation at each amino acid position. For example, a 50% mutation rate is attained by synthesizing oligonucleotides using a mixture containing 70% of the base found in the wild-type sequence and 10% each of the other three bases at each position (30). Many residues can be simultaneously randomized, leading to libraries having more combinations of covaried residues than in hard randomization. This gain is at the expense of less complete representation of the library, because there are still $20^n$ possible protein variants (where $n =$ number of positions mutated). Nevertheless, these libraries can provide valuable information and can rapidly improve binding affinity. For example, complete soft randomization of a 20-residue peptide lead from a naïve library resulted in the selection of a variant with about 150-fold improved binding to Factor VIIa (FVIIa). In addition, conservation of wild-type residues at various positions in the peptide indicated their importance for peptide structure and/or function. In a subsequent library, conserved residues from earlier selections were held constant, and the remaining positions were fully explored using hard randomization. This led to a variant with an additional 100-fold improvement in binding affinity for FVIIa (11).

## 4 Target presentation

Many methods of target presentation have been used for phage selection, including whole cells and even *in vivo* binding selection (see Chapter 10). Here we summarize two general approaches: immobilization of the target on a solid support, and binding in solution followed by capture of soluble target. A support should be chosen based on a low background of phage binding and on the target molecule used. The target concentration must be adequate to capture more specifically bound phage than background phage in order to achieve selective enrichment. Characteristics of the target molecule, such as stability, solubility, and availability of free thiols ($-SH$), free amines ($-NH_2$), or carbohydrates ($-CHO$) for covalent matrix attachment, are important considerations.

## 4.1 Direct immobilization

For direct immobilization, there are a number of commercially available supports which can be derivatized by the user (see *Table 1*). These include resins or beads of various dimensions (e.g. Sepharose or controlled-pore glass), 96-well plates, and even analytical biosensor chips. In the later case, dextran-coated chips (BIAcore, Inc.) can be derivatized for specific coupling using a variety of chemistries for direct presentation or for solution capture strategies (see, e.g. 31).

Targets can also be immobilized using affinity tags (see *Table 1*). Use of these techniques requires separate considerations to reduce background. For example, avidin capture of biotinylated targets can lead to the selection of biotin mimics (see, e.g. 32), targets displayed as immunoglobulin constant-domain (Fc) fusions

**Table 1** Some matrices for phage binding selections

| Capture matrix | Immobilization mechanism | Caveats |
|---|---|---|
| Maxisorp (Nunc) | Non-specific adhesion to polystyrene | Matrix binders |
| Covalink (Costar) | Crosslinking of –NH$_2$, –SH, or –CHO to an activated plate | Matrix binders; capacity[b] |
| Biosensor chip (Biacore) | Crosslinking of –NH$_2$, –SH, or –CHO to modified dextran | Matrix binders; capacity[b] |
| Oxirane acrylic beads (Sigma) | Crosslinking of –NH$_2$ or –SH via epoxide | Matrix binders[b] |
| Avidin-coated plates/beads (Promega)[a] | Capture of biotinylated target via avidin, streptavidin, or neutravidin (Pierce) | Peptide biotin mimics (32)[b] |
| Anti-Fc antibody plates/beads (Pierce)[a] | Capture of IgG–Fc fusion (12) via anti-Fc antibody | Fc-binders; target is fusion to Fc[b] |
| Ni$^{2+}$ or antibody plates/beads (Qiagen)[a] | (His)$_6$ fusion coordination with antibody or Ni$^{2+}$ | Ni$^{2+}$ or antibody binders; target is (His)$_6$ fusion[b] |

[a] These approaches are particularly useful for binding phage to target in solution, then capturing bound phage.
[b] Covalent modification of the target in these cases may interfere with target binding/activity.

can lead to the selection of immunoglobulin G (IgG) Fc binders (see e.g. 33), and targets immobilized to nickel-adsorbed plates through a histidine tag can result in the selection of Ni$^{2+}$ binding phage. These artifacts can be overcome by alternating reagents. For example, to prevent selection of avidin binding clones, alternate avidin and streptavidin for capture between rounds of selection. Clones specific for avidin will then tend to be lost during selection with streptavidin; and conversely, streptavidin binding clones will tend to be lost in subsequent rounds using avidin. Inclusion of biotin in the binding buffer may also reduce recovery of biotin mimics.

## 4.2 Solution binding

Phage binding to target can also be carried out in solution, with subsequent capture by an appropriate affinity matrix (see Table 1). This technique can reduce avidity effects because the phage:target interaction does not occur on a surface, where multivalent interactions are favored. A low concentration of soluble target can be used to drive the selection of higher affinity clones. Conditions must be chosen for capture of a significant fraction of the target.

## 4.3 Blocking

Remaining sites on the support are generally blocked with an unrelated protein such as BSA, gelatin, casein, ovalbumin, or powdered milk. Other useful blocking reagents include nonionic detergents and commercially prepared blocking solutions (e.g. SuperBlock; Pierce). If the background increases in later rounds of selection (see Section 5.6) due to the binding to the blocking protein, the blocking agent can be changed and the previous round of selection repeated.

#### 4.4 Pilot selection

A pilot selection using only wild-type displaying phage, or wild-type phage spiked into a solution of non-displaying control phage (e.g. pBluescript (Stratagene), stop-codon template or helper phage) is extremely useful for verifying protein display, controlling background, and establishing conditions for library sorting. *Protocol 2* can serve as a starting point for testing the initial construct and also as the starting point for library sorting using 96-well Maxisorp plates for target presentation.

## Protocol 2

# Target coating and phage binding

### Equipment and reagents

- Coating buffer (PBS buffer; see *Protocol 1*); alternatively, carbonate buffer (100 mM NaHCO$_3$, pH 9.3)
- Blocking buffer (PBS, 0.5% bovine serum albumin (BSA)); alternatively, PBS containing 0.5% ovalbumin, 0.5% casein, or 0.5% (when used as a blocking agent) Tween-20.
- Wash buffer (PBS (see *Protocol 1*), 0.05% Tween-20)

- Target protein or antibody
- Protein-displaying phage and control (non-displaying) phage
- Nunc Maxisorp 96-well plates (VWR); for target immobilization
- Nunc F96 96-well plates (VWR); for phage dilution
- Binding buffer (Blocking buffer, 0.05% Tween-20 (see text for alternative binding buffers))

### Method

1  Prepare a coating solution (0.1 ml per well) by diluting each target protein or antibody to a concentration of 2 μg/ml in coating buffer. Initially, test coating concentrations over a range of 100–0.1 μg/ml.

2  Aliquot 100 μl/well into the Maxisorp plate and cover to prevent evaporation. Incubate overnight at 4 °C. In some cases, incubation at room temperature for 1 h is sufficient. As a negative control, add 200 μ/well of blocking buffer to a separate plate or separate row on the same Maxisorp plate.

3  Empty the plate and blot it dry on stacked paper towels. Add 200 μl/well of blocking buffer and incubate the plate(s) at room temperature for 2 h or overnight at 4 °C.

4  Prepare six 10-fold serial dilutions of phage for both the wild-type construct and the non-displaying phage (stop-codon phage, helper phage, or other) in binding buffer starting from a concentration of $10^{11}$ phage/ml. This is conveniently performed in F96 plates.

5 Add 100 μl of each protein-displaying phage solution to each target-coated and to each control (blocked) well of the Maxisorp plate. Repeat for the non-displaying control phage.

6 Incubate for 1 h at room temperature. Higher or lower temperatures may be used if appropriate for the target protein. For example, lower temperatures may be used for unstable proteins.

7 Proceed with ELISA detection of phage binding as described in *Protocol 3*, or phage elution as described in *Protocol 4*.

An ELISA (see *Protocol 3*) can be used to quantitate the bound phage (see also *Figure 2*).

## Protocol 3

## ELISA detection of phage binding

### Equipment and reagents

- Wash buffer (see *Protocol 2*)
- Binding buffer (see *Protocol 2*)
- Horse radish peroxidase (HRP)/anti-M13 monoclonal antibody (mAb) (Amersham Biosciences)

- HRP substrate (e.g. OPD (*o*-phenylenediamine) (Sigma) or ABTS (2,2'-azinobis (3-ethylbenzothiazoline-6-sulfonic acid))) (Kinkegaard and Perry)
- A phage-bound plate prepared as in *Protocol 2* or *Protocol 6*

### Method

1 Wash each plate three times with wash buffer.

2 Add 100 μl of 1:5000 HRP/anti-M13 mAb in binding buffer; incubate at room temperature for 30 min.

3 Wash each plate three times with wash buffer.

4 Add 100 μl OPD, quench after no more than 15 min with 50 μl of 2.5 M $H_2SO_4$.

5 Measure the optical absorbance at 492 nm for each plate.

## 5 Selection

An analogy is often made between selection of phage-displayed protein variants in the laboratory and evolutionary selection of protein variants in nature. To continue this analogy, one expects more "fit" phage variants to be selected in each "generation" of binding to the target molecule, culminating in the emergence or enrichment (34) of a dominant clone or set of clones that are optimized for binding. For this to occur, one must ensure that the "selective pressure" is

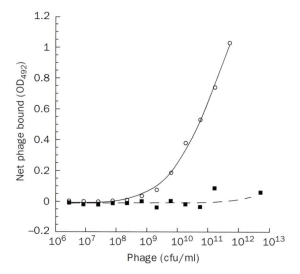

**Figure 2** Determination of protein display by target binding. As an example, kistrin, a protein isolated from snake venom and a known ligand (40) for the platelet receptor IIbIIIa, was displayed monovalently. Kistrin-phage particles (O) were tested for binding to immobilized IIbIIIa versus control wells coated with milk. A non-displaying phage (■) was included as a control. Net binding reflects the difference between signal observed on IIbIIIa and that observed on milk coated wells.

sufficient to result in a subpopulation of phage being retained at the expense of less-fit variants, then recovered and propagated for use in the next round.

The binding of phage to the target or matrix and washing away of nonbinding or weakly binding variants are the most critical steps in obtaining optimized variants. We discuss here considerations for selection conditions. However, it is important to modify the stringency conditions during selection based upon the initial affinity (i.e. the $K_d$ for the wild-type-protein:target interaction) and the affinity of optimized variants that emerge.

## 5.1 Binding buffer considerations

A buffer should be chosen which allows for the stability of the target and minimal non-specific binding. Inclusion of an irrelevant protein and a nonionic detergent as well as a physiological concentration of salt is helpful in reducing background. In addition some targets may, for example, require the addition of divalent cations or other cofactors. Protein cofactors can be co-immobilized with the target, added to the sorting buffer, or even used as a method of capture of the target (e.g. immobilized Tissue factor for the capture of FVIIa; 7, 9).

## 5.2 Stringency of selection

In the first round of selection the goal is to capture as many positive variants as possible while removing nonfunctionally displaying phage remaining from the library construction. Multiple wells of coated target, or larger amounts of soluble target should be used to favor the capture of large numbers of binding phage in order to maintain library diversity.

As enrichment increases (see below) in the later rounds of selection, the stringency of selection should be increased. By using multiple washings and increased washing times, clones with slower dissociation rate constants ($k_{off}$) are selected (see e.g. 3, 6, 16), resulting in improved $K_d$. Other techniques to increase stringency include addition of denaturants (e.g. acid (8), urea, or guanidine), organic solvents, increasing temperature (16), or increasing concentrations of a competing ligand or target (to prevent rebinding; see 6, 16) in the wash buffer.

### 5.3 Competitive selection

In addition to binding affinity, target selectivity may also be engineered into a displayed peptide or protein using competitive selection. This can be accomplished by selecting on an immobilized target while providing a competing ("decoy") target in solution (see *Figure 1*).

To establish the initial selection conditions, the desired target is immobilized, and dilutions of the competing target are added, followed by wild-type displaying phage. After an incubation to establish equilibrium, the plate is washed and developed by ELISA. The concentration of competitive target to be added in solution depends upon the affinity of the initial lead for the decoy target and can be determined empirically (*Figure 3*). The point at which 75–95% of the wild-type phage are prevented from binding to the immobilized target determines the starting concentration of competing target to add in the first round of selection. In subsequent rounds, as an increase in enrichment is observed, the stringency can be increased by increasing the concentration of competing target. Competitive selection can also be performed in solution with a tagged target and non-tagged decoy.

For example, competitive selection has been used to increase the selectivity of a general serine protease inhibitor for FVIIa (9) and to make receptor-selective atrial naturetic peptide (ANP) variants (4). In both cases, randomization of the

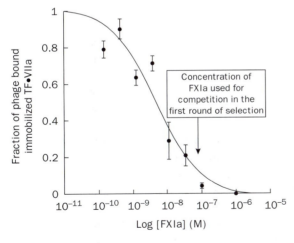

**Figure 3** Determination of Factor XIa (FXIa) concentration for competitive selection. A Kunitz domain (TF7I-C) phage was bound to TF•FVIIa in the presence of increasing concentrations of FXIa. Bound phage were quantitated using *Protocol 3*. The fraction of TF7I-C phage (bound versus added) is plotted as a function of FXIa concentration (reprinted from (9), with permission).

protein and selection for binding to immobilized target in the presence of competing targets in solution led to variants with more than 1000-fold selectivity. In each round of selection, as enrichment was observed, the concentration of soluble competing target was increased and more selective variants were selected.

## 5.4 Elution of bound phage

Phage remain infective after exposure to rather extreme conditions (see 34; also Chapter 1, *Table 3*) and may be eluted from the target using many methods. The eluent should be nontoxic to *E. coli* or able to be neutralized (e.g. acid–base), because eluted phage will subsequently be amplified by transfection of the eluted phage into *E. coli*.

A simple method for eluting bound phage from an immobilized target is to reduce the pH with 10–100 mM HCl, or 0.2 M glycine, to pH 2. After recovery of phage, the solution can be easily neutralized by the addition of 1 M Tris-HCl, pH 8.0 prior to infection of *E. coli* for amplification. Some tight-binding proteins may require rather long incubation times under acidic conditions for complete dissociation.

The addition of a competing ligand in solution is another method of elution. This requires very high concentrations of ligand in order to saturate any free sites that become available upon phage dissociation. This method of elution is not appropriate for polyvalent phage because of avidity effects described earlier.

Methods that specifically dissociate target from the matrix, and thus the phage bound to the target, are also acceptable. For example, the addition of thrombin to cleave an engineered thrombin site in a target fused to an IgG Fc domain was used for the selection of a peptide erythropoeitin (EPO) agonist (14).

A phage-ELISA of the target-coated plate after elution can be used to assess whether elution conditions are sufficient to displace a majority of bound phage. The plate should be reblocked with blocking buffer. Control wells, having no phage, should also be blocked to ensure that the detecting antibody does not adhere to the plate. Thereafter, the plate can be developed as described (see *Protocol 3*).

## Protocol 4

# Phage binding selection, washing, and elution

### Equipment and reagents

- Wash buffer (see *Protocol 2*)
- Binding buffer (see *Protocol 2*)
- Elution buffer (10 mM HCl, 500 mM KCl; or alternatively, 100 mM HCl for more stringent elution)
- Neutralization buffer (1 M Tris-HCl, pH 8.0)
- 2YT/Carb/Tet media (see *Protocol 1*)
- VCSM13 helper phage (see *Protocol 1*) or M13K07 helper phage (New England Biolabs)

## Method

1 Wash microtiter plate wells by pouring or pipeting Wash buffer across wells, 10 times. The use of an automated plate washer may introduce cross contamination between experiments and is not recommended.

2 For increased stringency, a dissociation period may be included. Add 200 µl/well binding buffer and incubate (see text, Section 5.2). In successive rounds, the incubation time might be increased from for example, 10 min to 1 h, for high-affinity interactions ($K_d < 10$ nM). Repeat step 1.

3 Add 200 µl per well of elution buffer. Incubate 10–30 min at room temperature.

4 Transfer to a clean tube containing 25–50 µl 1 M Tris-HCl, pH 8.0. Check for pH≈7 with pH paper. Phage can still be used for selection after storage at 4 °C for several days, or for propagation after similar storage for months.

5 For monitoring enrichment (colony counting), see *Protocol 5*.

6 To propagate of phage for a new round of selection, mix phagemid particles (e.g. $10^6$–$10^8$ colony forming units (cfu)) with 1 ml of a log-phase ($OD_{600} = 0.3$–$0.6$) culture of XL1-Blue cells, incubate with shaking for 1 h at 37 °C, and dilute into 25 ml of 2YT/carb/tet containing $10^9$ pfu/ml of helper phage for overnight growth with shaking (see *Protocol 1*). Note however, that additional selection rounds can sometimes be carried out without phage amplification (36; *Figure 1*).

## 5.5 Amplification

A typical round of selection may capture $10^6$–$10^9$ phage bound to the target (*Table 2*). These "positives" are generally amplified prior to the next round of selection by passage through *E. coli*.

After two or more cycles of selection and amplification, it is often possible to perform an additional selection step without amplification (36; Figure 1). Phage concentrations in the eluate will be greatly reduced; however, amplification artifacts, such as variants with increased expression but without improvement in affinity may be avoided.

**Table 2** Typical phage concentrations during sorting using Maxisorp plates

| Selection step | Phage concentration (cfu/ml) |
| --- | --- |
| Phage in culture media | $10^{12}$ |
| Input phage for selection | $10^{11}$ |
| Background phage binding | $10^3$–$10^6$ |
| Specifically bound phage | $10^4$–$10^9$ |

**Table 3** Troubleshooting suggestions

| Problem | Controls and suggestions |
|---|---|
| No phage produced in overnight culture | Use *E. coli* from a fresh plate<br>Confirm addition of helper-phage<br>Test helper-phage growth<br>Check media, antibiotics for contamination |
| No ELISA signal with wild-type phage | Confirm sequence of construct<br>Confirm target immobilization<br>Try an alternative vector/promoter<br>Test phage binding to antibodies (ELISA)<br>Check phage display by western blot<br>Test phage growth at 30 °C<br>Modify linker (longer, randomized)<br> or phage coat protein (see 42) |
| Binding of wild-type phage to negative-control plate is high | Test alternative blocking, binding buffers<br>Present target using another matrix (e.g. beads/<br> plates/in solution) |
| Background binding of negative-control phage to target plate is high | Present target using another matrix<br>Test alternative forms of target (e.g. domain<br> deletions) |
| Elution is incomplete; phage remain bound on plate | Increase elution time (50% may be maximum eluted)<br>Increase elution stringency |
| Recovery of phage from elution is low | Use less stringent wash conditions<br>Is elution complete (see above)?<br>Confirm target presentation<br>Sequence clones from initial library<br>Were sufficient phage added (*Table 2*) |
| No enrichment of wild-type phage | Is background high (see above)?<br>Use less stringent wash conditions<br>Is elution complete (see above)?<br> Present target using another matrix |
| No enrichment of library<br>*(Note: enrichment is rarely observed after the first round of selection.)* | Examine initial library. Is library theoretically complete?<br>Do wild-type phage enrich?<br>Sequence clones from library previous round of<br> selection<br>Is background high?(see above)<br>Test enrichment after 2–3 rounds of binding and<br> amplification |
| Library yields no consensus sequences | Perform additional rounds of selection<br>Increase washing stringency<br>Do randomized positions<br> influence binding? |
| Library yields clones with improved expression, but not improved target-binding affinity | Are phage monovalent?<br>Test noninducing conditions for promoter<br>Add one or more amber (TAG) insertions in<br> protein–phage linker<br>Test fusion to a low-expression protein on pIII |
| Enrichment is observed, but only wild-type phage are recovered | Is library pool contaminated? Compare DNA<br> sequence with starting library<br>Make fresh cells, helper phage, stock<br> solutions and media to avoid contamination<br>Clean work area and equipment with 0.1% SDS or<br> 10% bleach. |

76

## 5.6 Monitoring selection

Target-specific binding and successful phage propagation are critical for successful phage optimization. This is particularly true when increasing stringency is used during cycles of selection.

We use the term "background enrichment" to refer to the cfu ratio of phage recovered from the target versus those recovered from the non-target-coated or background matrix, and the term "marker enrichment" to refer to the cfu ratio of protein-displaying phage recovered from the target non-displaying phage. Both measures of enrichment can be important. In our experience, background enrichment is usually more critical in peptide selections, and marker enrichment is usually more critical in protein phage display.

### 5.6.1 Background enrichment

Background enrichment is measured to ensure that more phage are recovered from the target-coated matrix than from the control matrix alone. The plating protocol provided below enables titration of phage counts by colony counting. In addition to the determination of enrichment ratios, the determination of phage titers (cfu/ml) is an important test of whether sufficient numbers of viable phage are actually being recovered and propagated from round to round. Typically, $10^4$ phage are needed in order to obtain reasonable infection and growth in an overnight culture. In fact, growth of the infected culture can be used as a direct and semiquantitative readout of phage recovery. If $< 10^3$ are recovered, further amplification is unwarranted because essentially all clones can be individually evaluated.

---

## Protocol 5

## Monitoring enrichment

### Equipment and reagents

- XL1-Blue *E. coli* (Stratagene)
- LB/carb plates (see *Protocol 1*)
- 2YT/Tet media (2YT (see *Protocol 1*), 15 μg/ml tetracycline)
- Nunc 96F plates (VWR)

### Method

1  Pick a single colony of XL1-Blue *E. coli* from a fresh LB/Tet plate into 25 ml of 2YT/Tet and incubate with shaking at 37 °C for 6–10 h ($OD_{600}$ of 0.3–0.6).

2  For each phage sample to be titrated, aliquot 180 μl of 2YT media into 4–8 wells of a 96F plate.

3  To the first well, add 20 μl of phage solution. Mix using a pipette.

4  Perform 10-fold serial dilutions by transferring 20 μl from each well to the next, then mix using a pipette. Clean pipette tips must be used for each step.

**5** In a second 96F plate, aliquot 180 µl of fresh log phase XL1-Blue cells.

**6** Transfer 20 µl from each well of the phage plate to each well of the XL1-Blue plate. Mix using a pipette.

**7** Incubate for 1 h at 37 °C.

**8** Dry enough LB/carb plates to plate all dilutions (5 samples × 8 rows per plate) by incubation at 37 °C for 1 h with the petri dish lid ajar.

**9** Using a multichannel pipette, dispense 10 µl from each well of the XL1-Blue plate onto a pre-dried LB/carb plate. Allow the liquid to soak into the agar.

**10** Incubate the plate overnight at 37 °C.

**11** Count colonies at an appropriate dilution and calculate cfu/ml for each library sample or control.

### 5.6.2 Marker enrichment

Marker phage are useful for assessing nonspecific binding of phage to the affinity matrix or to nontarget coated matrix. These phage or phagemid vectors encode either a selectable marker (drug resistance, differing from the display phage, such as chloramphenicol or kanamycin resistance) or a reporter molecule (e.g. *β*-galactosidase *α*-peptide, to be used with a lacZ' *E. coli* strain such as XL1-Blue) that differs from the displaying phage. The marker phage can be non-displaying (wild-type phage proteins only), or may display a homologous protein for specificity testing. Marker phage particles are mixed with protein-displaying phage particles, and binding selection is carried out as described. In the case of drug resistance markers, the recovered phage are plated onto appropriate drug-containing LB-agar plates for colony counting. In the case of reporter genes, such as *β*-galactosidase, the recovered phage pool can be analyzed on a single IPTG/X-gal plate by scoring "blue" and "white" colonies (see (37)).

### 5.6.3 Evaluating enrichment factors

With well-established systems, we often proceed with 1–2 cycles of binding selection before beginning to determine enrichment factors. However, for new phage-display constructs or further rounds of selection, measurement of enrichment is strongly recommended. *Table 2* shows typical numbers of phage (measured as cfu) recovered from selections. Enrichments of >two-fold are usually significant; however >10-fold enrichment is ideal. Continuing selections when binding enrichment cannot be observed is rarely productive.

### 5.6.4 Monitoring by ELISA

A simple phage ELISA provides a rapid assessment of how many displaying or non-displaying phage are bound on target versus matrix. Using Maxisorp plates for selection, one row of wells is reserved for analysis, while the remaining wells are eluted for propagation. The analysis row is then developed as described (see *Protocol 3*).

## 6 Screening prior to analysis

Cycles of binding selection, or binding selection followed by amplification (see Figure 1) are continued until suitable variants are obtained, or until it can be deduced that no such clones can be recovered. A theoretical description of this process has been presented (38).

Once binding selections have progressed to the point of observing enrichment, individual colonies can be screened for qualitative binding properties prior to more complete analysis. *Protocol 6* is a rapid means of screening colonies in a 96-well format.

---

### Protocol 6

## Screening phage clones

### Equipment and reagents

- XL1-Blue *E. coli* (Stratagene)
- VCS-M13 phage (Stratagene)
- LB/carb plates (see *Protocol 1*)
- 2YT/carb media (see *Protocol 1*)
- 96 microtiter tube array (out patient services)
- Sterile toothpicks
- Binding buffer (see *Protocol 2*)
- Nunc Maxisorp plates (VWR): target-coated or blank, blocked as described (see *Protocol 2*)
- Centrifuge for spinning 96-well plates (e.g. Sigma-3K12, B. Braun Biotech International)

### Method

1 Infect log-phase ($OD_{600} = 0.3$–$0.6$) XL1-Blue cells with neutralized eluted phage. Incubate at 37 °C for 60 min, and plate onto LB/carb. Grow overnight at 37 °C.

2 Aliquot 400 μl of 2YT/carb containing VCSM13 ($10^9$ cfu/ml) into each well of a 96 microtiter tube array.

3 Pick individual colonies into each well by dipping a pipette tip or toothpick briefly into the media.

4 Secure the titer tube array to an orbital shaker, and grow at 37 °C overnight.

6 Spin titer tube rack at 2500 rpm for 10 min.

7 Transfer 50 μl of supernatant to both the target and blank plates along with 50 μl of binding buffer.

8 Wash the plate (see *Protocol 2*), and perform a standard ELISA (*Protocol 3*).

9 Phage from clones that demonstrate specific binding to target can be harvested from 96 microtiter tube array for DNA sequencing.

## 7 DNA sequence analysis

DNA sequence analysis is the primary means for identification of optimized variants from a phage-displayed protein or peptide library. It is useful to compare sequences at both the DNA and translated (amino acid sequence) level.

### 7.1 When to sequence

With modern automated DNA sequencing techniques, full-length (open-reading-frame) sequences can be obtained for many clones in a matter of hours. We typically begin sequencing clones after enrichment has reached a maximum. In monovalent phage display, this often occurs after 3–6 cycles of binding selection and amplification. A suggested strategy is to begin with sequencing 10–20 clones from the first round of selection that shows enrichment of 10-fold or more. Because 10 times more clones are recovered from target than from background, ≈90% of all clones are expected to have target-specific binding. Positive clones can be confirmed by single-well analysis (see *Protocol 6*).

The sequences should be aligned according to the window of mutagenesis either manually or by computer algorithm to evaluate what amino acids appear at the randomized positions. It is also important to confirm that clones from the selected library are generally intact, having an open reading frame throughout the displayed-protein gene and into the phage-coat fusion. It is possible for frame-shifted or deletion variants to dominate the phage pool. However, positive enrichment versus control matrix and versus negative-control phage are generally not observed in these cases. Often the library retains sequence diversity at this stage, and binding selections can be continued, perhaps with increasing stringency for selection of higher affinity clones (see above).

### 7.2 Sequence consensus and sibs

Two types of sequence convergence are often observed. "Sibs" occur when a single DNA clone dominates the pool of selected variants (see 37). For example, if NNS codon degeneracy is used, there are two possible codons for Val (GTC and GTG). Suppose the sequence GTC-GTC-GTC-GTC encoding $(Val)_4$ dominates the selected library. If none of the other 14 possible codon combinations for $(Val)_4$ are found (e.g. GTG-GTC-GTC-GTC), then the selected clone may have dominated the library. This may occur as a result of expression bias, unintended mutations elsewhere in the vector, or from the limited sampling size of the library. Further sequencing, especially from the preceding round of selection, is warranted. Note however, that some amino acid residues are encoded by only one codon (e.g. Trp or Met, using NNS codons), so that this test is not always meaningful. The occurrence of sibs does not preclude discovery of affinity-improved variants.

In the other type of sequence convergence, particular amino acid sequences dominate, but with a variety of codons at one or more positions. For example, ATG-CGG-GCC, ATG-AGG-GCC, ATG-CGC-GCG, and ATG-CGC-GCC, each encoding Met-Arg-Ala, may all be found. This type of outcome provides a high degree of confidence that the library was appropriately diverse, and that selection has occurred on the basis of binding properties of the amino acid sequence rather than on the basis of DNA dependent effects (e.g. expression level).

## 7.3 Evaluating the consensus

Except in cases where only one peptide sequence dominates, it will be necessary to evaluate to what extent the consensus sequence represents a significant departure from sequences sampled at random. Several approaches have been described.

We have made use of a scoring function (20), $S$, which accounts for the theoretical (or, ideally, experimentally determined) initial library diversity, codon degeneracy, and number of times each amino acid is observed. The function is expressed as $S = (P_f - P_e)/(P_e(1 - P_e)/n)^{1/2}$. $P_f$ is the observed frequency of a given amino acid residue; that is, the number of times it is observed divided by the total number of clones, $n$. $P_e$ is the expected frequency of each amino acid residue, based upon codon degeneracy, or ideally based upon sequencing (calculating $P_e$ as for $P_f$ above) of the library prior to selection. In our experience, scores of $S > 2$ have corresponded to functionally significant effects.

## 7.4 Evaluation of phage clones

The clones isolated from phage binding selections can be used in several ways for evaluation of affinity and specificity. Quantitative evaluation of protein binding interactions is often possible at the phage level, and this can be carried out with individual clones, or even pools of phage from different stages in the binding selections. Detailed protocols are provided in Chapter 5.

The effects of point mutations at different sites within proteins are often additive with respect to their contributions to the free energy of binding (39). A number of phage optimization efforts have made use of this principal to combine mutations that were selected from independent libraries each targeting only a few residues. Additivity and cumulative affinity improvements can be assessed using phage ELISA.

In cases where mutations are not additive, cumulative randomization and selection can be used. The strategy here is to select an optimized variant from one particular library, and randomize at new positions using this variant as template. Using this strategy, combined with additivity, a hormone variant having 400-fold improved affinity and mutated at 15 sites was identified (22). The maturation of peptide leads from naïve peptide phage libraries has also made use of this strategy (11, 14).

As described above, use of an amber stop codon in the same phagemid vector used for library selections can allow rapid production of soluble protein variants in E. coli. Peptide and protein variants can also be expressed as fusions with epitope tags for simplified purification (see e.g. 40). Because the details of soluble protein expression, recovery, and purification can be quite specific to the protein or peptide at hand, we do not include specific protocols here. The reader may wish to consult the review by Makrides (41) on high-level expression of proteins in E. coli.

## 8 Troubleshooting

Many factors involving protein expression, phage packaging, initial binding affinity, background binding effects, contamination, and the molecule-specific features of target molecules affect the outcome of phage optimization experiments. In *Table 3*, we summarize some common problems and suggest lines of investigation for solving them.

## References

1. Clackson, T. and Wells, J. A. (1994). *Trends Biotechnol.*, **12**, 173.
2. Lowman, H. B. (1997). *Annu. Rev. Biophys. Biomol. Struct.*, **26**, 401.
3. Lowman, H. B., Bass, S. H., Simpson, N., and Wells, J. A. (1991). *Biochemistry*, **30**, 10832.
4. Cunningham, B. C., Lowe, D. G., Li, B., Bennett, B. D., and Wells, J. A. (1994). *EMBO J.*, **13**, 2508.
5. Hawkins, R. E., Russell, S. J., and Winter, G. (1992). *J. Mol. Biol.*, **226**, 889.
6. Barbas, C. F. III, Hu, D., Dunlop, N., Sawyer, L., Cababa, D., Hendry, R. M., *et al.* (1994). *Proc. Natl. Acad. Sci. USA*, **91**, 3809.
7. Dennis, M. S. and Lazarus, R. A. (1994). *J. Biol. Chem.*, **269**, 22129.
8. Roberts, B. L., Markland, W., Ley, A. C., Kent, R. B., White, D. W., Guterman, S. K. *et al.* (1992). *Proc. Natl. Acad. Sci. USA*, **89**, 2429.
9. Dennis, M. S. and Lazarus, R. A. (1994). *J. Biol. Chem.*, **269**, 22137.
10. Dennis, M. S., Herzka, A., and Lazarus, R. A. (1995). *J. Biol. Chem.*, **270**, 25411.
11. Dennis, M. S., Eigenbrot, C., Skelton, N. J., Ultsch, M. H., Santell, L., Dwyer, M. A. *et al.* (2000). *Nature*, **404**, 465.
12. Rebar, E. J. and Pabo, C. O. (1994). *Science*, **263**, 671.
13. Jamieson, A. C., Kim, S. H., and Wells, J. A. (1994). *Biochemistry*, **33**, 5689.
14. Wrighton, N. C., Farrell, F. X., Chang, R., Kashyap, A. K., Barbone, F. P., Mulcahy, L. S. *et al.* (1996). *Science*, **273**, 458.
15. Foote, J. and Milstein, C. (1991). *Nature*, **352**, 530.
16. Chen, Y., Wiesmann, C., Fuh, G., Li, B., Christinger, H. W., McKay, P., *et al.* (1999). *J. Mol. Biol.*, **293**, 865.
17. Cwirla, S. E., Peters, E. A., Barrett, R. W., and Dower, W. J. (1990). *Proc. Natl. Acad. Sci. USA*, **87**, 6378.
18. Barbas, C. F. III, Kang, A. S., Lerner, R. A., and Benkovic, S. J. (1991). *Proc. Natl. Acad. Sci. USA*, **88**, 7978.
19. Müller, K. M., Arndt, K. M., and Plückthun, A. (1998). *Anal. Biochem.*, **261**, 149.
20. Bass, S., Greene, R., and Wells, J. A. (1990). *Proteins*, **8**, 309.
21. Wells, J. A. and Lowman, H. B. (1992). *Curr. Opin. Struct. Biol.*, **2**, 597.
22. Lowman, H. B. and Wells, J. A. (1993). *J. Mol. Biol.*, **234**, 564.
23. Crameri, R. and Suter, M. (1993). *Gene*, **137**, 69.
24. Fuh, G., Pisabarro, M. T., Li, Y., Quan, C., Lasky, L. A., and Sidhu, S. S. (2000). *J. Biol. Chem.*, **275**, 21486.
25. Bullock, W. O., Fernandez, J. M., and Short, J. M. (1987). *Biotechniques*, **5**, 376.
26. Muller, Y. A., Chen, Y., Christinger, H. W., Li, B., Cunningham, B. C., Lowman, H. B. *et al.* (1998). *Structure*, **6**, 1153.
27. Vierra, J. and Messing, J. (1987). In *Methods in enzymology*, Vol. 153, (ed. R. Wu and L. Grossman), p. 3, Academic Press, London.
28. Lowman, H. B. (1998). In *Methods in molecular biology*, Vol. 87, (ed. S. Cabilly), p. 249, Humana Press, Totowa, NJ.

29. Stemmer, W. C. (1994). *Nature*, **370**, 389.
30. Gallop, M. A., Barrett, R. W., Dower, W. J., Fodor, S. P. A., and Gordon, E. M. (1994). *J. Med. Chem.*, **37**, 1233.
31. Finucane, M. D., Tuna, M., Lees, J. H., and Woolfson, D. N. (1999). *Biochemistry*, **38**, 11604.
32. Giebel, L. B., Cass, R. T., Milligan, D. L., Young, D. C., Arze, R., and Johnson, C. R. (1995). *Biochemistry*, **34**, 15430.
33. DeLano, W. L., Ultsch, M. H., de Vos, A. M., and Wells, J. A. (2000). *Science*, **287**, 1279.
34. Smith, G. P. (1985). *Science*, **228**, 1315.
35. Marvin, D. and Hohn, B. (1969). *Bacteriol. Rev.*, **33**, 172.
36. Sparks, A. B., Adey, N. B., Quilliam, L. A., Thorn, J. M., and Kay, B. K. (1995). In *Methods in enzymology*, Vol. 255, (ed. W. E. Balch, C. J. Der, and A. Hall), p. 498, Academic Press, London.
37. Sambrook, I., Fritsch, E. F., and Maniatis, T. (ed.) (1989). *Molecular cloning: A laboratory manual*, 2nd edn., p. 1.8, Cold Spring Harbor Laboratory Press, New York.
38. Mandecki, W., Chen, Y. C., and Grihalde, N. (1995). *J. Theor. Biol.*, **176**, 523.
39. Wells, J. A. (1990). *Biochemistry*, **29**, 8509.
40. Dennis, M. S., Henzel, J. W., Pitti, R. M., Lipari, M. T., Napier, M. A., Deisher, T. A. *et al.* (1990). *Proc. Natl. Acad. Sci. USA*, **87**, 2471.
41. Makrides, S. C. (1996). *Microbiol. Rev.*, **60**, 512.
42. Weiss, G. and Sidhu, S. S. (2000). *J. Mol. Biol.*, **300**, 213.

Chapter 5

# Rapid screening of phage displayed protein binding affinities by phage ELISA

Warren L. DeLano and Brian C. Cunningham
Sunesis Pharmaceuticals Inc., 341 Oyster Point Boulevard,
South San Francisco, CA 94080, USA.

## 1 Introduction

Monovalent phage display enables a single protein (or peptide) molecule to be presented on the surface of a phage particle. This protein is translated from a phagemid vector as a fusion linked to the natural pIII coat protein in an *Escherichia coli* cell coinfected with helper phage. The display achieves monovalency since the fusion protein rarely substitutes for the native pIII protein from the helper phage genome. While display can be highly variable, typically just 0.05–5% of these phages actually display the protein of interest (1).

The key benefit of monovalent phage display is the elimination of avidity effects encountered in multivalent display systems (2). This allows for efficient selections that discriminate between variants that bind with very high affinity (i.e. $K_d \approx 1$ nM). Applications of this technology include the generation of polypeptides with enhanced binding affinities (3, 4) or with altered binding specificities (1). In practice, the most laborious part of the process is the characterization of selected mutant proteins.

Two methods have been used to speed up the analysis of binding affinities of selected proteins. One method incorporates an amber stop codon between the displayed protein and the anchoring pIII coat protein so that the free protein can be expressed in a non-suppressor strain (1). This permits the use of standard methods of binding analysis (5), which can be streamlined for higher throughput (6). Unfortunately, this approach suffers from the tendency of selected libraries to be overtaken by mutations that replace the amber codon with a sense codon (1). Such phages have an enormous selective advantage due to the elevated expression and display of their fusion protein. The alternative of omitting the termination codon and subcloning each gene into an expression vector is significantly more labor intensive.

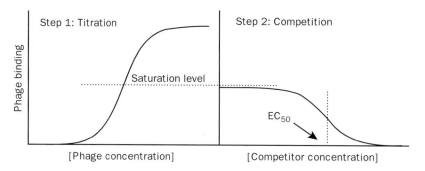

**Figure 1**

A second method, the focus of this chapter, measures the binding of the phage particle itself as a surrogate for the free protein (*Figure 1*). The phage binding assay follows the competitive displacement of phage binding to immobilize target protein by serial additions of the target protein in solution. The technique allows for the rapid analysis of mutants isolated from monovalent phage display selections and has also been used to estimate the affinities of directed mutants that are otherwise difficult to express. Deriving precise binding affinities using the phage binding assay is difficult because a key variable, the concentration of displayed protein, is unknown. Nevertheless, for a series of variants, it is possible to obtain ratios of $EC_{50}$ values that closely track the ratios of their dissociation constants (7). This chapter describes the important parameters to consider in designing phage binding assays and provides detailed sample protocols on how to perform these assays.

## 2 Parameters governing phage binding assays

The phage binding assay involves binding of a phage fusion protein (P) to an immobilized target protein (T) in the presence of a competing solution-phase target protein (C):

$$TP + C \longleftrightarrow T + P + C \longleftrightarrow T + PC \qquad (1)$$

The $EC_{50}$ is the concentration of competitor that results in a 50% reduction in binding, with fixed concentrations of phage and immobilized target:

$$EC_{50} = K_C(1/(1-f_0) + (2-f_0)T/(2K_T)) + f_0K_T/(2-f_0)/(1-f_0) + f_0T/2 \qquad (2)$$

where $f_0$ is the fractional phage saturation of the immobilized target in the absence of competitor, $T$ is the total target concentration, $K_T$ is the dissociation constant between target and phage, and $K_C$ is the dissociation constant between competitor and phage. Because the immobilized target and the competitor are

(a)

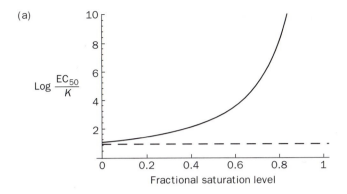

(b)

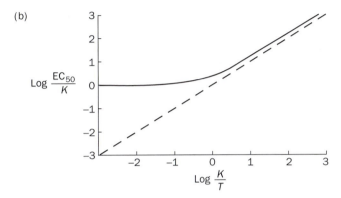

**Figure 2**

actually the same protein, we can simplify this equation by assuming that their affinities are equal ($K_C = K_T = K$) to give the following:

$$EC_{50} = 2K/(2 - f_0)/(1 - f_0) + T \qquad (3)$$

Equation [3] clearly shows that $EC_{50}$ determinations by phage binding assays depend upon the fractional saturation of the immobilized target ($f_0$) and the amount of immobilized target ($T$). This is shown graphically in *Figure 2*. As the concentration of phage is increased, the immobilized target approaches saturation, and the observed $EC_{50}$ values rise above the actual dissociation constant, approaching an asymptote (*Figure 2(a)*). On the other hand, as the concentration of immobilized target is raised, the ability to measure tighter affinities becomes increasingly limited (*Figure 2(b)*). These properties indicate that concentrations of both phage and immobilized target should be minimized in order to improve accuracy. However, since reasonable signal-to-noise ratios are also important for accurate measurements, the best results are usually obtained at a fractional saturation of 50%, using the minimum of immobilized target needed for a good signal.

When comparing $EC_{50}$ values across phage variants in order to estimate relative affinities, it is essential that their fractional saturation be uniform so that the following approximation is valid:

$$EC_{50}(mut)/EC_{50}(wt) = (cK(mut) + T)/(cK(wt) + T) \approx K(mut)/K(wt) \qquad (4)$$

where $c = 1/((2 - f_0)(1 - f_0))$. When the fractional saturation is constant at 50%, and the target concentration is 10-fold lower than the smaller dissociation constant, the ratio of the dissociation constants will vary from the ratio of $EC_{50}$ values by less than 7%. However, in practice, the errors will be larger because of variations in fractional saturation. For example, a 5% standard deviation in fractional saturation under these conditions will cause the $EC_{50}$ to deviate by 13%, and their ratios to deviate by about 18%.

# 3 Performing phage binding assays

## 3.1 Preparation of phage samples

Typically, after a full course of binding enrichments by monovalent phage display, single-stranded phagemid DNA from the surviving clones is isolated and sequenced. While the later rounds of selection may yield a consensus sequence, the corresponding polypeptide may not correspond to the variant of highest affinity because of other selective pressures. To assure that the best sequences are identified, the affinities of a large number of different variants from both earlier and later rounds of selection should be measured. To analyze these binding affinities using a phage binding assay, we start by preparing stock solutions of each phage (see Chapter 4, *Protocol 1*). A phage stock solution with titer of $3 \times 10^{14}$–$5 \times 10^{14}$ phage/ml can usually be obtained using carbenicillin resistance. Other antibiotic markers such as chloramphenicol tend to give less robust phage titers, but this can be substantially remedied by lowering the antibiotic concentration to a minimum. For example, our experience shows that 5 µg/ml of chloramphenicol is sufficient to maintain chloramphenicol-resistant phagemids, but gives substantially higher phage titers than cultures with higher concentrations of this antibiotic. We have found that phage titers can be conveniently estimated by measuring their optical density at 268 nm and assigning an extinction coefficient of 0.2 $nM^{-1}$ $cm^{-1}$ (see (8)). Optimal phage storage conditions vary depending on the displayed polypeptide and must be determined for each protein–phage fusion. We generally refrigerate our phagemid preparations at 4 °C and perform binding assays within 2 days of their preparation. Beyond this time, there is a progressive loss of binding competency and assay reliability. Generally, longer storage can be achieved by freezing aliquots of phage at −80 °C for single usage. Alternatively, some extension of storage time may be obtained by keeping phage stocks as 50% glycerol solutions at −20 °C.

## 3.2 Titration of target immobilization and phage addition

The basic conditions for immobilization of the target protein may already be established if the phage selection process was carried out on microtiter plates. Indirect immobilizations of target proteins to microplates coated through pre-coating reagents such as avidin or anti-tag antibodies are amenable to phagemid binding assays. However, immobilizations through anti-target antibodies should not be used since phagemid binding assays use solution-phase target as a binding competitor to generate binding curves. Below we describe a typical protocol for directly immobilizing target proteins to microtiter plates.

---

## Protocol 1

## Preparation of target coated microtiter plates

### Equipment and reagents

- 96-well microtiter plates (Nunc Maxisorp)
- 50 mM sodium carbonate buffer, pH 9.4
- Acetate plate sealers (Dynex)
- PBST: phosphate buffered saline pH 7.4, 0.05% Tween 20 (v/v)
- CBB: casein blocking buffer (Pierce)
- Eight channel P200 multichannel pipettor
- Automated microtiter plate washer (Skatron or similar)

### Method

1   Add 100 μl of the target protein (concentration established in *Protocol 2*) in sodium carbonate buffer to the bottom of each well of a Maxisorp microtiter plate. For no-target control wells, add 100 μl of sodium carbonate buffer only.

2   Cover the plate with sealing tape and gently transfer plates to a 4°C refrigerator without disturbing the coating solution. Incubate overnight.

3   Shake the solution out of the plate and stamp the plate upside down onto a stack of paper towels to remove the residual liquid.

4   Immediately block each well of the plate by adding 230 μl of CBB (no detergent) and incubate for 30 min at room temperature.

5   Aspirate the wells and wash the plate three times with PBST buffer using the plate washer just prior to use.

---

Before performing phage binding assays, both the concentration of target protein used to prepare the coated microtiter plates (*Protocol 1*) and the concentration of the phage addition to be used in the competition binding experiment (*Protocol 3*) must be determined. Initially, both of the processes can be performed simultaneously in a two-dimensional grid using a stock of the wild-type phage, as described below.

## Protocol 2

# Titration of the immobilized coating reagent and phage additions

### Equipment and reagents

- Plastic reagent reservoir (Costar)
- 96-well plates ("nonstick" Nunc F plates, or polypropylene plates)
- anti-M13 antibody horse radish peroxidase (HRP)-conjugate (Stratagene)
- Tetramethylbenzidine (TMB) substrate kit (Pierce)

- 2 M sulfuric acid
- CBB (see *Protocol 1*)
- Miniorbital shaker (Bellco)
- Microplate optical absorbance reader

### Method

1  Coat a Maxisorp microtiter plate with twofold serial dilutions (e.g. 16, 8, 4, 2, 1, 0.5, 0.25 µg/ml) of the target protein as described in *Protocol 1*. Distribute the highest concentration to row A, and successive dilutions to rows B through G. Do not add the target solution to row H, which will serve as a control to measure nonspecific phage binding.

2  In a separate, nonstick plate, make serial dilutions (twofold) of phage in CBB. Dilute so that column 1 contains the highest phage concentration and that successive dilutions occur in columns 2 through 12.

3  Shake and stamp the wash buffer out of the coated plate from step 1. Directly transfer 100 µl of the diluted phage solutions to the target immobilized plate.

4  Cover the plate with sealing tape and incubate at room temperature for 60 min.

5  Aspirate and wash the plate 12 times.

6  Add 100 µl of anti-M13 HRP conjugate, diluted 1/2000 into CBB, to each well. Incubate for 20 min at room temperature. Aspirate and wash the plate eight times. Shake and stamp the plate dry.

7  Add 100 µl of TMB substrate (equilibrated to room temperature) and shake vigorously on a miniorbital shaker for 10 min.

8  Immediately quench the substrate development by addition of 100 µl of 2.0 M sulfuric acid. Shake vigorously for 30 s and read the optical absorbance at 450 nm ($A_{450}$) with a microplate reader.

Plot the amount of phage bound as determined by the optical absorbance reading versus the dilution of the phage stock for each of the target coat concentrations tested (see *Figure 3*). This data generates a set of curves that permits the identification of the lowest target coat addition necessary to yield a signal of about 1.2 when 50% saturated with phage. The phage binding assays will use this condition for preparing target-coated plates.

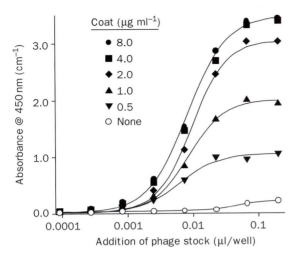

**Figure 3**

Next, binding titration curves for each stock solution of phage must be generated at the established target-coat conditions. Perform this experiment as described in *Protocol 2*, but with each row being a serial dilution of a different phage clone (including the wild-type phage) and the coat fixed as determined in *Protocol 2*. Afterwards, plot binding curves and calculate the dilution required for each phage stock to bind the target at 50% saturation. Make sure these determinations are accurate. This will be the concentration of each phage variant used in the phage binding assay.

### 3.3 Measurement of phagemid binding to target

Having worked out the conditions for target immobilization, and for adding a uniform 50% subsaturating level of each phage, we are ready to determine the relative binding affinities of each phage for the target protein.

## Protocol 3

## Phagemid binding assay

### Equipment and reagents

- 8-channel P10 multichannel pipettor (Eppendorf)
- TMB substrate kit (see *Protocol 1*)
- 96-well microtiter plates (Nunc Maxisorp, see *Protocol 1*)
- CBB buffer (see *Protocol 1*)

### Method

1 Coat columns 1–11 of a Maxisorp plate with the target protein at a concentration found optimal in *Protocol 2*. Leave column 12 empty as a control to measure nonspecific phage binding. Block and rinse the entire plate as described in *Protocol 1*.

**2** Make 10 twofold serial dilutions of the target protein in CBB in tubes. Adjust the concentration so that the target concentration is 1.11-fold higher than the desired final concentration in the assay. Be sure to change pipette tips between each transfer to prevent carryover error.

**3** Add the most concentrated target solution to a plastic trough and dispense 90 μl to column 1 of the target-coated plate. Repeat this process using fresh troughs to dispense the remaining dilutions to columns 2–10. Similarly, dispense 90 μl of CBB only to the control columns 11–12.

**4** Prepare 150 μl of each phage stock at 10-fold higher concentration than that required to give 50% saturated binding. Array these solutions in a column on a separate polypropylene (or F, nonstick) microtiter plate. In addition to the phage variant(s), be sure to include the wild-type phage.

**5** Using a P10 multichannel pipettor, transfer 10 μl of the phage stocks to each column starting from column 12 and ending at column 1 (i.e. moving from low to high competitor concentration). Vigorously mix the samples on a miniorbital shaker, remove and seal the plate with tape cover, and incubate the plate at room temperature for 60 min.

**6** Develop plates as described in *Protocol 2*, steps 5–8.

Frequently, it is necessary to run a preliminary assay to determine the best concentration from which to initiate the dilution series so that a complete displacement curve is obtained. Adjustments should be made so that the higher concentrations of competitor reduce the phage binding to near the level of nonspecific binding (column 12). Similarly, the lower concentrations of competitor should closely approach the phage binding of the no-competitor control (column 11). The above protocol makes use of common solutions of competitor to minimize the impact of dilution errors on determinations of relative affinity. However, if the phage variant affinities are dramatically different, greater accuracy can be obtained by making individual dilution sets tailored to give approximately the same number of serial dilutions before reaching the concentration corresponding to the $EC_{50}$.

## 3.4 Fitting of displacement curves

Displacement curves can be fit to a standard sigmoidal function (see 9) in order to calculate the $EC_{50}$ value:

$$f = (m3 - m4)/((EC_{50}/C)^{m2} + 1)) + m4 \qquad (6)$$

where $f$ is the signal from bound phage (i.e. absorbance reading), $m2$ is a slope constant, $m3$ is the maximal signal, $m4$ is the minimum signal, and $C$ is the total competitor concentration. This can be accomplished using the Kaleidagraph

(a)

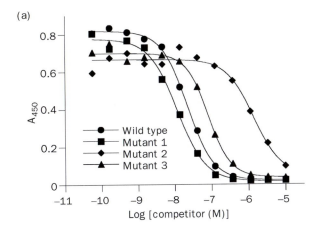

(b)

| | Wild type | Mutant 1 | Mutant 2 | Mutant 3 |
|---|---|---|---|---|
| $m1$ | −7.66 | −7.93 | −5.86 | −7.09 |
| $m2$ | −1.24 | −1.15 | −1.17 | −1.45 |
| $m3$ | 0.02 | 0.02 | 0.05 | 0.04 |
| $m4$ | 0.82 | 0.78 | 0.67 | 0.70 |
| $EC_{50}(nm)$ | 22 | 12 | 1366 | 80 |

**Figure 4**

(Synergy Software) least-squares fitting routine, by applying the following user-defined equation to a graph of signal versus competitor concentration:

$$((m4 - m3)/((m1/m0)^{\wedge}m2) + 1)) + m3; \tag{7}$$
$$m1 = 1e - 7; \quad m2 = -1.0; \quad m3 = 0.9; \quad m4 = 0.1$$

where $m0$ is the competitor concentration and $m1$ is the $EC_{50}$ concentration, both in molar units. The values that follow the equation are user-defined estimates that must sometimes be adjusted because of the poor convergence properties in fitting this equation. The initial estimate of the $EC_{50}$ value (100 nM in this example) is particularly important.

Frequently, it is desirable to fix $m3$ and $m4$ to the values obtained for the no-competitor and the nonspecific binding controls (see *Protocol 3*) to improve the curve fit. This is accomplished by simply replacing these equation variables with the appropriate values prior to fitting. Additionally, better convergence can be obtained by fitting the binding signal versus the logarithm of the competitor concentration. In this case, the user-defined equation should be modified to the following:

$$((m4 - m3)/(10^{\wedge}(m1 - m0) \times m2)) + 1) + m3; \tag{8}$$
$$m1 = -7; \quad m2 = -1.0; \quad m3 = 0.9; \quad m4 = 0.1$$

Here, $m1$ is equal to the logarithm of the $EC_{50}$ so both the initial estimate and the resulting fit will need to be converted appropriately. *Figure 4* shows an example of an actual fit of variant phage binding data using this approach.

## 4 Concluding remarks

Phage binding assays provide a rapid method for screening large numbers of phage and selecting the best candidates for further examination. As such, the assay is a useful tool for realizing the potential of phage display technology. However, phage binding assays do not provide definitive determinations of absolute binding affinities. Binding results can be influenced by a myriad of factors, including the level of display of the fusion protein on phage, unknown post-translational modifications, and avidity induced by increased protein display. It is essential that the optimized variants identified from these screens be verified and characterized by conventional binding analysis of the purified protein order to eliminate possible artifacts related to phage display.

## References

1. Cunningham, B. C., Lowe, D. G., Li, B., Bennett, B. D., and Wells, J. A. (1994). *EMBO J.*, **13**, 2508.
2. Lowman, H. B., Bass, S. H., Simpson, N., and Wells, J. A. (1991). *Biochemistry*, **30**, 10832.
3. Fairbrother, W. J., Christinger, H. W., Cochran, A. G., Fuh, G., Kennan, C. J., Quan, C. et al. (1999). *Biochemistry*, **37**, 17754.
4. DeLano, W. L., Ultsch, M. H., de Vos, A. M., and Wells, J. A. (2000). *Science*, **287**, 1279.
5. Englebienne, P. (ed.) (1999). *Immune and receptor assays in theory and practice*, CRC Press, Cleveland, OH.
6. Cunningham, B. C., Jhurani, P., Ng, P., and Wells, J. A. (1989). *Science*, **243**, 1330.
7. DeLano, W. L. (1999). *Mimicry and analysis of convergent protein interactions*, Ph.D. Thesis, University of California, San Francisco.
8. Day, L. A. and Berkowitz, S. A. (1977). *J. Mol. Biol.*, **116**, 603–6.
9. Peterman, J. H. (1991). In *Immunochemistry of solid-phase immunoassay* (ed. J. E. Butler), pp. 47–65, CRC Press, Boston.

# Chapter 6

# Identification of novel ligands for receptors using recombinant peptide libraries

## Steven E. Cwirla and William J. Dower
Xeno Port, Inc., 3410 Central Expressway, Santa Clara,
CA 95051, USA.

## Christian M. Gates
Amersham Biosciences, 800 Centennial Avenue, Piscataway,
NJ 08855, USA.

## Christopher R. Wagstrom and Peter J. Schatz
Affymax Inc., 4001 Miranda Avenue, Palo Alto,
CA 94304, USA.

## 1 Introduction

Large libraries containing billions of random peptide sequences have proven to be a rich source of ligands for a wide variety of protein targets. Examples include cytokine and growth factor receptor antagonists and agonists (1–4), protease inhibitors (5, 6), and modulators of ligand gated ion channels (7). In most cases peptide ligands were isolated by screening phage display libraries against an immobilized purified protein. Recent reports indicate that even more complex targets such as whole cells and tissues can be used to identify peptides that bind specifically to cell surface antigens (8, 9). High affinity ligands 10–20 amino acids in length have been identified from both linear and cyclic peptide libraries. The majority of these sequences bear no resemblance to known ligands for that particular target protein. Indeed the strength of this technology and other combinatorial approaches is based on the notion that ligands to receptors can be identified without prior knowledge of the nature of the interaction.

In this chapter we describe a multistep approach to identify high affinity peptide ligands for receptors. The first step involves constructing a random peptide library in a filamentous phagemid pVIII display vector. The library is then screened under conditions that allow the isolation of relatively weak binding sequences. The goal at this point is to identify one or more families of ligands that specifically bind to the target protein. Secondary libraries, which contain a large number of variants of the lead peptide sequences, can then be constructed in affinity selective display systems. A number of oligonucleotide mutagenesis strategies are employed at this point to fix certain residues while varying others, or to add random residues at each termini to probe for additional contact points with the receptor. The libraries are then subjected to selection methods that are designed to enrich for clones that display higher affinity peptides. Characterizing the isolated clones in an affinity sensitive enzyme linked immunoadsorbent assay completes the process.

While we have applied these techniques to the discovery of high-affinity ligands to cytokine and growth factor receptors, they are applicable to any target that can be produced in a soluble form and immobilized for screening.

## 2 Description of the peptide display systems

We employ several types of display systems to create large libraries of random peptides. The phage display system described in this chapter utilizes the phagemid vector p8V5 (*Figure 1*) in which the cloned oligonucleotide is inserted at the 5' end of gVIII. Peptides encoded by the oligonucleotides are displayed fused to a linker sequence at the N-terminus of the major coat protein pVIII. The expression of gVIII is under the control of the inducible arabinose promoter, which allows one to produce phage particles with ~100–500 copies of peptide-bearing coat protein relative to the wild-type protein supplied by helper phage. Induction conditions used for primary library construction are designed to result in the

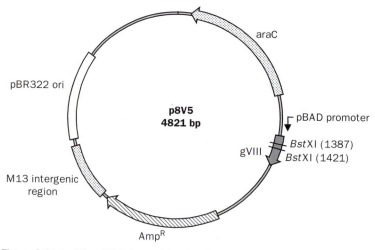

**Figure 1** Map of the p8V5 phagemid vector. The vector was constructed by inserting gene VIII of M13 into the multiple cloning site of the phagemid vector pBAD18 (10). The pBAD promoter of the arabinose operon and its regulatory gene, *ara*C, tightly regulate the expression of pVIII.

display of several hundred copies of peptide on each phage (polyvalent display). Displaying high levels of peptide per phage particle encourages multivalent attachment to the immobilized target, which increases the overall sensitivity of the primary screening process.

Another system that we routinely employ for random peptide library screening is called "peptides-on-plasmids" (11). This system utilizes the *Escherichia coli lac* repressor protein (LacI) where peptides are fused to the C-terminus of LacI rather than the N-terminus of the fusion partner as in phage display. The DNA binding activity of the repressor protein physically links the peptides to the plasmid encoding them by binding to the *lac* operator (*lacO*) sequences present on the plasmid. Complexes form intracellularly and are sufficiently stable to allow affinity selection on an immobilized target. To amplify the selected population the plasmid DNA is released from the bound complexes and reintroduced into *E. coli* by transformation. The complexes may form multivalent attachments with the immobilized target since the repressor normally exists as a tetramer. As with the pVIII display system this avidity effect is an asset in the initial rounds of selection.

Once lead peptides have been identified, secondary libraries constructed in lower valency display systems are used to select for higher affinity variants. The two systems we employ are based on those used for primary library screening. The first is a phagemid system in which the mutagenized oligonucleotide is inserted at the 5′ end of gene III. This places the expressed peptides on the N-terminus of the minor coat protein pIII, which is present at three to five copies on the tip of each phage particle. The amount of fusion protein that is displayed can be controlled to some degree by the induction conditions used during the amplification of the phage. In theory it is possible to produce a population of phage particles that display mostly the wild-type pIII protein supplied by helper phage with only a small proportion displaying one or more copies fused to peptide. Lower valency display favors the selection of clones based on the intrinsic affinity of the displayed peptide. We have found that it is quite difficult to produce a truly monovalent phage display system with peptide libraries, and we have often isolated clones with weak binding affinity despite using very low induction conditions.

A second lower valency display system that has proved to be quite useful in our hands is based on the LacI system and is called "headpiece dimer." The approximately 60 amino acid headpiece domain mediates DNA binding by the *E. coli lac* repressor. The dimer of the headpiece domains that binds to the *lac* operator is normally formed by association of the much larger (~300 amino acid) C-terminal domain. The headpiece dimer system utilizes an engineered DNA binding protein containing two headpiece domains connected via a short peptide linker (12). This protein can bind DNA with sufficient affinity to allow association of a peptide epitope displayed at the C-terminus of the headpiece dimer with the plasmid encoding that peptide. In theory this should produce plasmid complexes that display a single peptide per DNA molecule. We have determined, however, that the *lac* operator sequence is not necessary for plasmid binding by the headpiece dimer protein, so the actual number of peptides displayed per plasmid is unknown. In practice this system has consistently yielded higher affinity peptides than its LacI-based counterpart or even the pIII phagemid display system. Since a variety of pIII phage display systems are described in other chapters, we felt that it might be useful to include a detailed protocol for affinity selection in the headpiece dimer system for lead optimization (*Protocol 8*).

## 3 Constructing random peptide libraries in pVIII phagemid

*Protocol 1* describes the procedure used to prepare vector DNA for library construction. A one-liter culture of bacterial cells containing the p8V5 phagemid vector is grown to saturation and DNA is isolated using a commercially available kit. The expected yield of double-stranded DNA (dsDNA) is approximately 1 mg/L. The vector DNA is digested to completion with *Bst*XI to expose a pair of 3'-overhangs that are neither self-complementary nor complementary to one another. Efficient digestion as well as complete removal of the "stuffer" fragment is important for eliminating the background of nonrecombinant clones from the library. The ultracentrifugation step gradient described in *Protocol 2* is somewhat time consuming, but does allow one to purify large amounts of linearized vector DNA. The expected yield from the step gradient is approximately 50–70%. One advantage of this approach is that small amounts of undigested vector DNA, which could contribute a high level of nonrecombinant clones to a library, form a pellet in the bottom of the tube and can be eliminated from the preparation. Alternative methods of purification utilizing affinity matrix or size exclusion spin columns do not offer this advantage. To construct a library of $10^{10}$ recombinants requires approximately 20 µg of linearized p8V5.

The insert encoding the random peptide region is prepared from two oligonucleotides (*Figure 2*). The first oligonucleotide (ON-1) contains the variable oligocodon region flanked by sequence on the 5' end that reconstructs the C-terminal portion of the signal peptide in p8V5 and on the 3' end contains a region of overlap to anneal with the second oligonucleotide. A *Bsi*HKAI site is located at the 5' end of ON-1. The second oligonucleotide (ON-2) encodes a spacer region consisting of three $Gly_4Ser$ repeats and contains a *Bst*XI site at its 3' end. The variable region of ON-1 is created during synthesis of the DNA and is encoded using the codon scheme NNK, where N is an equimolar mixture of the nucleotides A, C, G, and T; and K is an equimolar mixture of G and T. This scheme encodes all 20 amino acids and one (amber) stop codon using 32 different codons.

The oligonucleotides are purified by polyacrylamide gel electrophoresis. Equimolar amounts of ON-1 and ON-2 are mixed and allowed to anneal followed by extending the 3' ends with T7 DNA polymerase to create a double stranded fragment. The insert DNA is then digested to completion with *Bsi*HKAI and

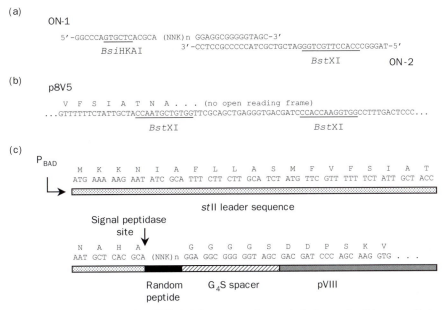

**Figure 2** Cloning scheme for p8V5. (a) Sequence of annealed oligonucleotides encoding a random peptide and Gly₄Ser spacer. Restriction sites used for cloning are underlined. (b) Sequence of the p8V5 cloning site comprised of two noncomplementary *Bst*XI sites. (c) Diagram of a random peptide insert cloned into p8V5. The heat-stable enterotoxin II (stII) leader sequence of *E. coli* directs the assembly of peptide-pVIII fusion protein into the inner membrane. Following signal peptide cleavage, peptides are displayed fused to a flexible linker (Gly₄Ser) at the N-terminus of mature pVIII.

*Bst*XI to create 3′ overhangs that are complementary to the two *Bst*XI overhangs found in p8V5. To assess the ligation efficiency of the prepared insert a series of test ligations should be performed prior to committing large amounts of linearized vector. The number of transformants obtained from the vector plus insert control should be 50–100 times greater than the sample that contains only vector.

The large-scale ligation described in *Protocol 2* is designed to produce libraries of up to $10^{10}$ individual transformants. The quality of electrocompetent cells used to transform the ligated DNA is perhaps the most important factor in producing large libraries. Electrocompetent MC1061 F′ cells are best prepared from cells cultured in a rich medium and harvested in early exponential growth phase. The washed cells should be resuspended in 10% glycerol at a minimum concentration of $5 \times 10^{10}$ cells per milliliter, frozen on dry ice, and stored at $-70°$C. A test transformation with uncut pUC18 DNA should yield $2-5 \times 10^{10}$ transformants per microgram. We typically observe approximately 10-fold lower transformation efficiencies for uncut p8V5.

Individual transformations of the large-scale ligation are pooled after a 1 h period of outgrowth, and added to a larger culture of medium containing antibiotic to select for the presence of phagemid, and glucose to repress the expression of the pVIII fusion protein. The cells are grown for 3–4 h (approximately 10-doublings) and are then harvested. During this period of growth no phage particles are produced since helper phage has not been added. One fifth of this culture, containing ~100 copies of each transformant, is used to produce a library of phage particles (*Protocol 4*) while the remaining cells are stored at −70°C for future library amplifications.

Phagemid library production requires the addition of helper phage to the culture. The cells are infected and then diluted fivefold into a culture containing antibiotics to select for the presence of phagemid and helper phage. The expression of pVIII fusion protein is under the control of the araB promoter in p8V5. In the presence of arabinose, transcription from the promoter is induced; in the presence of glucose it is strongly repressed. Arabinose and glucose are present at a final concentration of 0.2% in the amplification culture. We have found that this is the maximum level of induction that can be used without negatively affecting the overall yield of phage particles. Protein sequencing of intact phage particles from several clones produced using these conditions revealed that on average several hundred copies of recombinant pVIII were present on each phage. This number will vary greatly according to the individual peptide sequence that is displayed.

The phage library is then harvested by precipitation, resuspended in aliquots of about 100–1000 library equivalents, and stored at −20 or −70 °C. One library equivalent is considered to be the number of phage particles equal to the number of independent transformants that constitute the library. To assess library diversity, DNA from individual clones may be sequenced to deduce the primary structure of the displayed peptide. Sequencing twenty to thirty clones is usually sufficient to detect large biases in the library.

## Protocol 1

## pVIII phagemid library construction: vector preparation

### Equipment and reagents

- p8V5 phagemid vector (Affymax Inc.)
- Luria–Bertani (LB) amp medium (1% bacto-tryptone, 0.5% bacto-yeast extract, 0.5% NaCl, 100 µg/ml ampicillin)
- *Bst*XI (New England Biolabs)
- 10 × NEBuffer 3 (0.5 M Tris-HCl, pH 7.9 at 25 °C, 1 M NaCl, 100 mM MgCl$_2$, 10 mM dithiothreitol)
- Phenol : chloroform (1 : 1)
- TE buffer (10 mM Tris-HCl, 1 mM ethylenediaminetetraacetic acid (EDTA), pH 8.0)
- Potassium acetate solutions (10, 15, 20, and 25% w/v in 2 mM EDTA, 1 µg/ml ethidium bromide)
- Qiagen Plasmid Maxi Kit (Qiagen Inc.)
- 3 M NaOAc, pH 5.4 at 25 °C

## Method

### A. p8V5 DNA isolation and *Bst*XI digestion

1  Inoculate 5 ml of LB amp medium with a single colony of MC1061 p8V5 and grow at 37 °C with shaking for 6–8 h. Add culture to a 2.8 L Fernbach flask containing 1 L LB amp medium. Grow at 37 °C shaking vigorously overnight.

2  Isolate double-stranded phagemid DNA using Qiagen Plasmid Maxi columns according to the manufacturer's directions. The expected yield of double stranded p8V5 DNA is ~1 mg.

3  Combine 200 µg p8V5 DNA with 40 µl 10X NEBuffer 3 and 20 µl *Bst*XI (200 units) in a total volume of 400 µl. Incubate the restriction digest overnight at 55 °C.

4  Extract the reaction mixture twice with an equal volume of phenol : chloroform and once with chloroform.

5  Precipitate the DNA by adding 40 µl 3 M NaOAc solution and 880 µl ethanol. Spin in a microcentrifuge for 10 min at room temperature. Carefully remove the supernatant and resuspend the DNA pellet in 400 µl TE buffer.

### B. Potassium acetate step gradient centrifugation

1  Slowly layer 1 ml of each potassium acetate solution, in order of decreasing density, into two 13 × 55 mm polyallomer ultracentrifuge tubes. Carefully layer 200 µl of digested p8V5 on top of each gradient and ultracentrufuge in a SW50.1 rotor at 280,000$g$ (48,000 rpm) for 3 h at 20 °C.

2  Visualize the broad band of linerarized DNA contained in the lower part of each tube using a longwave hand-held UV lamp. Remove each band with a syringe and extract the solution several times with water-saturated butanol to remove the ethidium bromide.

3  Combine the aqueous fractions and add 0.1 volume 3 M sodium acetate solution and 2 volumes ethanol. Spin in a microcentrifuge for 10 min at room temperature to precipitate the DNA. Wash the pellet with 70% ethanol and resuspend in TE buffer. Determine the DNA concentration by measuring the absorbance at 260 nm ($A_{260}$).

## Protocol 2

## pVIII phagemid library construction: insert preparation

### Equipment and reagents

- TBE (0.089 M Tris-HCl, 0.089 M boric acid, 2 mM EDTA, pH 8.3)

- 8% polyacrylamide / 7 M urea / TBE gel

- 2X Prep TBE-urea sample buffer (Invitrogen)

- Fluorescent aluminum oxide TLC plate (J.T. Baker)

- Microspin Sephadex G-25 columns (Amersham Biosciences)
- *Bst*XI, *Bsi*HKAI (New England Biolabs)
- 5X Sequenase buffer (200 mM Tris-HCl, pH 7.5, 100 mM $MgCl_2$, 250 mM NaCl)

- Sequenase 2.0 (Amersham Biosciences)
- Deoxyribonucleotide triphosphate (dNTP) mix (25 mM of each dNTP)
- Microspin S-200 HR columns (Amersham Biosciences)

## Method

### A. Oligonucleotide synthesis and purification

1   Synthesize oligonucleotides with the following sequences:

ON-1: 5′-GGCCCAGTGCTCACGCA (NNK)*n* GGAGGCGGGGGTAGC-3′

N = A,G,C,T (equimolar)[a], K = G,T (equimolar), *n* is the desired number of random amino acids

ON-2: 5′-TAGGGCCCACCTTGCTGGGATCGTCGGAGCCACCTCCCCCACTTCCCCCACC-GCCGCTACCCCCGCCTCC-3′.

2   Elute each oligonucleotide from the synthesis column with 2 ml of 30% $NH_4OH$ and deprotect overnight at 55 °C. Lyophilize the crude deprotected oligonucleotide and resuspend in 400 μl water. Determine the DNA concentration by measuring the $A_{260}$.

3   Add 500 μg of each oligonucleotide to an equal volume of 2 × Prep TBE-urea sample buffer and load into a 2 cm well of an 8% polyacrylamide/7 M urea/TBE gel (24 × 15 cm, 1.5 mm spacers). Add sample buffer that contains xylene cyanol dye to one well.

4   Electrophorese at 500 V until the xylene cyanol dye reaches the bottom of the gel. Remove the gel and place it on plastic wrap on top of a fluorescent TLC plate. Visualize the dark oligonucleotide band by briefly shadowing with a shortwave hand-held UV lamp. Cut out the band, transfer to a microfuge tube and crush it into small pieces. Add 1 ml of water and elute the DNA overnight at room temperature.

5   Centrifuge briefly to pellet the polyacrylamide gel fragments. Transfer the solution to a new microfuge tube and reduce the volume to approximately 200 μl by evaporation in a Speed Vac. Desalt each oligonucleotide by loading onto a Microspin Sephadex G-25 spin column.

6   Collect the flow-through, dry each sample in a Speed Vac, and resuspend the pellets in water. Measure the $A_{260}$ to determine the DNA concentration.

### B. Fill-in reaction and restriction enzyme digest

1   Combine the following in a microcentrifuge tube:

- 100 pmol ON-1,
- 100 pmol ON-2,

- 10 μl 5 × Sequenase buffer, and
- Water to 45 μl total reaction volume.

2   Incubate at 70 °C for 5 min and cool slowly to room temperature to allow oligo-nucleotides to anneal.

**3** Add 2.5 µl 25 mM dNTP mix and 2.5 µl Sequenase 2.0 (32 units). Incubate at 37 °C for 1 h.

**4** Incubate at 70 °C for 10 min to inactivate the polymerase. Remove excess dNTPs from the fill-in reaction by gel filtration using Microspin S-200 HR columns.

**5** Combine 40 µl of the fill-in reaction with 20 µl NEBuffer 3. Add 10 µl *Bst*XI (200 U), 10 µl *Bsi*HKAI (200 U), and 120 µl of water. Incubate the restriction digest at 55 °C overnight in a covered water bath.

**6** Extract the reaction mixture twice with an equal volume of phenol : chloroform and once with chloroform. Purify the digested fragment by gel filtration using Microspin S-200 HR columns. Measure the $A_{260}$ to estimate the DNA concentration.[b]

[a] The phosphoramidites are pre-mixed in a single reagent bottle rather than allowing the synthesizer to deliver equal amounts from four bottles. This method consistently yields the expected distribution of bases in the random codons.

[b] A small amount of insert DNA (~10 µl) should be electrophoresed on a 10% polyacrylamide TBE gel to determine if the digest is complete.

## Protocol 3

# pVIII phagemid library construction: ligation of insert with *Bst*XI-digested p8V5

### Equipment and reagents

- 10X ligation buffer (500 mM Tris-HCl, pH 7.8, 100 mM MgCl$_2$, 100 mM DTT, 1 mM ATP, 500 µg/ml BSA)
- T4 DNA ligase (New England Biolabs)
- *E. coli* MC1061 F′ (Affymax Inc.)
- Electroporation apparatus (*E. coli* pulser, BioRad)
- Superbroth (3.5% bacto-tryptone, 2% bacto-yeast extract, 0.5% NaCl, pH 7.5 with NaOH)
- SOC medium (2% bacto-tryptone, 0.5% bacto-yeast extract, 10 mM NaCl, 2.5 mM

KCl, 10 mM MgCl$_2$, 10 mM MgSO$_4$, 20 mM glucose)
- LB amp medium (*Protocol 1*)
- 3 M sodium acetate
- SOP amp/glu medium (2% bacto-tryptone, 1% bacto-yeast extract, 100 mM NaCl, 15 mM K$_2$HPO$_4$, 5 mM MgSO$_4$, pH 7.2, 100 µg/ml ampicillin, 0.2% w/v glucose)
- 13 ml centrifuge tubes (Sarstedt)
- LB/15% glycerol (v/v)

### A. Ligation

**1** Combine the following in a 13 ml centrifuge tube:

- 20 µg p8V5 (*Bst*XI digested and purified)
- 1.25 µg insert DNA (*Bst*XI and *Bsi*HKAI digested and purified)
- 200 µl 10X ligation buffer
- 4 µl T4 DNA ligase (units)
- water to 2 ml total reaction volume

**2** Incubate at 14 °C overnight. Precipitate the DNA by adding 0.1 volume 3 M sodium acetate and 2 volumes ethanol. Centrifuge at 10,000 rpm (JA.20 Rotor) for 15 min. Remove the supernatant and resuspend the pellet in 400 µl of TE buffer. Transfer to a microcentrifuge tube and reprecipitate the DNA by adding 0.1 volume 3 M sodium acetate and 2 volumes ethanol. Wash the pellet with 70% ethanol, dry, and resuspend the ligated DNA in 20 µl TE buffer.

## B. Preparation of electrocompetent cells

**1** Prepare a culture of *E. coli* MC1061 F′ cells in 1 L of superbroth medium and harvest in early to mid log phase ($OD_{600} = 0.5$).

**2** Chill the growth flask on ice for 30 min. Centrifuge the cell suspension at 5,000 rpm (JA.10 Rotor) for 15 min at 4 °C.

**3** Resuspend the cell pellets in a total of 1 L of cold water. Repeat the centrifugation step and resuspend the cell pellets in a total of 0.5 L of cold water.

**4** Repeat the centrifugation step and resuspend the cells in 20 ml of cold 10% glycerol. Centrifuge the cell suspension at 6,000 rpm for 15 min at 4 °C.

**5** Resuspend the cell pellet in cold 10% glycerol to a final volume of 2 ml.[a] Freeze 200 µl aliquots in microcentrifuge tubes on dry ice and transfer to −70 °C for long term storage.

## C. Electrotransformation

**1** Perform four electrotransformations, each containing 200 µl of cells and 5 µl of ligated DNA in a 0.2 cm electroporation cuvette, by pulsing at 2.5 kV. Immediately transfer the cells from each cuvette to a culture tube containing 2 ml SOC medium and incubate without selection for 1 h at 37 °C.

**2** Combine the transformations and prepare 10-fold serial dilutions from a small aliquot. Plate 100 µl of the $10^{-4}$-$10^{-6}$ dilutions on LB amp agar plates to assess the efficiency of transformation. Add the remainder to 500 ml of SOP amp/glu medium and grow with vigorous shaking at 37 °C to an $OD_{600}$ of 0.8.

**3** Remove 100 ml of the culture and proceed to *Protocol 4* to prepare phage from the library. Centrifuge the remaining cell suspension at 6,000 rpm for 10 min.

**4** Resuspend the cell pellet in 10 mL LB/15% glycerol. Transfer 1 mL aliquots to Nunc freezer vials and store at −70 °C for future library amplifications.

[a] The cells should be resuspended at a concentration $5 \times 10^{10}$ to $10^{11}$ cells/ml to obtain the highest transformation efficiency.

## Protocol 4

# pVIII phagemid library construction: amplification and storage

### Equipment and reagents

- VCS-M13 helper phage (Stratagene; see Chapter 2, *Protocol 6* for preparation)
- 20% arabinose (w/v)
- 20 mg/mL kanamycin (Kan)
- PEG/NaCl (20% polyethylene glycol, average molecular weight 8000, 2.5 M NaCl)
- Phosphate buffered saline (PBS) (0.137 M NaCl, 4.3 mM $Na_2HPO_4$, 1.4 mM $KH_2PO_4$, 2.7 mM KCl, pH 7.4 at 25 °C)
- SOP amp/glu medium (*Protocol 3*)

### A. Helper phage infection and pVIII induction

1 Add VCS-M13 helper phage at a multiplicity of infection of 10:1 ($\sim 10^{12}$ plaque forming units (pfu)) to the 100 ml culture of transformed cells (*Protocol 3*). Incubate at 37 °C without shaking for 30 min.

2 Add the entire culture to a flask containing 500 ml SOP amp/glu medium. Add 0.6 ml of Kan solution and 6 ml of 20% arabinose. Incubate at 37 °C with shaking overnight.

### B. Harvest of amplified phagemid library

1 Centrifuge the overnight culture at 12,000g (8,000 rpm in a JA-10 rotor) for 15 min at 4 °C. Transfer the supernatant to a new bottle and repeat the spin.

2 Transfer the cleared supernatant to a new bottle and precipitate the phage by adding 0.2 volumes of PEG/NaCl. Mix well and hold on ice for 1 h.

3 Pellet the phage by centrifugation at 8,000 rpm for 15 min. Carefully remove the supernatant and resuspend the phage pellet in 10 ml of PBS. Titer the phage stock; the suspension may be stored for up to 24 h at 4 °C while the titer is determined. For long-term storage dispense the phage into aliquots of 1,000 library equivalents and store at −20 °C.

## 4 Screening pVIII phagemid libraries

Numerous methods have been devised to screen very large collections of peptides displayed on phage. In its simplest form, the target of interest is immobilized on a solid support by direct adsorption and exposed to a library. Phage that are not bound are removed by a series of washes, and binding clones are recovered by acid elution.

*Protocol 5* describes conditions that we routinely use to isolate peptides that bind to growth factor and cytokine receptors. These targets are often extracellular domains of single transmembrane receptors produced in a soluble form (13). A monoclonal antibody (mAb) (e.g. the proprietary Affymax Ab179) that binds with high affinity to a peptide epitope fused to the C-terminus of each of the target proteins, is used to capture and immobilize the target on the surface of enzyme-linked immunosorbant assay (ELISA)-type microtiter plates. We have used this generic method of immobilization to present a wide variety of receptors in an active form for library screening. This format may present a high density of active binding sites that enhances multivalent attachment of the peptides displayed on the phage.

There are many factors that can effect the efficiency of library screening. Some immobilization strategies can compromise the activity of the target. Prior to screening a library one should test the activity of the immobilized target with a known ligand if possible. The density of immobilized receptor should also be maximized to provide the highest number of contact points for target-binding phage. The enrichment of specific binding clones can be hindered by the recovery of phage that bind to things other than the target protein, such as proteins used to capture the target and even the solid support itself. Blocking with a solution of bovine serum albumin (BSA) is usually sufficient to prevent the isolation of clones that bind to plastic, and by including it in the binding buffer one can also eliminate clones that bind to albumin. We also include a high concentration of synthetic peptide ligand to Ab179 in the binding buffer to eliminate clones that bind to combining sites not occupied by receptor. Targets captured with streptavidin or protein A should be screened in the presence of excess biotin or IgG, respectively.

Normally 100–1,000 library equivalents of phage are exposed to the immobilized receptor. For a library of $10^{10}$ independent transformants this amounts to $10^{12}$–$10^{13}$ infectious phage particles. The incubations and washes are carried out in the cold to minimize dissociation of phage from the surface once bound. After completing this series of washes an acid elution solution is added to each well to recover the captured phage. For some targets multiple rounds of affinity selection yield only nonspecific binding phage and it is necessary to use specific elution techniques. Many of the soluble receptors we have screened contain a thrombin cleavage site between the extracellular domain of the receptor and the Ab179 epitope (2). Receptor bound phage can be released by adding thrombin to each well, which will cleave the receptor–phage complex from the immobilized antibody. It does not appear to be necessary to disrupt the receptor–phage complex prior to amplification, but this could be accomplished by simply heating the eluate at 70 °C for 10 min.

Amplifying the recovered phage completes a single round of affinity enrichment. An F' *E. coli* host is required for phage infection. A host that is also deficient for recombination (recA⁻) will reduce the frequency of deletions in the phagemid that occur during the amplification process. Exponentially growing cells are concentrated prior to infection such that the multiplicity of infection is between $10^{-2}$ and $10^{-3}$. After a short incubation helper phage are added to the culture and allowed to infect the cells. Antibiotics are added to select for the presence of phagemid and helper phage, and arabinose is added to induce the expression of the recombinant gene VIII. After overnight growth the phage are harvested, concentrated, and selected again on immobilized receptor.

## Protocol 5

## Screening pVIII phagemid libraries: affinity selection

### Equipment and reagents

- ELISA microplates (Dynatech Immulon-4) PBS (see *Protocol 4*)
- PBS/0.1% and 1% BSA
- Antibody blocking peptide
- Acid elution buffer (0.1 M HCl, pH 2.2 with glycine, 0.1% BSA)

### Method

1  Dilute the capture antibody to 100 μg/ml in PBS. Add 50 μl to twelve wells of a 96-well microplate and incubate at 37 °C for 1 h.

2  Wash the wells three times with PBS. Completely fill the wells with PBS/1% BSA and incubate for 1 h at 37 °C to block the plate.

3  Wash the plate with PBS, and add 100 μl of soluble receptor to each well. Incubate at room temperature for 1 h or alternatively the plate may be stored at 4 °C for up to 24 h.

4  Wash the receptor coated wells three times with PBS. Add 50 μl of a 100 μM solution of antibody blocking peptide to each well, and incubate at 4 °C for 30 min.

5  Dilute one aliquot of the phagemid peptide library to a final volume of 1.2 ml in PBS/0.1% BSA and add 100 μl to each receptor coated well. Incubate the plate at 4 °C for 2 h with shaking.

6  Wash the wells five times with cold PBS. Completely fill the wells with cold PBS/0.1% BSA and incubate at 4 °C for 30 min.

7  Wash the wells five times with cold PBS. Add 100 μl of acid elution buffer to each well and incubate at room temperature for 10 min. Transfer eluates to a microcentrifuge tube and neutralize with 70 μl of 2 M Tris base (pH unadjusted).

---

## Protocol 6

### Screening pVIII phagemid libraries: amplification of selected phage

#### Equipment and reagents

- *E. coli* ARI 292—derived from K91: Hfr-Cavalli thi recA::cat (Affymax Inc.)
- LB medium (1% bacto-tryptone, 0.5% bacto-yeast extract, 0.5% NaCl)
- PBS/0.1% BSA

#### Method

1 Inoculate 5 ml of LB medium with a single colony of ARI 292. Grow at 37 °C with shaking overnight.

2 Add 1 ml of the overnight culture to a flask containing 20 ml LB medium and grow with shaking to an $OD_{600}$ of 0.4. Centrifuge the culture at 6,000 rpm (JA.20 rotor) for 10 min at 4 °C. Resuspend the cell pellet in 2 ml LB medium.

3 Add 1 ml of eluate to 1 ml of ARI 292 cells and incubate at 37 °C for 20 min without shaking. Add the phage infected cells to a flask containing 20 ml SOP amp/glu medium and grow at 37 °C with shaking to an $OD_{600}$ of 0.4.

4 Add VCS-M13 helper phage ($\sim10^{10}$ pfu) to the culture and incubate at 37 °C for 20 min without shaking. Add 220 µl 20% arabinose and 22 µl 20 mg/ml kan solution to the culture and grow overnight at 37 °C with vigorous shaking.

5 Centrifuge the culture at 10,000 rpm for 15 min to pellet the cells. Transfer the supernatant to a new tube and add 0.2 volumes PEG/NaCl. Mix well and hold on ice for 1 h.

6 Pellet the phage by centrifugation at 10,000 rpm for 15 min. Remove the supernatant and resuspend the phage pellet in 1 ml of PBS/0.1% BSA.

## 5 Characterization of recovered clones

Three to four rounds of affinity selection should yield a population of phage that is enriched for receptor specific clones. The total number of phage recovered in later rounds usually increases, but this alone does not indicate that receptor binding peptides have been isolated. Clones that bind nonspecifically to the target or other components of the selection matrix can yield the same result. The phage ELISA described in *Protocol 7* is one of the most sensitive methods for characterizing the binding specificity of recovered clones. Individual wells are coated with receptor, as described in *Protocol 5*, and bound phage are detected with an horseradish peroxidase (HRP) conjugated anti-M13 antibody. Wells that contain only the capture antibody or BSA are included as negative controls. A known ligand competing for receptor binding can be added with the phage to distinguish competitive from non-competitive peptides. It is important to note that the conditions used in this assay are designed to maximize the sensitivity of detection rather than to distinguish clones based on the intrinsic affinity of the displayed peptide. Multivalent binding, differences in expression level, and protease susceptibility of different peptides can greatly influence the signal intensity.

Another method for identifying receptor specific clones from an enriched population is by probing colony filter lifts with receptor. This is done by infecting bacterial cells with selected phage and plating on LB amp/kan agar plates that contain glucose and arabinose at the same concentrations used for liquid amplification. The phage extruded from the bacterial colonies are transferred to nitrocellulose filters and probed with labeled (usually radiolabeled) target protein. The sensitivity of this method is lower than the phage ELISA, however it allows for the rapid analysis of a large number of clones and may be useful in identifying higher affinity sequences.

The peptide sequences displayed on individual positive clones are then deduced by sequencing the DNA insert in p8V5.

## Protocol 7

## Screening pVIII phagemid libraries: characterization of recovered clones by phage ELISA

### Reagents and equipment

- HRP-conjugated anti-M13 antibody (Amercham Biosciences)
- ABTS development buffer (50 mM citric acid, pH 4.0 with NaOH, 0.22 mg/ml 2, 2′-azino-bis 3-ethylbenzthiazoline-6-sulfonic acid; immediately before use add $H_2O_2$ to 0.05%)
- Qiagen Plasmid Mini Kit (Qiagen, Inc.)
- p8V5 reverse sequencing primer (5′-TGAGGCTTGCAGGGAGTC-3′)

### Method

1 Inoculate 5 ml of SOP amp/glu medium with a single colony from a titer plate and grow at 37 °C with shaking to an $OD_{600}$ of 0.5.

2 Add VCS-M13 helper phage ($1 \times 10^9$ pfu) and incubate at 37 °C for 30 min without shaking.

3 Add 50 μl 20% arabinose and 5 μl 20 mg/ml kan to the culture and grow at 37 °C overnight with vigorous shaking.

4 Pellet cells by centrifugation at 3,400 rpm (Tabletop Centrifuge) for 15 min and transfer the phage containing supernatant to a new tube.

5  Immobilize receptor in a 96-well microtiter plate as described in *Protocol 5*. Add 50 µl PBS/0.1% BSA and 50 µl phage containing supernatant to each well and incubate for 2 h at 4 °C.

6  Wash plate five times with PBS. Dilute HRP-conjugated anti-M13 antibody 1 : 5000 in PBS/0.1% BSA and add 50 µl to each well. Incubate the plate for 1 h at 4 °C.

7  Wash plate five times with PBS. Add 100 µl ABTS development buffer to each well and incubate at room temperature until color develops. Measure the $A_{405}$ with a microtiter plate reader.

8  Isolate dsDNA from cultures of ELISA positive clones with Qiagen Plasmid Mini Kit columns according to the manufacturer's directions. Sequence the DNA by the dideoxy method (14) using the p8V5 reverse primer.

## 6  Lead optimization strategies

Successful primary library screening results in the identification of one or more peptide sequences that bind specifically to receptor. The binding affinity of these isolated peptides is typically in the low micromolar range, which is often too low for measuring responses in cell based assays. To improve the affinity of these initial isolates, mutagenesis libraries that contain a large collection of variants are constructed in low valency display systems. The libraries are then screened using a variety of techniques that favor the recovery of higher affinity peptide ligands.

Mutagenesis of a sequence is accomplished through various oligonucleotide synthesis strategies. One typical strategy is to synthesize the first two positions of a codon with a phosphoramidite mixture containing 70% of the starting base and 10% of each of the other three nucleotides. The third position is encoded with an equimolar mixture of G and T. This "70 : 10 : 10 : 10" encoding strategy produces an amino acid change at the targeted position in approximately 50% of the sequences. The mutagenesis rate can be adjusted by increasing or decreasing the ratio of starting to mutant bases in the mixture. Choosing the appropriate degenerate codon scheme can also create subsets of amino acids. For example to examine the requirements for a cysteine in a certain position one can synthesize the codon TST, where S represents an equimolar mixture of G and C. This degenerate codon will encode a 50 : 50 mixture of serine and cysteine. Strongly conserved amino acids in a sequence family can be fixed while varying flanking residues. Additional random residues can also be placed at either end of a peptide with the addition of NNK codons. These strategies are often combined to create secondary libraries that contain as many different peptides as the fully random primary libraries.

The challenge of secondary library screening is to efficiently sort through the large number of variants to identify clones with improved binding affinity for the receptor. Using decreased valency display systems lowers the probability of selecting clones that bind through multivalent interactions with immobilized receptor. A pIII phagemid system similar to that described for pVIII is used in our laboratory for this purpose. The headpiece dimer system that is described in the next section has also proven even more useful for identifying high affinity peptides.

Libraries constructed in these vectors are normally screened for one initial, low stringency round against a high density of receptor to isolate a population of clones that bind to the target. The amplified clones are then subjected to subsequent rounds of panning with increased selection stringency. Lowering the density of immobilized receptor favors monovalent interactions between the displayed peptides and receptor. Dissociation of library members bound through low affinity interactions can be accelerated by the addition of a competing natural ligand or a synthetic peptide derived from the initial screens, and the low affinity clones are washed away prior to elution. The process is completed by characterizing clones from the enriched pool in an affinity sensitive ELISA.

## 7 Constructing secondary libraries in the "headpiece dimer" system

Construction and screening of headpiece dimer libraries is similar in many ways to those methods as first developed for LacI libraries (15), therefore only methods that differ substantially from those for LacI libraries will be described here.

The library cloning scheme for the headpiece dimer vector, pCMG42A, is identical to that used for LacI libraries (*Figures 3 and 4*). It uses two mutually incompatible *Sfi*I restriction sites in the vector ligated to a set of three complementary oligonucleotides. Procedures for vector preparation, library construction and library amplification are as previously described (15). The amplification of libraries is carried out in the *E. coli* strain ARI 814.

## 8 Affinity selection of headpiece dimer libraries

Headpiece dimer libraries are screened using a method modified from that used for LacI libraries, with several important differences. For example, the elution of plasmids from wells after washing is done with high salt and phenol instead of with isopropyl $\beta$-D-thiogalactoside (IPTG) because the headpiece dimer does not have the LacI IPTG-binding domain. Caution must be taken with both systems about ionic strength of buffers, as the affinity for many DNA-binding proteins is weakened by high salt. Buffers other than those described here should be tested with a model system before being used in panning of novel receptors.

The following procedure is for panning done in microtiter wells with a target receptor immobilized on a capture antibody such as Ab179. Many other panning formats are possible, including batch panning on beads, or the use of column chromatography materials coated with receptor. Whatever format is used, the receptor should be tested to ensure that it maintains appropriate activity after immobilization.

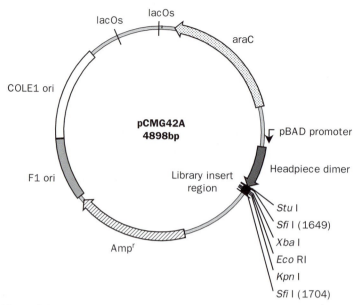

**Figure 3** Map of the pCMG42A vector. The vector is a derivative of pJS142 (15), with expression of the headpiece dimer gene being driven by the pBAD promoter of the arabinose operon, regulated by *araC* that is also present on the vector. Two *Sfi*I sites allow cloning of random oligonucleotides. Unique *Xba*I, *Eco*RI, or *Kpn*I sites allow "killer-cutting" of the vector during preparation of libraries to lower the background of clones without inserts.

(a)

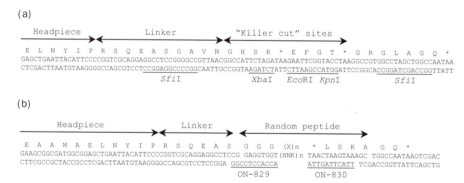

(b)

**Figure 4** Construction of headpiece dimer libraries in vector pCMG42A. (a) Sequence of the cloning region at the 3′ end of the headpiece dimer gene, including the *Sfi*I and killer cut sites used during library construction. (b) Ligation of annealed library oligonucleotide, ON-829, and ON-830 to *Sfi*I sites of pCMG42A to produce a library. Single spaces in the sequence indicate sites of ligation.

## Protocol 8

# Affinity selection of peptide complexes displayed in the headpiece dimer system

### Equipment and reagents

- BSA, protease free (USB)
- Bulk DNA, sonicated, phenol extracted (Sigma)
- Glycogen, molecular biology grade (Roche Molecular Biochemicals)
- Lysozyme from hen egg white (Roche Molecular Biochemicals)
- Microtiter plate (Dynatech Immulon-4)
- PBS without $MgCl_2$ (Sigma)
- Phenol, equilibrated with Tris buffer (USB)
- Phenol : chloroform : isoamyl alcohol (25 : 24 : 1) (USB)
- 0.1 M PMSF in isopropanol
- HE buffer (35 mM HEPES, 0.1 mM EDTA, pH 7.5 with KOH)

- HEK buffer (35 mM N-[2-hydroxyethyl] piperazine-N′-[2-ethanesulfonic acid (HEPES), 0.1 mM EDTA, 50 mM KCl, pH 7.5 with KOH)
- WTEK buffer (50 mM Tris-HCl, pH 7.5, 10 mM EDTA, 100 mM KCl)
- TEK buffer (10 mM Tris-HCl, pH 7.5, 0.1 mM EDTA, 100 mM KCl)
- HEK buffer/1% BSA (fraction V)
- Lysis buffer (4.2 ml HE buffer, 1 ml 50% glycerol, 0.75 ml 10mg/ml BSA in HE buffer, 10μl 0.5 M DTT, 12.5 μl 0.1 M PMSF added immediately before use)
- Elution buffer (10 mM Tris-HCl, 1 mM EDTA, 0.5M NaCl)

### Method

1  Dilute the capture antibody to 100 μg/ml in PBS. Add 100 μl to the appropriate number of wells of a 96-well microtiter plate and incubate at 37 °C for 1h.[a]

2  Wash the wells four times with PBS. Completely fill the wells with PBS/1% BSA and incubate at 37 °C for 1 h to block the plate.[b]

3  Wash the plate four times with PBS. Add 100 μl of soluble receptor to each well and incubate at 4 °C for 1 h to overnight with shaking.

4  Wash the plate four times with cold HEK/1% BSA to remove excess salts that could interfere with headpiece dimer binding.

5  Prepare 1 ml of 10 mg/ml lysozyme in cold HE buffer. Thaw a library aliquot (2 ml) and transfer to a 13 ml Sarstedt screw cap tube that contains 6 ml lysis buffer.

6  Add 150 μl lysozyme solution to the library and invert gently to mix. Incubate for approximately 3 min on ice.

7  As soon as lysis occurs, as evidenced by an increase in viscosity upon gentle inversion, add 200 μl 2M KCl and gently invert to mix. Immediately centrifuge the tubes at 27,000g for 15 min at 4 °C.

8  Gently pipet or pour the clear supernatant to a new tube. (The lysis incubation was too long if the cell debris pellet is very loose.)

9  Add 50–250 μl of crude lysate to each well and incubate at 4 °C for 1 h with shaking.[c]

10  Wash the plate four times with cold HEK/1% BSA followed by 2 times with cold HEK buffer to reduce the amount of residual BSA in the wells.[d]

11  Add 50 µl of elution buffer to each well and mix on a plate shaker for 5 min. Add 50 µl of phenol to each well and mix by shaking. Transfer the eluates to microcentrifuge tubes and spin for 5 min at maximum speed.

12  Transfer the aqueous phase to new tubes and add an equal volume of phenol : chloroform. Mix well by vortexing and spin for 5 min in a microcentrifuge.

13  Transfer the aqueous phase to new tubes, add 1 µl of 20 mg/ml glycogen and mix by vortexing. Add 1 volume of isopropanol to each tube and mix well.

14  Spin in a microcentrifuge for 10 min at maximum speed. Carefully remove the supernatants and wash the DNA pellets with 70% ethanol. Repeat the spin and remove all of the supernatant. Resuspend the DNA pellets in a total volume of 4 µl of water.

15  Perform two electrotransformations, each containing 40 µl of electrocompetent ARI 814 cells and 2 µl of recovered plasmid DNA in a 0.1 cm cuvet, by pulsing at 1.8 kV. Immediately transfer the cells from each cuvet to a culture tube containing 2 ml SOC medium and incubate for 1 h at 37 °C with shaking.

16  Add the cultures to a flask containing 200 ml LB amp medium. Remove a small aliquot and prepare 10-fold serial dilutions. Plate 100 µl of the undiluted to $10^{-3}$ dilution on LB amp agar plates to determine the number of plasmids that were recovered in that round of selection.

17  Incubate the culture with shaking at 37 °C to an $OD_{600}$ of 0.5–0.8.

18  Chill the culture flask in an ice-water bath for at least 10 min. Centrifuge the cell suspension at 6,000g for 5 min at 4 °C.

19  Resuspend the cell pellets in 100 ml cold WTEK buffer. Centrifuge at 6,000g for 5 min at 4 °C.

20  Resuspend the cell pellets in 50 ml cold TEK buffer. Repeat the centrifugation step.

21  Resuspend the cell pellet in 4 ml HEK buffer. Transfer 2 ml aliquots to Nunc freezer vials and store at −70 °C. One aliquot is used for the next round of selection.

[a] In the first round of selection we normally use 24 receptor coated wells. In subsequent rounds 6 wells are coated with receptor while 6 wells containing only antibody serve as a negative control.

[b] Blocking with PBS/1% nonfat milk in later rounds is recommended to reduce the enrichment of BSA specific clones.

[c] If the receptor has requirements for certain ions (e.g. $Mg^{2+}$, $Ca^{2+}$) they can be included at this point, although significant increases in the ionic strength of the lysis buffer should be avoided to prevent disruption of the plasmid complexes.

[d] Competition with a peptide or known ligand, and additional washes can be added in the third round to increase the selection stringency.

## 9 Transfer to MBP for analysis

As described in our previous paper (15), we frequently transfer sequences from late rounds of panning to vectors that fuse those sequences in frame with the gene encoding the *E. coli* maltose binding protein (MBP). This allows examination in an ELISA where the signal generally correlates with the relative affinity of the peptides encoded by each clone. It also permits easy purification of milligram amounts of fusion protein so that many peptides can undergo preliminary analysis of their properties before incurring the expense of chemical peptide synthesis.

Transfer to the MBP vector is accomplished by digesting the enriched plasmid population with *Bsp*EI and *Sca*I and transfering the fragment into one of two MBP vectors, pELM3 or pELM15 (15). To aid in the purification of the library fragment, the digest also includes *Nhe*I to cleave a fragment that would otherwise comigrate with the fragment containing the random coding sequences. After purification of a ~900 bp library fragment, it is ligated to a ~5.6 kb *Age*I-to-*Sca*I fragment from pELM3 or pELM15. Transformation of ARI 814 produces individual clones for analysis by ELISA or by colony lifts with labeled receptor.

The MBP ELISA method is identical to that described previously (15), with the exception of the lysis procedure to release the MBP-peptide fusions from the *E. coli* cells. We have found that lysis of cells can be more easily and efficiently accomplished using a commercial detergent (B-PER extraction reagent from Pierce) rather than using lysozyme. After pelleting induced cells from a 3 ml culture, they are resuspended in 1 ml of B-PER containing 0.1 mM EDTA. After 20 min at room temperature, the cell debris is removed by centrifugation for 5 min followed by transfer of the supernatant to a new tube. The lysate is stored at $-70\,^{\circ}$C and is diluted $1:50$ for use in the ELISA.

In several projects, we have found that high signal in the MBP ELISA is predictive of higher affinity of the peptides for the receptor under study. Of course, eventual chemical synthesis of the peptides is necessary to allow rigorous analysis of their activities. For screening of a larger numbers of clones, colony lifts using standard procedures (14) can allow the isolation of a smaller set of clones for further analysis by ELISA.


## 10 Conclusions

We have found that there is a strong synergy between use of phage-display vectors and *LacI*/headpiece dimer display for the discovery of high affinity peptide ligands. Very often, the two methods identify similar families of peptides (3). In some cases, we have found that one or the other method works better to find initial peptides specific for a particular receptor or to select for the highest affinity variants from a sequence family. This may be due to differences in the orientation of the displayed peptides, or differing biological constraints imposed by the two methods. With either system it is clear that the most efficient approach for identifying high affinity ligands is to isolate at least one sequence family that binds to the target, and to then screen, under increased stringency, additional libraries that contain many variants of the lead peptide. By this multi-step approach one can effectively probe areas of sequence space that could not be represented in a single large library.

## References

1. Yanofsky, S. D., Baldwin, D. N., Butler, J. H., Holden, F. R., Jacobs, J. W., Balasubramanian, P. *et al.* (1996). *Proc. Natl. Acad. Sci. USA*, **93**, 7381.

2. Wrighton, N. C., Farrell, F. X., Chang, R., Kashyap, A. K., Barbone, F. P., Mulcahy, L. S. *et al.* (1996). *Science*, **273**, 458.

3. Cwirla, S. E., Balasubramanian, P., Duffin, D. J., Wagstrom, C. R., Gates, C. M., Singer, S .C. *et al.* (1997). *Science*, **276**, 1696.

4. Lowman, H. B., Chen, Y. N., Skelton, N. J., Mortensen, D. L., Tomlinson, E. E., Sadick, M. D. *et al.* (1998). *Biochemistry*, **37**, 8870.

5. Smith, M. W., Shi, L., and Navre, M. (1994). *J. Biol. Chem.*, **270**, 6440.

6. Koivunen, E., Arap, W., Valtanen, H., Rainisalo, A., Medina, O. P., Heikkila, P. *et al.* (1999). *Nat. Biotech.*, **17**, 768.

7. Li, M., Yu, W., Chen, C. H., Cwirla, S., Whitehorn, E., Tate, E. *et al.* (1996). *Nat. Biotech.*, **14**, 986.

8. Rajotte, D., Arap, W., Hagedorn, M., Koivunen, E., Pasqualini, R., and Ruoslahti, E. (1998). *J. Clin. Invest.*, **102**, 430.

9. Barry, M. A., Dower, W. J., and Johnston, S. A. (1996). *Nat. Med.*, **3**, 299.

10. Guzman, L.-M., Belin, D., Carson, M. J., and Beckwith, J. (1995). *J. Bacteriol.*, **177**, 4129.

11. Cull, M. G., Miller, J. F., and Schatz, P. J. (1992). *Proc. Natl. Acad. Sci.*, **89**, 1865.

12. Gates, C. M., Stemmer, W. P. C., Kaptein, R., and Schatz, P. J. (1996). *J. Mol. Biol.*, **255**, 373.

13. Whitehorn, E. A., Tate, E. T., Yanofsky, S. D., Kochersperger, L. K., Davis, A., Mortensen, R. B. *et al.* (1995). *Biotechnology*, **13**, 1215.

14. Sambrook, J., Fritsch, E. F., and Maniatis, T. (ed.) (1989). *Molecular cloning: a laboratory manual*, 2nd edn., Cold Spring Harbor Laboratory Press, NY.

15. Schatz, P. J., Cull, M. G., Martin, E. L., and Gates, C. M. (1996). *Meth. Enzymol.*, **267**, 171.

# Chapter 7

# Substrate phage display

## David J. Matthews
Exelixis Inc., 170 Harbor Way, South San Francisco, CA 94083, USA.

## Marcus D. Ballinger
Sunesis Pharmaceuticals Inc., 341 Oyster Point Boulevard, South
San Francisco, CA 94080, USA.

## 1 Introduction

Substrate phage is a phage display technique that enables one to rapidly survey
the substrate specificity of enzymes that recognize peptide substrates. The basic
requirements of a substrate phage experiment are as follows:

(a) a suitable library of random peptide sequences displayed on the surface of
  phage particles.
(b) a method for *in vitro* modification of the peptide substrate sequences.
(c) a method for separation of modified peptide phage from unmodified peptide
  phage.

As with other phage display methods, the power of the technique lies in the
ability to repeat the process through many cycles, allowing a small subset of good
substrates to be isolated from a vast library of potential substrates. The first
substrate phage experiments focused on determining substrate specificity for
various proteases (1). In these examples, the peptide modification event is proteolytic cleavage. By inserting the randomized peptide library between the phage
capsid and an affinity domain, it is possible to separate the modified (cleaved)
from unmodified (uncleaved) peptides by capture of the affinity domain (*Figure 1*).
The process has been shown to be applicable to a wide range of different proteases (*Table 1*). It is also worth noting that the substrate phage technique may be
used to isolate peptide sequences that are resistant to a given protease or set of
proteases (*Figure 1*). This "non-substrate phage" approach has been described as a
side product of substrate phage selection (1–3), and also forms the basis of a
method for isolation of substrates that are selective for one protease relative to
another ((10); see Section 5). Variations of this resistance selection have also been
used to isolate protein domains that are stably folded, based on the assumption
that properly folded proteins are less susceptible to proteolysis than unfolded
proteins ((13–15); see Chapter 8).

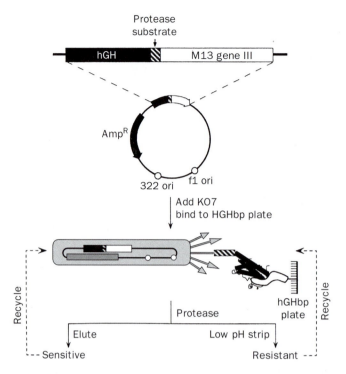

**Figure 1** Scheme for protease substrate phage selection. In this example, phagemid particles display a fusion protein comprising an affinity domain (in this case, a high affinity variant of human growth hormone (hGH)), a randomized substrate linker, and the gene III protein from M13 phage (pIII). Phagemids are captured by binding to immobilized hGH binding protein, then treated with the protease of interest. The phagemids that display good substrates are released and can be subjected to further rounds of selection. Similarly, protease-resistant phagemids may be isolated and iteratively enriched. (Reprinted with permission from (1); Copyright 1993 American Association for the Advancement of Science.)

In this chapter, we will focus on the application of substrate phage to study protease substrate specificity. However, it should be noted that several variations of the method have been described for studying non-proteolytic enzymes, particularly protein tyrosine kinases. These will be discussed in Section 5.

## 2 Preparation of substrate library constructs

The first step in preparation of a substrate phage experiment is construction of the substrate library display vector and subsequent generation of the substrate library. Both monovalent and polyvalent pIII phage display systems may be used for substrate phage (*Table 1*). In the case of monovalent display, only a small fraction of phage particles will bear a single substrate sequence and affinity domain on the phage surface. This means that phage must be captured on a solid support via the affinity domain prior to selection. Polyvalent substrate

**Table 1** Proteases studied using substrate phage

| Enzyme | Display method | Affinity tag | Reference |
| --- | --- | --- | --- |
| Subtilisin mutant S24C/H64A/E156S/ G166A/G169A/Y217L | Monovalent | hGH/hGH binding protein | 1 |
| Subtilisin mutant N62D/G166D | Monovalent | hGH/hGH binding protein | 2 |
| Furin | Monvalent | hGH/hGH binding protein | 3 |
| Factor Xa | Monovalent | hGH/hGH binding protein | 1 |
| HSV-1 protease | Monovalent | BPTI/anhydrotrypsin | 4 |
| MCP-7 | Monovalent | FLAG peptide/ anti-Flag | 5 |
| Granzyme B mutants | Monovalent | His-tag/Ni-NTA resin | 6 |
| Stromelysin | Polyvalent | YGGFL/mAb 3E7 and ACLEPYTACD/mAb 179 | 7 |
| Matrilysin | Polyvalent | YGGFL/mAb 3E7 and ACLEPYTACD/mAb 179 | 7 |
| Urokinase | Polyvalent | YGGFL/mAb 3E7 | 8, 10 |
| Tissue-plasminogen activator | Polyvalent | YGGFL/mAb 3E7 | 9, 10 |
| Prostate specific antigen | Polyvalent | YGGFL/mAb3E7 | 11 |
| Elastase mutant H57A | Monovalent | hGH/hGH binding protein | 12 |

phage particles are each expected to contain more than one copy of the affinity domain and substrate, so multiple substrates may have to be cleaved to select a given phage particle. Assuming that all polyvalent phage capsids display substrates, selection may be performed in solution with subsequent capture of the affinity domain. This offers advantages in terms of scalability; however, efficient selection requires that all non-cleaved phage are removed by capture of the affinity domain. In practice, there may be some nonreactive phage clones with defective epitopes that will be selected as substrates by this procedure. Also, a given population of polyvalent phage is likely to contain many phage that undergo degradation, and thus the valency of individual virions may be variable. Polyvalent phage are also required to have a full-length, intact pIII protein for infectivity. This may be used as a selective advantage if cleaved phage show greater infectivity than uncleaved phage. However, some library sequences may have higher infectivity than others even in the absence of enzymatic clea-vage, thus biasing the distribution of the starting library and subsequent rounds of selection.

Despite the above caveats, both monovalent and polyvalent systems have been successfully used to identify a wide range of protease substrates. In this chapter, we will focus on a pIII phagemid/monovalent display system, although many of the features described will apply to polyvalent substrate phage as well. A typical vector for substrate phagemid display is shown in *Figure 1*. Important features to note include:

(a) a tripartite fusion protein gene, comprising an affinity domain gene, a substrate-coding sequence and gene III from M13;

(b) a promoter sequence (not illustrated) for constitutive expression of the fusion protein in *Escherichia coli.*;

(c) bacterial and f1 origins of replication, plus an antibiotic resistance gene.

More detailed descriptions of phage display vectors can be found in Chapters 1 and 4. Here, we focus on some aspects of vector design that are specific to substrate phage applications.

## 2.1 Choice of binding pair

A critical component of protease substrate phage is the affinity domain, which allows separation of substrate from non-substrate sequences. As shown in *Table 1*, several different systems have been described. These include protein–protein interactions, protein–peptide interactions, and binding of polyhistidine to a nickel chelate matrix (the use of which is described in Chapter 8). The one underlying design factor uniting these different systems is that they are all based on high affinity interactions. Typically, the dissociation constant ($K_D$) for the binding pair should be in the low nM range or better, such that the dissociation rate of the complex is much slower than the rate of proteolysis in the experiment. For example, our previous work in this area has utilized a high affinity variant of hGH which binds its receptor with a $K_D$ of ~13 pM. However, in the case of protease substrate phage it is also useful to choose an interaction that can be readily disrupted by suitable buffer conditions (e.g. low or high pH). This allows for the selective enrichment of non-substrate sequences (which remain bound to the capturing solid support) as well as optimal substrate sequences (which are released by proteolysis). Another factor to consider is that the target protease (or any other protease encountered during synthesis of the phage capsid in *E. coli*) must not appreciably cleave the affinity domain. For this reason, the affinity domain should be either a stable globular protein or a short, non-labile peptide.

## 2.2 Design and construction of substrate cassettes

For substrate phage to be successful, it is important that cleavage should only occur between the affinity (capture) domain and the phage capsid. Fortunately, filamentous phage are remarkably resistant to proteolysis: amongst commonly known proteases, only subtilisin is known to cause significant degradation of phage coat proteins (16). To promote accessibility of the substrate region to the protease, it is prudent to design some segmental flexibility into the surrounding linker sequence. For example, the substrate sequence GPGG(X)$_n$GGPG may be introduced between the affinity domain and pIII, wherein a flexible Gly-Pro linker flanks the randomized region. In our hands, a truncated form of pIII has proved to be effective: however, others have reported success with the full-length pIII. In the case of polyvalent substrate phage display, use of the full-length pIII is essential for infectivity to occur.

Several other important factors govern the design of library cassettes, including number of randomized codons, choice of codons, and library size. These are reviewed in detail in Chapter 2. In many cases, it may not be possible to exhaustively explore the entire substrate sequence space, which may extend over eight or more residues. Nevertheless, it is possible to obtain useful information about substrate specificity even when the library represents only a fraction of available sequences. For example, initial exploration of a library with five randomized codons may lead to a clear consensus sequence for two positions within the sequence. This information may be used to design a secondary library with these two positions fixed, allowing for a more complete exploration of the remaining three positions and/or screening an extended substrate sequence (3). Alternatively, a more focused library may be designed from the outset, incorporating known specificity determinants and probing those that are less clearly defined (4, 6). Suitable protocols for construction of randomized substrate libraries are given in Chapter 2.

## 3  Substrate selection procedure

A detailed procedure for sorting substrate sequences is given in *Protocol 1*. In general, the substrate concentrations will be much less than both their respective $K_M$ values and the enzyme concentration, and sorting might be expected to occur based on the relative $k_{cat}/K_M$ of the substrates (1). However, synthetic peptide substrates derived from substrates selected in the same substrate phage experiment can exhibit a wide variation in $k_{cat}$, $K_M$, and $k_{cat}/K_M$ values (9, 11). Studies on reaction kinetics with polyvalent substrate phage have suggested that phage substrate cleavage is consistent with a single exponential event, with substrate binding being the rate-limiting step (17).

Due to the empirical nature of substrate phage selection, it is important and instructive to perform a series of pilot experiments prior to sorting an actual substrate phage library. Such experiments can be used to verify that specific substrate cleavage can actually occur within the context of the library linker and under the chosen assay conditions. Further optimization can then be performed to choose optimal enzyme concentrations and incubation times for the selection procedure. Typically, these pilot experiments will involve two substrate phage constructs, one containing a known substrate sequence and one containing a known non-substrate sequence. Equal numbers of substrate and non-substrate phage are (separately) immobilized via the affinity domain and treated with the protease of interest (*Protocol 1A*). By titrating the number of phage released from each sample, the degree of enrichment at each round of selection may be estimated. An alternative means of optimizing the enrichment procedure is to perform a mock selection experiment by "spiking" a small number of substrate phage (i.e. from $1:10^3$–$1:10^6$) into a preparation of non-substrate phage. By monitoring the ratio of substrate to non-substrate phage recovered at each round of sorting, one may predict the rate of convergence on a consensus substrate population.

## Protocol 1
# Substrate phage selection

### Equipment and reagents

- ~$10^{12}$ colony forming units (cfu) of positive and negative control substrate phage stocks (see text)
- ~$10^{12}$ cfu of substrate phagemid library stock at a concentration of >$10^{12}$ cfu/ml (see Chapter 2, *Protocols 3 and 4*)
- ~$10^{12}$ plaque forming units (pfu) of helper phage, for example, M13K07 (New England Biolabs), or VCSM13 (Stratagene; see Chapter 2, *Protocol 6* for preparation)
- ~$10$ μg capture reagent (e.g., hGH binding protein, anti-peptide antibody)
- High-binding 96-well microtiter plates such as Nunc Maxisorp plates (Nalge Nunc)
- A fresh (<1 week) streaked plate of F′ *E. coli*, such as XL-1 blue (Stratagene)

- Immobilization buffer for capture reagent (e.g. 50 mM carbonate buffer, pH 9.6)
- Phosphate buffered saline (PBS) containing 0.05% Tween 20 (PBST)
- Dilution and reaction buffers suitable for the protease of interest
- 2% skim milk in PBS
- 0.2 M glycine buffer, pH 2.0
- 1 M Tris base
- LB (Luria Broth) + 100 μg/ml ampicillin or LB + 50 μg/ml carbenicillin agar plates and liquid media
- 2YT medium
- Low- or non-protein binding microtiter plates (e.g. Nalge Nunc 96F polypropylene plates)

### Method

### A. Titration of protease activity versus positive and negative control substrate phagemids

1  Coat two columns of wells of a high-binding microtiter plate with 1 μg/ml of the capture reagent in immobilization buffer overnight at 4°C or for 2 h at room temperature.

2  Aspirate the coating solution and add 200 μl of skim milk/PBS per well to block the plate. Incubate at room temperature for 2 h with mild shaking.

**3** Wash the plate three times with PBST and dry the plate by slapping it face down into a paper towel on a bench top. Add 100 μl/well of phagemid stocks ($10^{12}$ cfu/ml) in PBST as follows: Positive control in column 1, negative control in column 2. Incubate for 2 h at room temperature with mild shaking.

**4** During the incubation period, make threefold serial dilutions of protease (at five times the desired final concentrations) in dilution buffer down one column of a separate, nonbinding microtiter plate. Leave the last well blank (buffer only). A reasonable (final) concentration to start from is 10- to 100-fold above what might be used in a typical chromogenic substrate hydrolysis assay. For example, for an enzyme having a $k_{cat}/K_M$ of $10^5$ M$^{-1}$s$^{-1}$ towards the positive control substrate, 50 nM would be a reasonable maximum concentration.

**5** Aspirate the phagemid solutions. Wash the wells four times with PBST. Leave the PBST from the final wash in the wells and incubate the plate for 5 min with mild shaking. Then wash again four times.

**6** Rinse the wells twice with the protease reaction buffer, and add 80 μl of the same buffer per well. Add 20 μl of protease dilutions from step 4 to both columns of the plate. Incubate at room temperature for 10 min with mild shaking.

**7** Transfer the protease-eluted phagemid solutions to the first columns of two separate nonbinding plates (one each for the positive and negative controls). These plates will be used for cfu titrations of the phagemid samples (see step 10 below).

**8** Rinse the wells of the binding plate three times with PBST, and add 100 μl of 0.2 M glycine, pH 2.0 to each well. Incubate for 5 min at room temperature.

**9** Transfer the glycine-eluted phagemid solutions to the first columns of two separate nonbinding plates. Add 10 μl of 1 M Tris base to each well to neutralize the acid.

**10** Make 10-fold serial dilutions of the phagemid elutions from steps 7 and 9 (across the nonbinding plates) and perform cfu titrations by infecting log-phase F' *E. coli* (see Chapter 4 *Protocol 5*).

**11** Calculate the number of eluted phagemids in each sample and determine a suitable concentration of enzyme for substrate phagemid library selection. Ideally, a large difference between the positive and negative control elutions will be obtained. It is also preferable to use a protease concentration that cleaves only a fraction (<50%) of the total bound positive control phagemids, to keep the selection stringent. Pick a concentration where the positive control is enriched by at least 10-fold relative to the well without enzyme, shows negligible enrichment towards the negative control well, and yet does not cleave the majority of the total bound phagemids. The extent of phagemid cleavage can also be optimized by varying the reaction time with a fixed concentration of enzyme. However, it is more difficult to cover a wide range of total activity by varying time as opposed to enzyme concentration. Furthermore, it is important that the reaction time should be short relative to the dissociation rate of the affinity tag from the capture molecule.

## B. Substrate phagemid library selection

1  Coat one well of a high-binding microtiter plate for positive selection (enzyme treatment) and one well for background (no enzyme). If a large library ($>10^9$ independent clones) is used, more wells (or large wells, with corresponding changes in the volumes of the selection procedure) may be used for positive selection in the first round to allow for more phage binding and preservation of the full diversity of clones. Roughly $10^9$ phagemids can be bound to one well of a 96-well plate, and in the first round it is preferable to bind at least 10-fold more phagemids than the number of clones in the library. In subsequent rounds, the diversity of the library is likely to be reduced substantially due to the selection, and one positive well should be sufficient.

2  Aspirate, block, wash, and dry the plates as described above in *Protocol 1A*, steps 2 and 3.

3  Apply $10^{12}$ cfu of the substrate phagemid library stock in PBST to each well. Incubate for 2 h at room temperature. Wash as described above in *Protocol 1A*, step 5.

4  Rinse the wells twice with protease reaction buffer. Add 100 μl of protease in reaction buffer (at the concentration predetermined in *Protocol 1A* above) to the enzyme positive well(s), and add 100 μl of buffer only to the background well. Incubate the plate at room temperature for 10 min (or optimized incubation time) with mild shaking.

5  Transfer the protease-eluted and background phagemid solutions to microfuge tubes.

6  *Propagation of protease-sensitive phagemids:* For the protease-treated sample, save a small aliquot (~25 μl) for titration and use the remainder to propagate phage for a new round of selection (see Chapter 4, *Protocol 4*, step 6).

7  *Propagation of protease-resistant phagemids:* Rinse the wells three times with PBST, add 100 μl of 0.2 M glycine, pH 2.0 to each well, and incubate with shaking for 5 min.

8  Transfer the glycine-eluted phagemids to microfuge tubes and neutralize with 10 μl of 1 M Tris base.

9  The glycine-eluted sample may be propagated as described for the protease-sensitive phagemids in step 6 and further selected for protease resistance by sorting in a separate pool. If this is to be carried out, it may be desirable to increase the protease concentration and/or incubation time in subsequent rounds to more stringently select for cleavage-resistant sequences.

10  *Phagemid titrations:* Perform cfu titrations on the starting phagemid stocks, the protease-eluted phagemids, and acid-eluted phagemid samples by infecting log-phase F′ *E. coli* (see Chapter 4, *Protocol 5*).

11  On the following day, prepare enriched phagemid stock(s) (see Chapter 4, *Protocol 1*) and repeat the selection procedure. Always include the background (no protease) control well to monitor the enrichment during the selection.

**12** *Analysis of titration results:* If the selection is strong, positive enrichment [(cfu eluted with protease)/(cfu eluted without protease) >1] may be observed after 2–3 rounds. Sequence ~20 clones every 2–3 rounds to see if a consensus is developing. If several clones of the same sequence (siblings) are found, the selection is complete or nearly complete and it may be of little value to select further (see Chapter 4, Section 7). More information will be gleaned from a pool of different but related sequences than from a library that has been selected down to a single clone. If, at any time, a high proportion (>50%) of the total bound phage are undergoing cleavage, the protease concentration should be reduced to increase selective pressure. Although monitoring enrichment by phage titration can be a useful guide to the progression of the substrate selection process, it can give variable results and should not be relied upon as a surrogate measurement of sequence convergence. In fact, good consensus sequences can sometimes be obtained even when no significant enrichment of phage titer is observed during the sorting.

# 4 Characterization of selected sequences

## 4.1 Initial sequence analysis

Following sorting of the substrate library, it is necessary to sequence individual phage clones to determine whether a consensus sequence has emerged. In some cases, the evidence for a common consensus sequence is clear: however, the unifying theme in a set of selected sequences is frequently less obvious. Identification of the consensus sequence is further complicated by imbalances in codon degeneracy, resulting in a nonuniform distribution of amino acids in the starting library. For example, the NNS codon set contains three arginine codons but only one lysine codon, so that even in the absence of selective pressure one would expect to find 3 times the number of arginine residues compared to lysine. A more rigorous method of probing the overrepresentation of specific amino acids in the selected library is to compare the number of empirically observed amino acids to the predicted number. If $P(x)$ is the probability of observing a given amino acid $(x)$, and $Obs(x)$ is the actual number of times that $x$ is observed in $n$ sequences, then

$$\Delta\sigma = \frac{Obs(x) - nP(x)}{\{nP(x)[1 - P(x)]\}^{1/2}},$$

where the value $\Delta\sigma$ denotes the difference of the observed number from the expected number in terms of standard deviations (3). $P(x)$ is obtained by assuming a binomial distribution of amino acids, which may be approximated by the Poisson distribution for most phage library applications.

## 4.2 "Expression editing" and other potential complications

It should be noted that several factors other than *in vitro* enzymatic selection might affect the observed distribution of amino acids found in a selected library.

125

One such factor is "expression editing"—the observation that certain amino acids are underrepresented in the library due to reasons of poor expression and/or display on the phage surface. For example, cysteine is frequently found to be underrepresented in phage libraries, possibly due to inappropriate intramolecular and intermolecular disulfide bond formation. Another possible cause of expression editing is the effect of endogenous *E. coli* proteases on the substrate library. This is unlikely to be a major problem considering the diverse set of enzymes that have been successfully screened to date. However, it may limit the utility of the technique if the protease of interest is closely related to an endogenous *E. coli* protease. Finally, if the enzyme of interest has very broad specificity, it may be difficult to distinguish between a *bona fide* set of selected substrates and a set of random sequences resulting from an ineffective selection procedure.

## 4.3 Detailed characterization of substrates

Once a substrate consensus sequence has been identified, it is usually necessary to evaluate the activity of the enzyme against selected substrate sequences from the enriched library. Depending on the goals of the experiment, this may range from qualitative comparison between several different sequences to rigorous enzymological determinations of $k_{cat}$ and $K_M$ for each substrate. In addition, the identity of the scissile bond may be unknown or may be ambiguous. Enzymological parameters, as well as the identity of the scissile bond, may be identified using synthetic peptide substrates. Such studies require the costly and time-consuming synthesis of peptides and typically involve low-throughput high performance liquid chromatography (HPLC) assays (combined with mass spectrometry for definitive identification of the cleavage site). However, they provide high quality, rigorous data to support the validity of the phage sorting procedure.

Several simpler approaches may be used to rapidly compare and characterize selected substrate sequences. One such technique is the dot blot assay (7). The substrate phage are treated with protease, spotted onto nitrocellulose filters and probed with an antibody to the affinity domain. This may be used as a rapid, low-resolution method for determining whether a given clone is a good or bad substrate, since bad substrates (where the affinity domain is still attached) will give a signal and good substrates (where the affinity domain has been cleaved) will not. A phage enzyme-linked immunosorbant assay (ELISA) method for monitoring substrate phage proteolysis has also been described (17).

In *Protocol 2*, we present an alternative method which allows for a relatively rapid, qualitative comparison of relative cleavage efficiencies for different substrates (1–3, 12). The method involves presentation of the substrate sequence in the context of a fusion protein, similar to that used in the selection procedure (*Figure 2*). Furthermore, isolation of cleavage products from this fusion protein

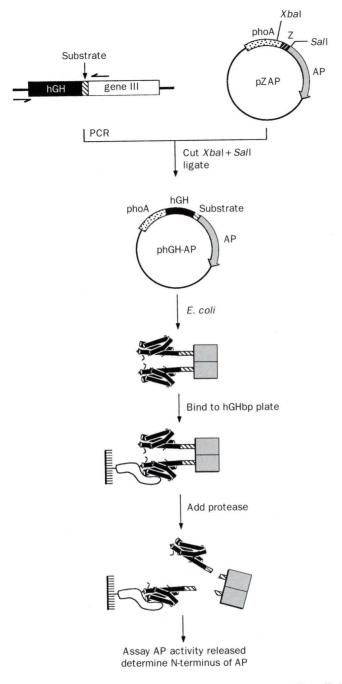

**Figure 2** Alkaline phosphatase (AP) fusion protein assay. The affinity domain (hGH) and substrate sequence genes are isolated from selected phagemids by PCR and fused to an alkaline phosphatase gene in a periplasmic expression vector, driven by an alkaline phosphatase promoter (phoA). Fusion proteins are expressed, quantitated and captured by binding to immobilized hGH binding protein. On treatment with protease, the AP moiety is released and may be quantitated, enabling the rate of substrate cleavage to be monitored. The exact cleavage site may be determined by NH$_2$-terminal sequencing of the AP product. (Reprinted with the permission of Cambridge University Press.)

enables the scissile bond to be determined by N-terminal sequence analysis. We describe the construction of alkaline phosphatase (AP) fusion proteins as an example: however, the method may be used with other reporter enzymes such as luciferase or horse radish peroxidase (HRP).

---

## Protocol 2

## Construction and assay of AP fusion proteins

### Equipment and reagents

- An AP fusion protein periplasmic expression vector, containing restriction sites suitable for cloning affinity domain—substrate linker polymerase chain reaction (PCR) products derived from selected clones as N-terminal fusions to AP. An example construct, with hGH as the affinity domain, is shown in *Figure 2*.

- Double-stranded or single-stranded phagemid DNA, prepared from the clones selected for protease sensitivity (or resistance, for a negative control)

- *E. coli* strain suitable for secreted protein expression, such as KS330 (18)

- LB + 100 µg/ml ampicillin or LB + 50 µg/ml carbenicillin agar plates and liquid media

- Anti-tag affinity resin (e.g. for hGH affinity domain, hGHR affinity resin—see *Table 1*)

- TE buffer (10 mM Tris-HCl (pH 7.6), 1 mM EDTA) containing 1 mM PMSF (phenyl methyl sulfonyl fluoride)

- Reaction buffer suitable for the protease of interest

- Purified AP for use as a standard (Sigma)

- AP colorimetric substrate, such as *p*-nitrophenol phosphate (*p*NPP) (Sigma)

- AP substrate buffer (supplied with *p*NPP substrate kit)

- Nonbinding microtiter plates, such as Nunc 96F plates (Nalge Nunc)

### Method

### A. Subcloning and expression of AP fusion proteins

1 Set up PCR reactions using template DNA from the selected phagemid clones and oligonucleotide primers containing restriction sites corresponding to the cloning sites for the AP fusion expression vector (*Figure 2*).

2 Analyze the reactions by agarose gel electrophoresis to verify the presence of correct-sized amplification products. Purify the PCR products by method of choice.

3 Digest the PCR products with enzymes corresponding to the cloning sites, and ligate them into the digested AP fusion vector using standard techniques.

4 Transform the ligation products into *E. coli* by standard methods, and plate onto LB agar plates containing 100 µg/ml ampicillin. Prepare individual clones and screen for insert by a standard method such as restriction digestion or PCR.

**5** Grow small (~5 ml) overnight cultures of positive clones in LB + 100 µg/ml ampicillin. Make 1/100 inoculations into expression media suitable for the induction system employed.

**6** Grow the cells and induce expression as appropriate for the system employed. Harvest the cells by centrifugation at 7500g and remove the supernatant. Freeze the cell pellets at − 20°C for > 4 h or on alcohol/dry ice for 10 min.

**7** Resuspend the pellets thoroughly in TE + PMSF buffer to release periplasmic proteins. Pellet the cells by centrifugation (7500g) and collect the supernatant.

**8** The fusion proteins may be further purified by a method specific to the tag employed. This will facilitate quantitation of the fusion proteins and remove contaminating proteins that could interfere with the cleavage assay. However, if the expression level of the AP fusion is high, the crude periplasmic fraction may be used directly, since the assay effectively contains an affinity purification step.

**9** Analyze the periplasmic proteins by SDS-polyacrylamide gel electrophoresis (PAGE) alongside a set of concentration standards of purified AP. Quantitate the levels of AP fusion protein present in the samples by gel densitometry.

## B. Cleavage assay

**1** In microfuge tubes, add a constant concentration (1–5 µg) of AP fusion proteins to an amount of anti-tag affinity resin sufficient to bind all of the material. Incubate for 2 h with slow tumbling at room temperature.

**2** Centrifuge the tubes at 5000g for 1 min, and aspirate the supernatant. Resuspend the beads in 1 ml of TE and centrifuge again. Wash the resin once more with TE and once with protease reaction buffer.

**3** Resuspend the washed resin in 200 µl of protease in protease reaction buffer. The amount of protease required to obtain a suitable reaction rate may have to be determined empirically. For an empirical determination, positive and negative control substrate fusions should be reacted with a range of concentrations of protease (see *Protocol 1A*, step 4).

**4** At various time intervals, withdraw 20 µl aliquots of supernatants from the reactions and transfer them to fresh tubes or a microtiter plate. Distribute the intervals such that widely varying rates may be monitored. For example, start with 2 min intervals for the first 10 min of reaction, then 5 min intervals for the next 20 min, then 15 min intervals for the next 60 min.

**5** Add 10 µl of the recovered AP samples to 90 µl AP substrate solution (prepared according to the manufacturer's instructions) in a separate microtiter plate. Include a standard curve of known concentrations of purified AP. Monitor the reactions at a wavelength appropriate for the product, preferably using a microplate reader capable of taking kinetic readings.

6 Determine the initial rates of the reactions, and calculate the concentrations of AP released in the samples by comparison to the standard curve. Plot the release of AP by protease cleavage over time. If the substrate concentration is much lower than its $K_M$ value, the initial rates should be proportional to $k_{cat}/K_M$ for cleavage of the substrate linkers.

## C. Determination of the point of cleavage of the substrate linker

1 Set up cleavage reactions as in *Protocol 2B* above, but in smaller volumes (50 μl) and using more AP fusion protein (5–10 μg) with corresponding increases in the amount of capture resin. Add enough protease and incubate the reactions long enough to completely digest the substrate linker, using the cleavage kinetics results as a guide.

2 Separate the reaction mixtures using SDS-PAGE. For some proteases, it is advisable to inactivate the enzyme using an appropriate reagent prior to gel sample preparation, since proteases such as subtilisin can rapidly digest proteins in a nonspecific manner during denaturation and heating.

3 Electroblot the proteins from the gel onto membranes, stain to reveal the cleaved AP products, and subject them to $NH_2$-terminal sequencing using standard techniques.

## 5 Variations and future directions

Since the inception of the substrate phage method, a number of novel modifications to the basic method have been implemented. In this final section, we review some of these systems and describe their utility in exploring a range of scientific problems.

### 5.1 Substrate subtraction libraries

An elegant refinement of the substrate phage technique has been described which enables the discrimination of substrate specificity between two different proteases (10). The substrate library is first sorted for several rounds with the first protease, and the resulting pool of good substrates isolated. Next, this pool is sorted with the second protease and the *non-substrate* sequences isolated. These sequences are good substrates for the first protease but poor substrates for the second protease. This method was used to discriminate between the specificities of the plasminogen activators t-PA and u-PA (10). Although both proteases preferentially cleave substrates with the consensus sequence GR in the P2 and P1 positions, it was found that u-PA prefers serine in the P3 position whilst t-PA prefers bulkier amino acids such as tyrosine and arginine.

### 5.2 Substrate specificity of a peptide ligase

A variant of the substrate phage technique has been used to elucidate the substrate specificity of an engineered peptide ligase (19). A subtilisin mutant, in which the active site serine 221 was converted to cysteine and Pro 225 was converted to alanine, was found to ligate esterified peptides onto the N-termini of other peptides or proteins. A substrate phage library was prepared by randomization of the first three codons of phage-displayed hGH. This was incubated with the subtilisin mutant in the presence of a biotinylated peptide ester (iminobiotin-KGAAPF-glc-F-amide), and ligated substrate phage were isolated by capture on avidin–agarose beads. Thus, library peptides that were good nucleophiles for the ligation reaction were selectively enriched. The results corresponded well with those obtained using synthetic substrates, especially in the P1′ position.

## 5.3 Substrate specificity of protein kinases

The substrate phage concept has been successfully extended to the study of sequence–specific protein phosphorylation. In this case, the separation step depends on a capture reagent that separates phosphorylated substrate from non-phosphorylated non-substrate sequences. Although many anti-phosphopeptide antibodies exist, great care must be taken to ensure that the capture step involves recognition of the phosphorylated amino acid alone and does not in itself impart any selection to the sequence. Such "capture editing" of the substrate phage library may be more of a problem for the isolation of phosphoserine and phosphothreonine residues than for the more bulky phosphotyrosine. There is at least one report of screening of a phage library for substrates containing phosphoserine and phosphothreonine (20). However, in this case the capture antibody was known to show specificity for M-phase phosphoproteins and the selection procedure thus only isolated kinase substrates which also contained epitopes for this antibody.

Several protein tyrosine kinases have been studied, including c-Src, Blk, Lyn and Syk (21), and Fyn (22). Two different phage display systems have been described: a monovalent system utilizing a substrate library fused to full-length gene III (21) and a polyvalent system wherein the substrate library was fused to gene VIII (22). As with protease substrate phage, before proceeding with library screening it is important to carry out pilot experiments to optimize selection conditions and demonstrate the feasibility of specifically enriching for good substrates. In the case of peptide phosphorylation, it is theoretically possible for phage coat proteins to be modified along with the substrate library, which would render specific selections difficult. In addition, selection of substrates using polyvalent display systems may be complicated if the kinase also contains phosphopeptide binding (e.g. SH2) domains that could interact with phosphorylated phage peptides and lead to selection of inferior kinase substrates (22). Nevertheless, with careful attention to these issues, substrate phage display can be successfully used to define subtle differences in substrate specificity between closely related protein tyrosine kinases (21).

## 5.4 *In vivo* selection with retroviral display vectors

Buchholz *et al.* (23) have described an alternative to bacteriophage substrate display that employs retroviral vectors to present the substrate library and enables substrate selection to be performed using intact mammalian cells. The selection process relies on the observation that fusion of epidermal growth factor (EGF) to one of the viral coat proteins greatly reduces the infectivity of the virus on EGF receptor-rich cells, and proteolytic cleavage of EGF restores full infectivity. By inserting a random substrate sequence between EGF and the coatprotein and screening for cleavage/infectivity, the authors were able to demonstrate enrichment of substrates containing consensus sites for the furin/prohormone convertase family of cellular proteases.

# 6 Conclusions

Since its inception, substrate phage has been shown to be a powerful tool in probing substrate specificity. Several different laboratories have created a variety of substrate phage display vectors and used them to discover substrates for a variety of enzymes, principally proteases and kinases. The potential applications are, however, much broader; many other peptide posttranslational modifications may also be amenable to the substrate phage approach. Finally, several ingenious new applications of phage display are beginning to emerge (for a review, see (24)). These include methods for probing enzyme–substrate interactions in a more general sense, where both enzyme and substrate can be displayed on the same phage particle.

## References

1. Matthews, D. J. and Wells, J. A. (1993). *Science*, **260**, 1113.
2. Ballinger, M. D., Tom, J., and Wells, J. A. (1995). *Biochemistry*, **34**, 13312.
3. Matthews, D. J., Goodman, L. J., Gorman, C., and Wells, J. A. (1994). *Protein Sci.*, **3**, 1197.
4. O'Boyle D. R., II, Pokornowski, K. A., McCann P. J., III, and Weinheimer, S. P. (1997). *Virology*, **236**, 338.
5. Huang, C., Wong, G. W., Ghildyal, N., Gurish, M. F., Sali, A., Matsumoto, R. *et al.* (1997). *J. Biol. Chem.*, **272**, 31885.
6. Harris, J. L., Peterson, E. P., Hudig, D., Thornberry, N. A., and Craik, C. S. (1998). *J. Biol. Chem.*, **273**, 27364.
7. Smith, M. M., Shi, L., and Navre, M. (1995). *J. Biol. Chem.*, **270**, 6440.
8. Ke, S.-H., Coombs, G. S., Tachias, K., Corey, D. R., and Madison, E. L. (1997). *J. Biol. Chem.*, **272**, 20456.
9. Ding, L., Coombs, G. S., Strandberg, L., Navre, M., Corey, D. R., and Madison, E. L. (1995). *Proc. Natl. Acad. Sci. USA*, **92**, 7627.
10. Ke, S.-H., Coombs, G. S., Tachias, K., Navre, M., Corey, D. R., and Madison, E. L. (1997). *J. Biol. Chem.*, 272, 16603.
11. Coombs, G. S., Bergstrom, R. C., Pellequer, J.-L., Baker, S. I., Navre, M., Smith, M. M. *et al.* (1998). *Chem. Biol.*, **5**, 475.
12. Dall'Acqua, W., Halin, C., Rodrigues, M. L., and Carter, P. (1999). *Protein Eng.*, **12**, 981.
13. Kristensen, P. and Winter, G. (1998). *Fold. Des.*, **3**, 321.
14. Sieber, V. and Plückthun, A. (1998). *Nat. Biotechnol.*, **16**, 955.
15. Finucane, M. D., Tuna, M., Lees, J., and Woolfson, D. N. (1999). *Biochemistry*, **38**, 11604.
16. Schwind, P., Kramer, H., Kremser, A., Ramsberger, U., and Rasched, I. (1992). *Eur. J. Biochem.*, **210**, 431.

17. Sharkov, N. A., Davis, R. M., Reidhaar-Olson, J. F., Navre, M., and Cai, D. (2001). *J. Biol. Chem.*, **276**, 10788–93.

18. Strauch, K. L. and Beckwith, J. (1988). *Proc. Natl. Acad. Sci. USA*, **85**, 1576.

19. Chang, T. K., Jackson, D. Y., Burnier, J. P., and Wells, J. A. (1994). *Proc. Natl. Acad. Sci. USA*, **91**, 12544.

20. Westendorf, J. M., Rao, P. N., and Gerace, L. (1994). *Proc. Natl. Acad. Sci. USA*, **91**, 714.

21. Schmitz, R., Baumann, G., and Gram, H. (1996). *J. Mol. Biol.*, **260**, 664.

22. Dente, L., Vetriani, C., Zucconi, A., Pelicci, G., Lanfrancone, L., Pelicci, P. G. *et al.* (1997). *J. Mol. Biol.*, **269**, 694.

23. Buchholz, C. J., Peng, K.-W., Morling, F. J., Zhang, J., Cosset, F.-L., and Russel, S. J. (1998). *Nat. Biotechnol.*, **16**, 951.

24. Forrer, P., Jung, S., and Plückthun, A. (1999). *Curr. Opin. Struct. Biol.*, **9**, 514.

# Chapter 8

# Protease-based selection of stably folded proteins and protein domains from phage display libraries

Mihriban Tuna, Michael D. Finucane, Natalie G. M. Vlachakis, and Derek N. Woolfson
Centre for Biomolecular Design and Drug Development,
School of Biological Sciences, University of Sussex, Falmer,
Brighton BN1 9QG, UK.

## 1 Introduction

Until recently, selection from phage display libraries has relied on binding of the displayed proteins to targeted ligands. While a wide variety of targets from small molecules to nucleic acids and proteins have been employed, these traditional bio-panning methods limit potential applications of phage display technology. Here we describe methods for selecting proteins from libraries based on their stability/structural integrity alone. This may be desirable in cases where no selectable protein function is available, for example in *de novo* design or proteomics. Three groups have presented different solutions to this problem: those of Plückthun/Schmid (1), Winter (2) and our own group (3). We will focus on our own studies with the aim of teaching how to select stably folded proteins from a variety of phage display libraries.

## 2 The methodology

We have combined phage display and proteolysis to select stably folded mutants from protein–phage mixtures and libraries as shown in *Figure 1*. The premise of the method is that unstable and/or partially folded proteins should be more susceptible to cleavage by proteases than stably folded proteins (4). This is combined with phage display to achieve selection as follows. The target proteins are made with N-terminal histidine tags and are then expressed in a phagemid vector as C-terminal fusions to the minor coat protein of bacteriophage, pIII, resulting

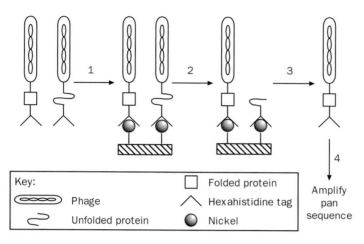

**Figure 1** The principle of the protease selection method developed in our laboratory (3). Protein-displaying phage are captured onto nickel-derivatized supports via N-terminal
histidine tags (Step 1). Bound phage are challenged with protease. Those phage that
display proteins that are susceptible to proteolysis are severed from the support and washed away (Step 2). Protein-displaying phage that resist proteolysis and, thus, remain bound are then eluted from the nickel support (Step 3) and amplified for further rounds of selection or DNA sequencing (Step 4).

in monovalent display (see Chapter 1). Phage displaying the fusion proteins are then immobilized onto nickel supports via the histidine tag and challenged with a protease. In this procedure, unstable protein linkers should be cleaved, and thus their associated phage may be removed simply by washing. Phage displaying proteins that resist proteolysis may then be eluted from the support by washing with imidazole or ethylenediaminetetraaceticacid (EDTA). Phage selected in this way may then be amplified for further rounds of selection and/or DNA sequencing by reinfection into *Escherichia coli* cells as usual. Filamentous phage are known to resist a broad range of proteases (5); thus this selection approach specifically targets the stability of the displayed proteins.

In some respects, the procedure is similar in principle to substrate phage (5) (see Chapter 7), in which libraries (in this case, of linear peptides) are also displayed between an affinity tag and pIII to allow immobilization on solid supports. However, treatment with specific proteases in this case is used to map substrate specificities; peptides that are cleaved are thereby identified as substrates for the enzymes. In contrast, our method was devised to select proteins that are protease resistant, despite containing protease substrate sites, by virtue of their foldedness (3, 6).

As described earlier, two other groups have also developed methods for selecting stable proteins using phage display and proteolysis (1, 2). In these cases, affinity tags are not required, but proteolysis of the displayed protein and phage infectivity are directly linked. The target proteins are inserted in the

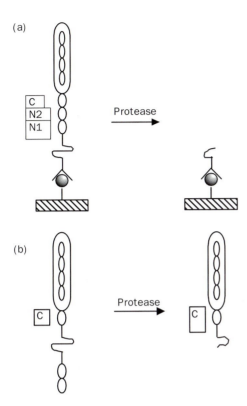

**Figure 2** A comparison of the two methods for protease selection using phage display. (a) The affinity-based approach developed by our group (3). (b) The approach that links proteolysis to phage infectivity described by the Plückthun/Schmid (1) and Winter (2) groups.

region between the N-terminal infectivity domains (N1, N2) and phage-anchoring C-terminal domain of pIII (see Chapter 1). Since the N-terminal domains are required for phage infection into *E. coli*, removal of these domains by proteolysis of the inserted linker leads to loss of infectivity. Thus after protease selection, only phage that display stable proteins that resist proteolysis will be propagated. This approach and our affinity-based procedure are compared schematically in *Figure 2*.

One potential drawback of the approach of the Plückthun/Schmid and Winter groups is that they are not compatible with conventional monovalent display. This is because the presence of wild-type pIII on the display phage would result in their being selected and propagated regardless of whether or not the pIII-fusion protein is cleaved, leading to false positives. Sieber *et al.* (1) deal with this problem by using a phage system in which all copies of pIII are fused to the target protein (i.e. polyvalent display). In turn, loss of phage infectivity during selection requires cleavage of all the copies of the displayed protein. Kristensen and Winter (2) surmount the problem elegantly by engineering a helper phage to provide protease-sensitive pIII. Thus, in their system the displaying phage particles only remain infective if the protein-pIII fusions resist proteolysis. As with our method, this approach avoids polyvalent display.

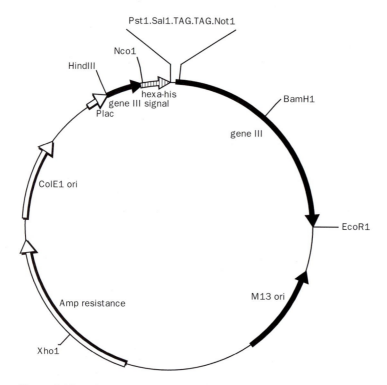

**Figure 3** Map of the phage display vector pCANTAB B. The salient features of the vector are described in the text.

## 3 General procedures and a specific example—rescue of stable ubiquitin variants from a library of hydrophobic core mutants

We originally developed our own methodology using ubiquitin as a model protein. Our goal was to redesign the hydrophobic core of ubiquitin by mutating it to generate a library of protein sequences with various combinations of hydrophobic residues, and then using protease selection to identify stably folded variants.

The ubiquitin gene was amplified from a rat complementary DNA (cDNA) library and cloned into an in-house vector, pCANTAB B (*Figure 3*) (3). This vector was constructed from the phage display vector pCANTAB 5 (Pharmacia) by adding the coding sequence for a hexahistidine tag to the 5′ end of the gIII sequence. The ubiquitin gene was inserted into pCANTAB B to generate the following protein fusion: hexahistidine tag-LQG-ubiquitin-GLDQQ-pIII. This construct also includes two suppressible amber (TAG) stop codons preceding gIII, allowing expression of hexahistidine-tagged proteins by using non-suppressor strains of *E. coli* (see also Chapter 4). Other protein domains could be similarly cloned into pCANTAB B to perform stability-based selections.

The procedure for library generation and propogation is similar to general phage display protocols (see Chapter 4). Libraries of appropriate variants can be created using a variety of mutational strategies, including oligonucleotide-directed mutagenesis (Chapter 2) and DNA shuffling (Chapter 3). In our specific use to create stable variants of ubiquitin, we created a hydrophobic core library by mutating residues at positions 1, 3, 5, 13, 15, 17, 26, and 30 of ubiquitin to combinations of hydrophobic residues. These residues were identified as contributors to the hydrophobic core by inspection of the high-resolution crystal structure of ubiquitin. Combinations of methionine, leucine, isoleucine, valine, and phenylalanine were introduced at these positions by incorporating the degenerate codon DTS (D is A, T, or G and S is G or C), which includes one codon each for phenylalanine, leucine, isoleucine and methionine, and two for valine. Thus, the resulting library was expected to contain 390,625 mutants—that is, eight positions mutated to combinations of the five amino acid residues—including the wild-type sequence. Since these residues are dispersed throughout the coding sequence, we used a recursive polymerase chain reaction (PCR) procedure (7) to assemble the library, in which overlapping oligonucleotides specifying the mutated gene are annealed and amplified with an excess of flanking "outer" primers.

An important issue when performing stability selections on a library is to ensure that the library is not contaminated with the wild-type sequence, which would probably be efficiently selected over variant sequences. For example, ubiquitin is an exceptionally stable protein (8) and most variants are likely to be less stable than the wild type. To solve this problem, it is important to first create a DNA template for mutagenesis that encodes an inactive variant, and to use this to create the library (see Chapter 2). In the case of ubiquitin, we first created a destabilized version of the protein, called $AL_7$, by replacing methionine-1 with alanine and the remainder of the aforementioned residues with leucine, using classical site-directed mutagenesis (9). Subsequently, DNA for this variant was used as a template for sticky feet-directed mutagenesis (10) to construct the library, using the recursive PCR product as primer. Some of the $AL_7$ parent remained in the final library used in the selection studies.

The protocols below describe procedures that can be used to generate and then select libraries of protein variants to identify those of highest stability. The procedures are based on those used in our ubiquitin work (3, 6) and are illustrated throughout by examples from that work.

## 3.1 Construction of a phagemid library of hydrophobic core mutants

### 3.1.1 Assembling the mutagenic oligonucleotide by recursive PCR

Recursive PCR (7) is an efficient way to target mutations to distant but discrete sites in a coding sequence. In the case of ubiquitin, four overlapping primers containing the degenerate codon DTS at specific sites were assembled to give a

128 bp product. The primer sequences are listed in *Table 1*. *Protocol 1* gives the procedure for assembly of a mutagenic primer using a recursive PCR procedure.

## Protocol 1

## Recursive PCR to assemble a double-stranded degenerate mutagenic oligonucleotide

### Reagents

- Deep Vent DNA polymerase (New England Biolabs)

- $10 \times$ Deep Vent DNA polymerase buffer (New England Biolabs)

- 100 mM MgSO$_4$ (New England Biolabs)

- Equimolar mix of deoxyribonucleotide triphosphate (dNTPs), 25 mM in each of G, A, T, and C (New England Biolabs)

- Sterile ultrapure water

- Thermal cycler

- Mineral oil (Sigma)[a]

- Apparatus and reagents for preparative agarose gel electrophoresis

- Spin-column kits (Stratagene–Strataprep DNA gel extraction kit, Cat. number 400766)

### Method

1  Design the primers for the recursive PCR procedure, based on the principles illustrated by the example of the ubiquitin library primers listed in *Table 1*. The overlaps between internal primers should be approximately 20 nucleotides, and to give melting temperatures of 55 to 60 °C. Overlaps should be targeted to avoid mutagenized positions if possible.

2  Prepare stock solutions of the primers in 1 ml volumes at concentrations of 10 pmol/ μl in sterile deionized water.

3  Prepare serial dilutions in 1 ml volumes of the "inner" primers (primers 2 and 3 in the case of the ubiquitin example; see *Table 1*) to give concentrations of 1, 0.1, and 0.01 pmol/μl. These will be needed to set up a series of PCR reactions with varying ratios of outer to inner primers (e.g. 1 : 1, 1 : 0.1, 1 : 0.01, and/or 1 : 0.001).

4  Set up the series of PCRs by mixing the following in order:

- Ultrapure H$_2$O — 67 μl
- $10 \times$ Deep Vent DNA polymerase buffer — 10 μl
- 100 mM MgSO$_4$ — 6 μl
- dNTPs — 5 μl
- Primer 1 (10 pmol/μl) — 2.5 μl
- Primer 2 (appropriate dilution) — 2.5 μl
- Primer 3 (appropriate dilution) — 2.5 μl
- Primer 4 (10 pmol/μl) — 2.5 μl
- Deep Vent DNA polymerase (2 U/μl) — 2 μl

5  Centrifuge the tubes briefly, overlay with mineral oil[a] and place the tubes in a thermal cycler.

6  Incubate the reactions for 25 to 30 cycles of 2 min at 94 °C, 2 min at the melting temperature of primer overlapping regions (in the ubiquitin example this was 56 °C) and 1 min at 72 °C. At the end, incubate the reactions at 72 °C for a further 10 min.

**7** Purify the PCR products using a low-melting-point agarose gel of appropriate percentage (e.g. 1–2% for the 128 bp product targeted here), cut out the appropriately sized band and extract the DNA using spin-column kits or alternative standard procedures and pool the successful reactions.

[a] Mineral oil is not required if a thermal cycler with a heated lid is used.

**Table 1** Primers used in recursive PCR to create ubiquitin hydrophobic core library[a, b]

| |
| --- |
| Primer 1: 5′-CTG CAG GGA DTS CAG DTS TTC DTS AAG ACC CTG ACC GGC AAG ACC-3′ |
| Primer 2: 5′-CTC GAT GGT GTC ACT GGG CTC SAH CTC SAH GGT SAH GGT CTT GCC GGT CAG G-3′ |
| Primer 3: 5′-CC AGT GAC ACC ATC GAG AAC DTS AAG GCC AAG DTS CAG GAT AAA GAG GGC ATC C-3′ |
| Primer 4: 5′-TGA TCA GGG GGG ATG CCC TCT TTA TCC TG-3′ |

[a] Degeneracy codes: D = A, T or G; S = G or C; H = the complement of D, that is, A, T or C.
[b] The overlapping regions between the primers were designed to have melting temperatures of around 56 °C.

### 3.1.2 Sticky-feet mutagenesis to generate a randomized library

The product from the recursive PCR (section 3.1.1) is then used as a mutagenic oligonucleotide and annealed to a single-stranded DNA (ssDNA) template encoding a destabilized protein (pCANTAB B-AL$_7$ in the case of the ubiquitin example) by "sticky-feet-directed mutagenesis" (10) as outlined in *Protocol 2*. In this procedure the double-stranded primer is first denatured, then the temperature is slowly lowered in the presence of the DNA template and *Taq* polymerase, allowing the primer to anneal to the template and be extended. This procedure increases the complementarity between the extending primer and the template ssDNA, so that the appropriate strand of the recursive PCR product should preferentially bind to the ssDNA template rather than to the complementary synthetic oligonucleotide strand. After annealing, a procedure similar to that described by Kunkel is followed to complete mutagenesis (9).

### 3.1.3 Rescue of the phagemid library

After the preparation of the library DNA, the DNA is electroporated into *E. coli* using standard methods, transformants are recovered and archived, and phage are then prepared for analysis and selection. We typically grow the transformed cells further to log phase in liquid media, as described in *Protocol 3*. After this amplification in liquid culture, some of the culture is used to prepare glycerol stocks and the remainder is used to prepare phage. Alternatively, the transformants may be recovered on solid media directly by plating out the whole transformation reaction on agar with selective antibiotics (e.g. see Chapter 13, *Protocol 10*). This approach is believed to prevent biased propagation of particular clones during the rescue procedure.

## Protocol 2

## Sticky-feet-directed mutagenesis to generate a randomized library

### Reagents

- 1 M Tris-HCl, pH 8.0
- 0.2 M MgCl$_2$
- 0.1 M dithiothreitol (DTT)
- 10 mM ATP (New England Biolabs)
- 10 × PCR buffer (0.1 M Tris-HCl pH 8.3, 0.5 M KCl, 15 mM MgCl$_2$)
- 10 × T7 polymerase buffer: (0.4 M Tris-HCl pH 7.4, 0.2 M MgCl$_2$, 0.32 M NaCl, 50 mM DTT)
- Equimolar mix of dNTPs, 25 mM in each of G, A, T, and C (New England Biolabs)
- T4 polynucleotide kinase (New England Biolabs)
- T4 DNA ligase (New England Biolabs)
- T7 DNA polymerase (New England Biolabs)

- *Taq* DNA polymerase (Promega)
- 0.5 M EDTA
- Ultrapure water
- Thermal cycler
- Mineral oil
- Gel-purified recursive PCR product from *Protocol 1*
- Reagents and equipment for preparation of uracil-containing single-stranded DNA (dU-ssDNA) template for mutagenesis (see Chapter 2, *Protocol 1*)
- Phage display vector DNA encoding a destabilized protein variant (e.g. pCANTAB B-AL$_7$ in the case of the ubiquitin example)

### Method

1 Prepare dU-ssDNA template of the phage display vector encoding the destabilized protein variant, as described in Chapter 2, *Protocol 1*. Using a uracil-containing template, prepared from a *dut ung* strain of *E. coli* such as CJ236, allows selection against template strand by transformation of the annealing products into a *dut*$^+$ *ung*$^+$ strain (9, 10).

2 Phosphorylate the double-stranded mutagenic primer (the recursive PCR product from *Protocol 1*) by making the following reaction mix, and incubating it at 37 °C for 45 min, and then at 65 °C for 10 min:

- H$_2$O              to give a total volume of 20 µl
- Recursive PCR product      to give a final concentration of 6 pmol/µl
- 1 M Tris-HCl, pH 8.0       2 µl
- 0.2 M MgCl$_2$            1 µl
- 0.1 M DTT               1 µl
- 10 mM ATP              1.3 µl
- T4 polynucleotide kinase (10 U/µl)    0.5 µl

3 Anneal the phosphorylated mutagenic primer (from step 2) to the ssDNA template (from step 1) in the following reaction mixture. It is important to adjust the concentrations of phosphorylated primer and ssDNA so that the primer is in excess (typically, we use a 25-times molar excess).

- H$_2$O to give a
  total volume of          10 µl
- 10 × PCR buffer          1 µl
- dNTPs                    0.5 µl
- ssDNA template
  from step 1              to 0.24 pmol

- Phosphorylated primer from
  Step 2 (6 pmol/µl)       1 µl
- *Taq* DNA polymerase
  (5 U/µl)                 0.5 µl

Denature the DNA by heating the reaction mixture under mineral oil at 92 °C for 2 min on the thermocycler. Then incubate the mixture at 67 °C for 30 s and then at 37 °C for 1 min, using 2-min ramps between each step.

4  During the annealing step, prepare the following mixture for the extension/ligation reaction, and store it on ice:
- H$_2$O                           5.5 µl
- 10 × T7 polymerase buffer        1 µl
- 5 mM ATP                         1 µl
- 0.1 M DTT                        1 µl
- T4 DNA ligase (400 U/µl)         1 µl
- T7 DNA polymerase (10 U/µl)      0.5 µl

5  Add this mixture to the template/primer mixture from step 3 immediately after the annealing step, and incubate the mixture at 37 °C for 20 min.

6  Stop the reaction by adding 1 µl 0.5 M EDTA and 30 µl H$_2$O. Store the reaction mix on ice or at −20 °C.

# Protocol 3

## Electroporation and rescue of phagemid libraries, preparation of glycerol stocks and preparation of phage for selections

### Reagents

- 2TY/Carb/Tet[a] agar plates (31 g/L 2TY, 15 gr/L agar, 100 µg/ml carbenicillin and 10 µg/ml) tetracycline. Add 31 g of 2TY, 15 g of agar and water to 1 L, then autoclave the solution. Allow the solution to cool down to 40 °C, then add 1 ml of carbenicillin (100 mg/ml) and 1 ml of tetracycline (10 mg/ml), then mix well and pour agar into petri dishes.

- Electrocompetent *E. coli* cells; for example, electrocompetent XL1-Blue cells (Stratagene)

- Reagents and equipment for electroporation of DNA into *E. coli* (see Chapter 2, *Protocol 3*).

- 2TY and 2TY/Glu medium (31 g/L 2TY broth with or without 2% glucose). Add 31 g of 2TY to 1 L of water, stir the solution to dissolve 2TY and autoclave. Allow the solution to cool, separate 100 ml to a sterile bottle and add 10 ml of 20% w/v glucose (sterile-filtered).

- 10 mg/ml sterile tetracycline in ultrapure water[a]

- 100 mg/ml carbenicillin in ultrapure water

- 100 mg/ml kanamycin in ultrapure water

- Sterile 50% glycerol in H$_2$O

- Kanamycin-resistant VCS-M13 helper phage stock (10$^{10}$ pfu/ml) (Stratagene; see Chapter 2, *Protocol 6* for preparation)

**Protocol 3** continued

- 20% PEG in 2.5 M NaCl (PEG-NaCl)
- Phosphate buffered saline (PBS) (2 mM $NaH_2PO_4$, 16 mM $Na_2HPO_4$, 150 mM NaCl, pH 7.4–7.6). To prepare PBS, dissolve 8 g of NaCl, 0.2 g of KCl, 1.44 g of $Na_2HPO_4$, and 0.24 g of $KH_2PO_4$ in 800 ml of de-ionized water, adjust the pH to 7.4 with HCl, and add water to 1 L.

- Library DNA (from *Protocol 2*, Step 4)
- 40 ml Oakridge centrifugation tubes (Nalge Nunc)
- Sorvall RC *5C plus* centrifuge and SS-34 rotor or equivalent

## Method

1  Electroporate 10 µl of the mixture from *Protocol 2*, step 6 into a *dut⁺ ung⁺* strain of *E. coli* (e.g. XL1-blue). Several electroporations may be required for large libraries. A protocol for high-efficiency electroporation is given in Chapter 2, *Protocol 3*. In our work with the ubiquitin system, we have electroporated libraries into electro-competent XL1-Blue cells purchased from Stratagene. Alternative strains include SS320 (see Chapter 2), or TG1.[b]

2  Plate out appropriate dilutions of the electroporated cells on 2TY/Carb/Tet[a] agar plates to determine the number of transformants. Colonies from the plates can be used later for testing the library by PCR, restriction enzyme digests and sequencing.

3  Dilute the electroporated cell culture (from step 1) into 40 ml 2TY/Glu containing 12 µl 100 mg/ml carbenicillin and 20 µl 10 mg/ml tetracycline[a] in 250 ml growth flasks.

4  Incubate the culture at 37 °C with shaking to log phase until optical density $(OD_{600}) = 0.5$, this usually takes approximately 4–5 h. Add 12 µl 100 mg/ml carbenicillin and 20 µl 10 mg/ml tetracycline[a] to the flask after 1 h.

5  Remove 4 ml of the cell culture, add 1.5 ml of 50% glycerol and store the frozen cells in 1 ml aliquots at −80 °C for future use.

6  To the remaining cells add 4 ml of VCSM13 helper phage (at approximately $10^{10}$ pfu/ml) to obtain approximately a 10-fold ratio of phage to *E. coli* cells. Incubate the flask at 37 °C for 15–30 min without shaking.

7  Add one quarter of this culture, approximately 10 ml, to each of four 250 ml growth flasks containing 100 ml 2TY/Glu, 75 µl 100mg/ml carbenicillin and 100 µl 10 mg/ml tetracycline. Incubate the flasks at 37 °C with shaking until the cell culture starts to cloud slightly (around 2 h).

8  Add 70 µl of 100 mg/ml kanamycin and incubate the flasks overnight at 37 °C with shaking.

9  The next morning, divide the overnight cultures into three Oakridge tubes each, and centrifuge using a SS-34 rotor in a Sorvall RC *5C plus* centrifuge at 12,000 rpm for 15 min at 4 °C.

10  Remove 30 ml of the supernatant from each tube and mix with 7.5 ml of PEG-NaCl to precipitate the phage particles. Leave the tubes on ice with occasional mixing for 1 h.

**11**  Centrifuge the tubes at 12,000 rpm for 15 min as described in step 9. Remove and discard the supernatant.

**12**  Resuspend the phage pellet in 1 ml PBS per tube. Pool the resuspended pellets in a new tube and centrifuge the tube briefly at 10,000 rpm in a microfuge to pellet insoluble debris. Transfer the cleared supernatant to a new tube and store at 4 °C.

[a] This protocol includes the use of tetracycline to select for maintenance of F-pilus expression in *E. coli* strain XL1-Blue. For libraries electroporated into other *E. coli* strains, supplement media and plates with alternative antibiotics as appropriate for the background strain.

[b] A procedure for preparation and electroporation of electrocompetent *E. coli* TG1 cells is provided in Chapter 13, *Protocols 9* and *10*.

## 3.2  Characterization of the phagemid library

Characterization of the size and diversity of the library is carried out using procedures common to other phage display approaches (e.g. see Chapter 4). To extract sequence-to-structure relationships after selection, it is important to ensure a reasonable level of redundancy in the phagemid library to be confident that all clones are sampled during the selection process (see Chapters 2 and 4). We usually aim for an around 100-fold excess of library size compared with theoretical protein sequence diversity. The size of a phagemid library may be estimated by plating out serial dilutions of post-electroporation cells (*Protocol 3*, step 2). Subsequently, individual clones may be screened by colony PCR (10), or by restriction-enzyme digests if restriction sites are introduced or deleted during the mutagenesis in order to determine the presence of library inserts. For example, if the parental clone (template) has a unique restriction site that is destroyed upon mutagenesis, this can be used to determine the proportion of parental clones in the library. For a targeted library containing several partially randomized positions, such as that produced by *Protocols 1* and *2*, diversity is best assessed by sequencing of DNA from single clones. Sequence analysis will reveal whether there are any biases toward particular clones in the initial library. Statistical methods may be used to determine the minimum number of clones that need to be sequenced for a defined library size in order to gain confidence that no one clone is present in vast excess over the rest (11).

## 3.3  Protease selection of stably folded ubiquitin variants

Protease selection experiments may be performed using a range of proteases with different specificities for different protein sequence features. With the exception of subtilisin, which has been reported to cleave pIII (12), filamentous phages generally resist most site-specific proteases. In our experiments, protein-displaying phage immobilized on solid supports via their N-terminal histidine tags were treated with chymotrypsin or trypsin. Two different nickel-derivatized supports have been used: Ni-derivatized NTA sensor chips for use in Biacore instruments (Biacore AB), and stand-alone, Ni-NTA agarose beads (Qiagen). We have found that surface plasmon resonance (SPR) allows quick and direct visualization and monitoring of the binding and proteolysis steps, as well as screening of individual clones or phage populations before and after selection. By contrast, Ni-NTA immobilizations are more labor-intensive and time-consuming due to the extensive washing and plating out that is required. However, scalability and a larger binding capacity mean that Ni–NTA agarose beads are well suited to use in the selections themselves.

### 3.3.1 Screening of individual clones and phage populations using SPR in BIAcore instruments

SPR technology is useful for monitoring biomolecular interactions in real time. The method works by detecting changes in refractive index of solutions above specific surfaces, which accompany binding events to these surfaces (13). Typically, one of the interactants is immobilized on the surface in a micro-flowcell and a sample containing the other interactant(s) is introduced to the system. Binding and subsequent dissociation events are detected as an SPR signal measured in resonance units (RU). With the Biacore instrument, 1000 resonance units RU generally approximates to 1 ng of bound material per 1 $mm^2$ of chip surface. The change in response (RU) with time gives a record of the interactions as they occur at the surface of the sensor chip in the form of a sensogram (see *Figure 4*).

*Figure 4* shows the full sensogram of a typical SPR binding and protease-cleavage experiment performed in a Biacore 2000 instrument. In stability selection studies, SPR can be used to establish the stabilities of protein phage displaying wild-type and reduced-stability variants, in order to optimize the proteolytic step to be used in selections (*Protocol 4*). It can then be used to analyze the stabilities of selectants. For example, in our work with ubiquitin, we used SPR in a Biacore instrument to monitor binding of protein-displaying phage and subsequent cleavage by chymotrypsin. We demonstrated that wild-type ubiquitin-phage (WT-UBQ-phage), various mutants and the protein-phage library bound specifically to a nickel-coated NTA sensor chip via the hexahistidine tag. Cleavage by chymotrypsin was found to be specific for the protein linker, and resistance to proteolysis was directly linked to the stability of the displayed protein (3, 6). In the experiment shown in *Figure 5*, we compared the chymotrypsin susceptibility of WT-UBQ-phage and phage displaying a flexible Gly-Ser-based peptide linker that harbored a single tyrosine cleavage site (FLEXI-phage). When the two protein-phage were exposed to 10 μM chymotrypsin in parallel on a Biacore sensor chip virtually all of FLEXI-phage was cleaved and washed away, whereas WT-UBQ-phage resisted proteolysis and only a small fraction (around 20%) was removed from the chip surface in the same time.

(a)

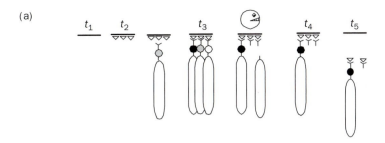

(b)

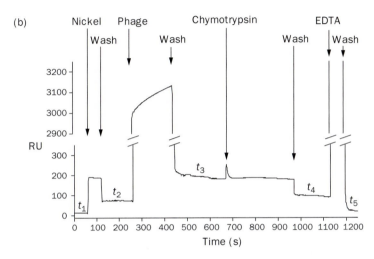

**Figure 4** Following protease selection by SPR in Biacore. (a) A schematic depiction of the method. Phage displaying proteins of different stabilities (represented by circles with different shading) and tagged with hexahistidine tags (Y-shaped tags in the figure) are bound to a Ni-coated chip surface ($\triangledown$). Subsequent treatment with protease cleaves the less stable proteins ($\bigcirc$, $\bigcirc$) and releases their associated phage from the surface. Any remaining phage, which should display more stable proteins ($\bullet$) are then eluted and retrieved. Each of these steps is labeled with a time ($t_1 - t_5$), which corresponds to an event in the SPR experiment shown in panel B. (b) Sensorgram trace of a typical SPR experiment carried out in Biacore. At time $t_1$, the resonance intensity (measured in RU) is at a minimum, and the intensity can be normalized to zero. As buffers containing additional components such as Nickel ions, phage, or enzymes are passed over the chip the resonance intensity changes. This is because of associated changes in refractive index. $Ni^{2+}$ is added first to derivatize the chip. When washing with normal running buffer (HPS) is resumed ($t_2$) the resonance intensity drops, but to a new equilibrium value higher than the starting value ($t_1$), thus, indicating that material ($Ni^{2+}$) has been bound. Phage are then added, which bind through their displayed hexahistidine tags. Resumption of the HPS wash sees another increase in the baseline resonance intensity, indicating that phage have bound. Addition of the chymotrypsin-containing solution is followed by a sharp decline in signal as phage are stripped from the surface by proteolysis. After resumption of the HPS wash, the new equilibrium value $t_4$ is lower than $t_3$ indicating a loss of phage from the surface. Finally, an EDTA solution is applied to strip all $Ni^{2+}$ ions from the surface and, with them, any remaining phage. This restores the resonance intensity ($t_5$) to approximately the initial value ($t_1$). (Reprinted with permission from ref. 3. Copyright 1999 American Chemical Society.)

147

(a)

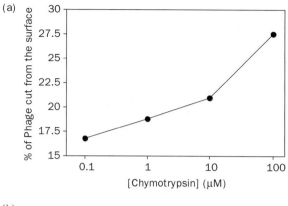

(b)

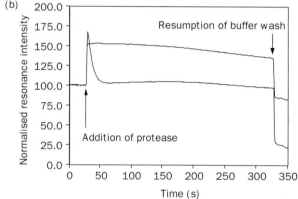

**Figure 5** Demonstrating the specificity of the protease selection method.
(a) The proportion of phage cleaved from the $Ni^{2+}$ surface in SPR is a function of the protease concentration used. The percentages cleaved were calculated as $(t_3 - t_4)/(t_3 - t_2) \times 100$ from experiments similar to that described in *Figure 4*, but performed with WT-UBQ-phage. (b) Target-sequence dependence of the protease cleavage profile as judged by changes in resonance intensity with two different phage fusions: the solid trace is for WT-UBQ-phage, whereas the dashed line is for FLEXI-phage. The data are normalized such that 100 on the *Y*-axis is equivalent to 100% of the phage bound immediately before proteolysis. After 300 s of exposure to chymotrypsin, 83% of the WT-UBQ-phage remained bound to the BIACORE chip surface; in contrast, only 23% of the FLEXI-phage survived. (Reprinted with permission from ref. 3. Copyright 1999 American Chemical Society.)

As described in *Protocol 4* and shown in *Figures 4* and *5*, SPR can be applied successfully to assess the relative stabilities of single clones. Such protease cleavage experiments may also be carried out for protein-phage recovered from protease challenges either within, or outside Biacore and after amplification via *E. coli*. However, there is one limitation of the SPR method—the surface area available for capturing protein phage is only 1 mm$^2$ per flowcell. We calculated, and have confirmed experimentally, that this limits the number of phage that may be captured onto the chip to around $10^6$ per flowcell (3). Therefore, at present it is not possible to pan large libraries directly on a Biacore sensor chip. Nonetheless, the power of the SPR method is in quickly optimizing selection conditions and studying single clones either independently or after protease selection from libraries.

---

## Protocol 4

## Monitoring protease cleavage of single protein-phage clones by SPR

### Equipment and reagents

- Biacore 2000 instrument (Biacore AB)
- NTA sensor chip (Biacore AB)
- HPS running buffer (Biacore AB) (10 mM HEPES, 0.15 M NaCl, 50 μM EDTA, 0.005% surfactant P20, pH 7.4)
- Nickel solution: 500 μM NiSO$_4$ in HPS running buffer

- Regeneration buffer (10 mM HEPES, 0.15 M NaCl, 0.35 M EDTA, 0.005% surfactant P20 (Biacore AB), pH 8.3)
- 10 μM chymotrypsin[a] (type II from bovine pancreas, Sigma) in HPS running buffer
- Concentrated display phage stocks (e.g. from *Protocol 3*)

### Method

1  Dock an NTA sensor chip into the Biacore instrument and prime it with HPS running buffer and set the temperature to 25 °C. This and all subsequent Biacore operations should be performed in accordance with the manufacturer's instructions.

2  Set the flow rate to 20 μl/min and choose the flowcell(s) to be used in the experiment.[b] Up to four distinct flowcells may be used simultaneously in Biacore 2000 and 3000 instruments. The procedure described here involves use of all four flowcells, for making sample and reference measurements.

3  Inject 20 μl of regeneration buffer into all four flowcells and record the baseline responses.

4  Inject 20 μl of the nickel solution into all flow cells and record the new baseline with nickel bound. Usually a response of 40 RU is indicative of the flowcell being fully saturated with nickel ions (about $10^{10}$ ions).

5  Change the flow rate to 10 μl/min.

6  Choose flowcell 1 and inject 50 μl of a test single-clone protein-phage solution (e.g. the wild-type construct).

7  Inject 50 μl of a control protein-phage sample into flowcell 2 (e.g. a destabilized construct).

8  Leave flowcell 3 as a control to follow changes associated with the chymotrypsin injection, and flowcell 4 as a control to monitor the baseline throughout the experiment.

9  After phage binding has equilibrated, record the resonance response.

**10** Change the flow rate to 30 μl/min and inject 50 μl of 10 μM chymotrypsin into flowcells 1,2 and 3.[c] This gives a 100s contact time with protease, but the volume and flow rate can be changed to vary the contact time.[b]

**11** Allow the HPS running buffer to wash the flowcells and record the responses after the traces have equilibrated.

**12** Change the flow rate to 10 μl/min. Inject 30 μl of regeneration buffer into all four flowcells to elute any protease-resistant phage that remain bound to the chip surface.[c] At this point, the trace should return to the level of the initial baseline recorded after step 3, since everything else on the chip surface, including the nickel ions, should have been eluted by the EDTA in the regeneration buffer.

[a] Other proteases, or even protease cocktails could be used depending on the sequence composition of the displayed protein and the precise application. For instance, we have also used trypsin.

[b] Injection volumes, flow rates, and temperatures should be viewed as guides: these are variables that may be optimized for individual experiments.

[c] If desired, the recovery option in the injection menu can be used to collect fractions that are being released from the sensor chip surface in steps 10 and/or 12.

## 3.3.2 Protease selection using Ni-NTA agarose beads

Because of the limitations posed by the small surface area available in Biacore instruments, we perform protease selections using Ni-NTA agarose beads, where binding capacity is not so limited (*Protocol 5*). This should allow panning of larger libraries simply by increasing the amount of Ni-NTA beads used for binding and selection. Our current selections using Ni-NTA agarose beads have been optimized for the ubiquitin-phage library, which contains approximately $4 \times 10^5$ mutants and a 20-fold redundancy. The proteolysis, washing, and elution conditions in this protocol were determined empirically from our work with ubiquitin selections, and optimal parameters may change for different proteins. According to the manufacturer, 1 ml of Ni-NTA agarose beads is sufficient for binding 5–10 mg of an average 20 kDa histidine tagged protein. On this basis, 1 ml of beads should have a binding capacity of approximately $10^{14}$ phage (with an approximate molecular weight $1.5 \times 10^7$).

Usually, we propagate phage through three to four rounds of protease selection. Successive rounds of selection may be performed using either identical conditions, or by varying them. For example, if the aim is to select super-stable variants, the stringency of selection may be increased in successive rounds of panning, for example, by gradually adding denaturants, increasing the temperature, or otherwise altering the selection conditions (14, 15). Alternatively, for experiments aimed at probing subtle, specific links between sequence and structure/stability, then the protease selection may be made less stringent in order to propagate many clones to provide data for sequence analysis (6).

## Protocol 5

# Protease-based phage display selection using Ni-NTA agarose beads

### Reagents

- Binding buffer (50 mM $Na_2HPO_4$, 300 mM NaCl, 10 mM imidazole, pH 7.8)
- Nickel-NTA agarose beads supplied as a 50% suspension (Qiagen)
- Microcentrifuge tube filter (10 μm cut-off) (Whatman, Catalog Number 6838-0002)
- 1 mM chymotrypsin (type II from bovine pancreas, Sigma)

- Elution buffer (50 mM $Na_2HPO_4$, 300 mM NaCl, 250 mM imidazole, pH7.8)
- 2TY/Carb/Tet agar plates (see Protocol 2)[a]
- Concentrated display phage stocks (e.g. from *Protocol 3*)

### Method

1 Mix 0.1 ml of library phage (approximately $4 \times 10^8$ phage) with 0.8 ml binding buffer and 0.2 ml Ni-NTA agarose beads and allow the protein-phage to bind for 10 min with gentle shaking.

2 Transfer the mixture to a microcentrifuge tube filter (10 μm cut-off) to collect the beads in the filter, and wash the beads 14 times with 0.75 ml aliquots of binding buffer. Remove the supernatant by centrifugation at each stage and discard it.

3 Resuspend the beads in 1.25 ml of binding buffer and transfer 1 ml of the suspension to a new microcentrifuge tube preheated to 37 °C. Add 40 μl of 1 mM chymotrypsin and incubate the tube for 100 s at 37 °C. Here, conditions may be varied—for example, protease concentrations, temperature, added denaturant—to change the stringency of the selection.

4 Wash the beads 4 times as described in Step 2.

5 Recover bead-bound phage particles by incubating the 0.2 ml washed beads in 0.2 ml elution buffer followed by a brief microcentrifugation and incubate for 10–15 min.

6 Infect aliquots from each step to log phase *E. coli* XL1-Blue cells and plate out relevant dilutions on 2TY/Carb/Tet agar plated to determine the number of phage and, thereby monitor binding and protease cleavage throughout the experiment.

7 Amplify the selected phage for a new round of selection by reinfecting *E. coli* (e.g. using the protocol given in Chapter 4, *Protocol 4*, Step 6).

[a] Use 2TY/Carb/Tet plates if the library was electroporated into XL1-Blue cells. The tetracycline selects for maintenance of the F-pilus expression required for phage infection. For other *E. coli* strains, use 2TY/Carb plates supplemented if necessary with additional antibiotics appropriate for the background strain.

### 3.3.4 Characterization of selected clones

For initial characterization of selected clones, DNA from individual phagemids can be sequenced to identify the displayed proteins using standard protocols (see Chapter 4, Section 7). Using pCANTAB B or similar vectors, selected proteins can also be expressed readily and free from pIII simply by changing to a non-suppressor strain of *E. coli* such as HB2151, since the vector has two amber (TAG) suppressible stop codons between the target protein and pIII (*Figure 3*). Because the histidine tag is N-terminal to the target protein, purification of the resulting proteins may be performed simply using standard Ni-NTA based methods. In our experiments with ubiquitin, we then used circular dichroism (CD) and nuclear magnetic resonance (NMR) spectroscopy to show that the selected clones were fully folded with structures and stabilities similar to the wild-type protein (6).

## 4 Potential new applications for protease-based selection in phage display

The phage display stability selection strategy that we describe here has potential applications in several areas. One general application is improving the stabilities of natural and designer proteins. For instance, in protein design studies a semi-rational approach can be envisaged in which basic target protein frameworks are derived, say computationally, and then stably folded structures for these are selected from pools of hydrophobic-core mutants (6) or other variants by protease selection (1). Second, the method allows sequence-to-structure/stability relation-ships to be probed directly by mutagenesis and selection, without the need to maintain, or to select for, a specific protein function (3, 6, 14, 15). Third, Reichmann and Winter describe a method for potentially selecting novel folded protein domains using protease selection from a chimeric polypeptide library constructed by replacing the gene encoding of the C-terminal half of CspA with randomly generated fragments of *E. coli* DNA (16). We propose that the method may provide a solution to a bottleneck commonly encountered in proteomics; namely, the selection of constructs for stably folded domains from large, or intransigent genes of interest and/or from cDNA and genomic DNA libraries. The stability selection approach should be tractable and valuable because many proteins comprise a number of independently folded domains that are possible to isolate as stable fragments. Such subunits are often associated with specific functions and their isolation carries advantages for detailed biochemical and structural characterization. Without any structural information it is not yet possible to know, and it is difficult to predict, the precise boundaries of protein domains. We propose that by using restriction enzymes to randomly digest a gene, or even a whole genome of interest, displaying appropriately sized encoded polypeptide fragments on phage and selecting for stably folded proteins, it should be possible to identify individual protein domains (*Figure 6*).

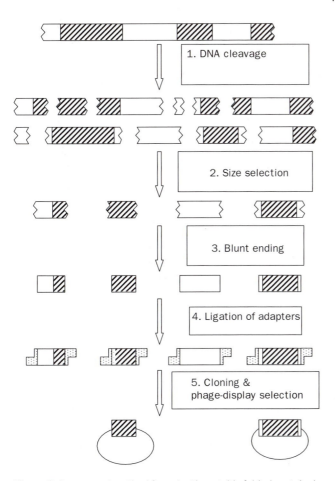

**Figure 6** A proposed method for selecting stably folded protein domains from cDNA and genomic DNA. The libraries could be made by randomly cleaving the starting DNA. Fragments that potentially encode protein domains could be selected by sizing on agarose gels. After cloning the resulting fragments into the hexahistidine-tagged phage display system, protease selection could be applied to remove incomplete and unfolded/unstable protein domain fragments.

## References

1. Sieber, V., Plückthun A., and Schmid, F. X. (1998). *Nat. Biotechnol.*, **16**, 955.
2. Kristensen, P. and Winter, G. (1998). *Folding & Design*, **3**, 321.
3. Finucane, M. D., Tuna, M., Lees, J. H., and Woolfson, D. N. (1999). *Biochemistry*, **38**, 11604.
4. Fontana, A., deLaureto, P. P., DeFilippis, V., Scaramella, E., and Zambonin, M. (1997). *Folding & Design*, **2**, R17.
5. Matthews, D. J. and Wells, J. A. (1993). *Science*, **260**, 1113.
6. Finucane, M. D. and Woolfson, D. N. (1999). *Biochemistry*, **38**, 11613.
7. Prodromou, C. and Pearl, L. H. (1992). *Protein Eng.*, **5**, 827.
8. Woolfson, D. N., Cooper, A., Harding, M. M., Williams, D. H., and Evans, P. A. (1993). *J. Mol. Biol.*, **229**, 502.
9. Kunkel, T. A. (1985). *Proc. Natl. Acad. Sci. USA*, **82**, 488.

10. Clackson, T. and Winter, G. (1989). *Nucl. Acids Res.*, **17**, 10163.
11. Levitan, B. (1998). *J. Mol. Biol.*, **277**, 893.
12. Gray, C. W., Brown, R. S., and Marvin D. A. (1981). *J. Mol. Biol.*, **146**, 621.
13. Rich, R. L. (2000). *Curr. Opin. Biotechnol.*, **11**, 54.
14. Jung, S., Honegger, A., and Plückthun, A. (1999). *J. Mol. Biol.*, **294**, 163.
15. Martin, A., Sieber, V., and Schmid, F. X. (2001). *J. Mol. Biol.*, **309**, 717.
16. Riechmann, L. and Winter, G. (2000). *Proc. Natl. Acad. Sci. USA*, **97**, 10068.

# Chapter 9

# Phage display of zinc fingers and other nucleic acid-binding motifs

Mark Isalan and Yen Choo

Medical Research Council Laboratory of Molecular Biology,
Hills Road, Cambridge CB2 2QH, UK.

## 1 Introduction

DNA-binding proteins play important roles in DNA replication, recombination, restriction, modification, repair, packaging, and the control of DNA transcription into RNA. Selecting DNA-binding proteins from pools of combinatorial variants should be a valuable method for the study of these proteins, and furthermore in the design of variants with new DNA sequence-specificity or effector function. In recent years, one key step in this direction has been the selection of zinc finger motifs with predetermined DNA-binding specificity by phage display (1–11), and their use to engineer customized transcription factors for use in research and medicine (reviewed in (12)). Although zinc finger domains are ideally suited to phage display studies, it is very likely that the lessons emerging from this work will be applicable to many other DNA- and RNA-binding motifs. To this end, the present chapter is intended to bring out these principles in addition to describing the practical aspects of creating and selecting zinc finger phage display libraries.

## 2 Preliminary considerations in creating a DNA-binding protein phage display library

### 2.1 Is phage display appropriate?

In phage display using filamentous bacteriophage, the DNA-binding domain of interest is typically expressed as a fusion to the minor coat protein pIII, translocated through the bacterial membranes and assembled into the tip of the particle that encloses the viral genome (13) *(Figure 1(a))*. The first phage display experiments of peptide domains were performed using secreted proteins such as antibodies, hormones, or extracellular enzymes which naturally traverse biological membranes and are stable in oxidizing environments; however, we now know

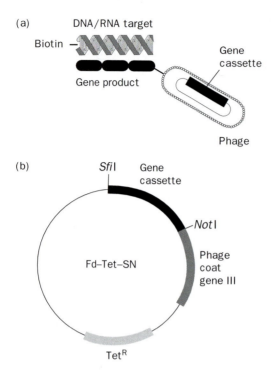

**Figure 1** Phage display of nucleic acid-binding domains. (a) The gene coding for a nucleic acid-binding domain is cloned into a phage vector so that its product is displayed (polyvalently—not shown) as a fusion protein on the surface of phage. This ensures genotype–phenotype linkage so that novel peptides, and their associated genes, can be simultaneously captured by screening against a nucleic acid target. The target is itself recovered by immobilization using biotin linkage to a streptavidin-coated matrix. (b) Schematic map of a typical phage display vector, Fd–Tet–SN (3). Key features include the *Sfi*I and *Not*I cloning sites that allow N-terminal fusion of a nucleic acid-binding domain, in frame with phage minor coat gene III. Tet$^R$ allows positive selection of phage-transformed bacteria.

that many intracellular domains, such as DNA-binding domains, can be successfully displayed on phage (see Chapter 1). Even so, protein translocation across the bacterial membrane occasionally presents an obstacle to phage display of certain peptides, particularly those that are highly charged. Nucleic acid-binding domains are generally positively charged in order to interact with the net negative charge of the phosphodiester backbone, and we have previously noted that certain highly basic domains, such as the BIV Tat minimal domain, will not be displayed on phage (Yen Choo, unpublished data).

In addition, at least two other factors should be considered before embarking on phage display. First, although there appears to be no general limit on the size of the domain that can be displayed (see Chapter 1), the technique may be best suited to studying small motifs (around 10 kDa)—of which there are plenty in nature. Second, a more serious limitation is that the majority of natural motifs have evolved to bind as non-covalent dimers to palindromic or near-symmetric DNA sequences. Although it is possible to adapt phage display to the selection of such motifs, other techniques, such as the bacterial "one-hybrid" method (14) might be better suited to this task.

## 2.2 Is the DNA-binding mode of the protein understood?

One of the important lessons from phage display of zinc fingers is that a preliminary understanding of the mode of protein-DNA interaction can be invaluable. The greatest aid in this regard is doubtless a high-resolution protein-DNA structure showing the mode of DNA binding and the contacts underlying intermolecular recognition. One obvious consideration, alluded to in Section 2.1 above, is whether the motif of interest binds DNA as a monomer or a dimer. Zinc fingers are particularly well suited to phage display as they bind DNA as covalently linked concatamers, allowing recognition of asymmetric DNA sequences.

Another consideration is whether the DNA in the complex is in any way distorted. While a severely distorted complex is not an indication that phage display experiments will be impossible, it may be that selections can only be performed against a limited subset of DNA sequences and/or that the resulting DNA-binding domains may bind nonspecifically. Induced fit protein-DNA interactions are usually highly specialized in the sense that very few DNA sequences will be able to adopt the structure required for binding. Hence intermolecular recognition in these instances is often heavily dependent on contacts to the DNA backbone, which does not carry direct sequence information. From the crystal structures of Zif268 (15, 16) and related proteins (17, 18), it can be seen that zinc fingers are geometrically adapted to binding DNA without any major perturbations of the double helix.

## 2.3 Which regions of the protein should be randomized?

Randomizing the appropriate regions of a DNA-binding domain can be the key difference between success and failure in phage display. The choice of residues will depend on the objective of the experiment, for instance whether one wishes to test conservative changes in DNA-binding specificity (e.g. by altering a few nucleotides of the DNA binding site), or if one is selecting proteins to bind radically different DNA sequences (e.g. engineering a homeodomain to bind a bacterial operator). If the protein residues that make phosphate contacts are likely to remain the same in complexes with other DNA sequences, for example, if it is clear that the protein will always prefer to bind B-form DNA, then these residues can be saved from randomization. The zinc finger is a particularly simple class of DNA-binding motif in which all the DNA-recognition residues can be readily localized to positions on one α-helix [reviewed in (19)]. In early experiments, we therefore randomized all the residues in this region (3), revealing that the residues most involved in general zinc finger-DNA recognition are the three α-helical positions observed to mediate base contacts in the crystal structure of Zif268 (15). However, we later used phage display to study the importance of a fourth α-helical position, and to illustrate how phage display libraries can be compromised by failing to randomize the full complement of DNA-binding residues in a protein domain (7, 20).

## 2.4 Polyvalent or monovalent display?

Proteins and peptides can be displayed either polyvalently or monovalently on phage (see Chapter 1). Monovalent display is generally considered to allow selective enrichment of higher-affinity displayed protein variants, and most other groups have used monovalent display to isolate zinc fingers with predetermined specificities (1, 2, 5). However, we have found that polyvalent libraries are highly effective starting points for selection. The protocols in this chapter describe the use of polyvalent pIII libraries displayed on phage (as opposed to phagemid), although they are readily adapted to monovalent libraries.

## 2.5 Choice of protein scaffold and selection strategy

Phage display of DNA-binding domains can be used to select proteins with novel DNA specificity for use in biotechnology and from the emerging publications it seems that this application, rather than basic research into the fundamental mechanism of DNA recognition, is becoming the driving force behind experiments (4, 11, 21–26). For these purposes, several potential DNA-binding domains and selection strategies might be used. The zinc finger motif was chosen as a subject for phage display by a number of groups because it was clear that single domains could be used as building blocks to construct multi-finger DNA-binding domains (27, 28). Since only a single zinc finger is randomized in any one phage display library, the diversity of the library can be kept low, greatly simplifying the construction of the library. Hence the zinc finger, with as few as four base recognition positions per DNA binding unit, is an ideal scaffold for phage display. More complicated domains, which make far more direct contacts with the nucleotides they recognize, would require massive libraries that are difficult to clone. At least three different strategies of creating DNA-binding domains comprising zinc fingers have been proposed, and each requires a unique phage display library design (4, 6, 9, 29). The early strategies assume zinc fingers are highly modular and feature either parallel (4, 9) or serial (6) selections of individual zinc fingers. The more advanced strategy recognizes that zinc fingers within a DNA-binding domain interact, and is based on a less modular approach that emphasizes selection of two adjacent zinc fingers (29).

## 3 Constructing a phage library cassette

Before initiating a project to select variant nucleic acid binding proteins from a library, it is worthwhile to carry out a pilot experiment to determine whether a single (i.e. wild type) domain retains its binding properties when displayed on phage. Certain motifs will prevent functional phage from being secreted from the bacterial host, while others may misfold when fused with pIII. The gene for the wild-type domain can be cloned into the display vector using the general procedures described for libraries in *Protocols 1* and *2*, but using a single gene in place of the oligonucleotide-derived randomized sequence. Carrying out a phage ELISA (described in *Protocol 4*) on phage derived from this wild-type display vector rapidly establishes whether particular domains displayed on phage will bind their desired nucleic acid targets under standard selection conditions. Moreover, such assays give a useful estimate of the affinity with which interactions occur, and help formulate specific selection conditions (*Protocol 3*).

Having confirmed that the protein in question can be functionally displayed, the next step is to construct a library of mutants. The choice of how many positions to mutate, and how diversely to mutate them, depends on standard considerations of the library size versus theoretical diversity (see Chapter 2 for a discussion). In our work, we have used the structure of Zif 268 bound to DNA, and have randomized those positions which make direct contacts with the edges of base pairs (alpha helical positions − 1, 2, 3, and 6). In addition, we have also randomized positions which are adjacent (or lie between) these, as their identity could well influence the fine structure of the DNA recognition helix. We have also partially randomized amino acids which may mediate inter-finger contacts, in order to allow some flexibility in domain positioning and stabilization. In general we have not mutated those amino acids which make contact with the DNA backbone, as the crystal structure of Zif 268 suggests that they do not play a major role in sequence recognition (although they are undoubtedly important in binding).

Construction of a gene cassette can be achieved by end-to-end ligation of synthetic oligonucleotides, as described in *Protocol 1*.

## Protocol 1

## Library cassette construction

### Equipment and reagents

- Designed library and guide DNA oligonucleotides (see Step 1 below)
- T4 polynucleotide kinase (10 U/μl) and buffer (New England Biolabs)
- T4 DNA ligase (400 U/μl) and buffer (New England Biolabs)
- 10 mM ATP (New England Biolabs)
- Polymerase chain reaction (PCR) Thermocycler block
- Sterile distilled water
- *Taq* DNA polymerase (Promega)
- 10 × *Taq* DNA polymerase buffer containing 15 mM MgCl$_2$ (Promega)
- Equimolar mix of deoxyribonucleotide triphosphates (dNTPs), 25 mM in each of G, A, T, and C (New England Biolabs)
- Two PCR primers that anneal at the ends of the library cassette, each at 10 pmol/μl (see Step 5 below)

- Mineral oil (Sigma)
- *Sfi*I 20,000 U/ml, NEBuffer 2 (New England Biolabs)
- *Not*I 10,000 U/ml, NEBuffer 3 (New England Biolabs)
- TAE buffer (50 × stock solution: Tris base, 242 g/L; glacial acetic acid, 57.1 ml/L; Na$_2$EDTA.2H$_2$O, 37.2 g/L)
- Buffer-saturated phenol (e.g. Sigma, Cat. Number P4557)
- Chloroform/isoamyl alcohol (IAA) 24:1 (e.g. Sigma, Cat. Number C0549)
- PCR purification kit (e.g. QIAquick, Qiagen)
- Agarase I and manufacturer's recommended buffer (Sigma)

## Method

1  Design and synthesize a set of oligonucleotides, each up to about 100 bases in length, that will code for the protein domain of interest when ligated end-to-end. Randomized codons can be introduced at appropriate positions to generate library diversity (see Chapter 2 for a discussion of diversity numerical limits), although bear in mind that practical cloning limits using phage (as opposed to phagemid) vectors are around $10^6$–$10^7$ transformants. Also construct a set of non-coding "guide" oligonucleotides, which anneal to the template strand, overlapping by ~20 bases in each direction (see *Figure 2*).

2  Place 10 µl of each coding-strand oligonucleotide (at 10 pmol/µl) in a 1.5 ml Eppendorf tube and add 1.2 µl of 10 × T4 kinase buffer, containing 1 mM ATP. Add 1 µl (10 U) of T4 polynucleotide kinase and incubate at 37 °C for 1 h. Heat inactivate the enzyme at 65 °C for 10 min.

3  Mix the kinase reactions with 10 µl of each non-coding strand guide oligonucleotide (from 10 pmol/µl stocks in 1 × T4 DNA ligase buffer). Denature the DNA by heating the tube to 94 °C for 1 min, then allow the DNA strands to anneal by cooling to 4 °C at a rate of 2 °C per min, on a thermocycler block.

4  Spin the tube briefly to collect condensation, then add an equal volume of 1 × T4 DNA ligase buffer with 8 µl (3200 U) of T4 DNA ligase per 100 µl. Carry out the ligations by incubating at 16 °C for 8–24 h.

5  Generate a full length double-stranded cassette by PCR amplification of the ligation mix, using appropriate primers that anneal at the ends of the library cassette and with terminal restriction sites complementary to those in the phage vector. Set up the PCR reaction by mixing the following in order:

- Distilled water            235 µl
- 10 × Taq buffer            30 µl
- dNTPs (25 mM)              15 µl
- Primer 1 (10 pmol/µl)      7.5 µl
- Primer 2 (10 pmol/µl)                            7.5 µl
- Annealed, ligated library (from Step 4)          3 µl
- *Taq* DNA polymerase                             2 µl

Distribute 50 µl into each of six PCR tubes, centrifuge the tubes briefly, overlay the solutions with mineral oil[a] and place the tubes in a thermal cycler.

6  Incubate the reactions for 25 cycles of 1 min at 94 °C, 1 min at the melting temperature of flanking primers (e.g. 60 °C) and 1 min at 72 °C. At the end, incubate the reactions at 72 °C for a further 10 min.

7  Verify that the PCR product is of the correct size by agarose gel electrophoresis. PCR conditions may need to be optimized in each individual case. If the PCR product is not of the predicted size, you may need to repeat the PCR using gel-purified DNA as template. This may be facilitated by $^{32}$P end-labeling of the 5'-terminal oligonucleotide used in constructing the gene cassette, allowing the position in the gel of the desired product to be identified by autoradiography.

**8** Purify 300 µl PCR product by phenol/chloroform extraction: extract twice with buffer-saturated phenol, then twice with chloroform/IAA, followed by ethanol precipitation. Alternatively, use a PCR purification kit (e.g. QIAquick)

**9** Digest the purified DNA cassette very thoroughly using the appropriate restriction enzymes. When cloning into phage vector Fd-Tet-SN, *Sfi*I and *Not*I are used to produce cohesive termini. Digestions with *Sfi*I and *Not*I are performed sequentially. To digest with *Sfi*I, resuspend the PCR product in 460 µl of $1 \times$ NEBuffer 2, containing 100 µg/ml BSA, and add 10 µl (200 U) of *Sfi*I. Overlay with mineral oil and incubate at 50 °C, supplementing the reaction with 10 µl (200 U) of *Sfi*I every 2 h, for a total incubation time of 8 h.

**10** Prepare the DNA for the second digestion (with *Not*I) by extracting once with phenol and once with chloroform, followed by ethanol precipitation. Resuspend the DNA in 460 µl of $1 \times$ NEBuffer 3 containing 100 µg/ml BSA. Add 10 µl (100 U) of *Not*I and incubate at 37 °C. Supplement the reaction with 10 µl (200 U) of *Not*I every 2 h for a total incubation time of 8 h.

**11** Remove the small terminal restriction fragments and prepare the digested DNA cassette for cloning by using agarose gel purification: run the digested cassette on a 2% low melting point agarose gel in $1 \times$ TAE containing 0.5 µg/ml ethidium bromide. Excise the band of interest from the gel and extract the DNA by any suitable method. We typically use *Agarase*I digestion: dissolve the agarose gel slices in $1 \times$ *Agarase*I buffer by incubating at 65 °C for 15 min. Add two units of *Agarase*I per 100 mg of gel slice, and incubate at 45 °C for 1 h. Pellet the resulting oligosaccharides by centrifuging at 13,000 g for 10 min at 4 °C. Recover the DNA from the supernatant by ethanol precipitation.

**12** Quantitate the DNA yield by absorbance at 260 nm ($A_{260}$), and store the cassette in distilled water at $-20$ °C. This protocol typically yields 1–3 µg of library cassette, of which less than 0.5 µg is typically required to generate a library of 5 million transformants.

[a] Mineral oil is not required if a thermal cycler with a heated lid is used.

## 4 Phage vector preparation and library construction

Phage vectors such as Fd–Tet–SN (3) allow the polyvalent surface display of proteins, such as zinc finger peptides, as in-frame fusions with phage minor coat protein (pIII; *Figure 1(b)*). This vector carries a tetracycline resistance (Tet[R]) gene, which allows positive selection of bacterial transformants. To clone large libraries in phage vectors such as Fd–Tet–SN, it is essential that vector DNA be properly prepared. Only double-stranded phage DNA purified on a caesium chloride gradient is suitable for library construction—the library size that may be

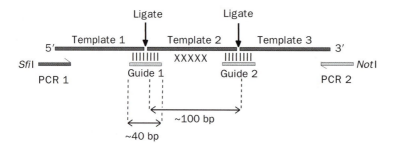

**Figure 2** Schematic representation of gene cassette construction by end-to-end ligation. Guide oligonucleotides anneal to the template (coding) strand, thereby directing the desired ligation reactions. Randomized codons (XXXXX) are introduced at appropriate positions in the coding strand to generate library diversity. The full-length, double-stranded cassette is recovered by PCR using primers that introduce appropriate restriction sites for cloning into the phage vector.

cloned is severely reduced by lower quality vector. Vector purification and subsequent library cloning steps are described in *Protocol 2*.

## Protocol 2

# Preparing the phage vector DNA and cloning the binding protein library

### Equipment and reagents

- Phage vector (e.g. Fd–Tet–SN, (3)).
- F' strain of *E. coli* (e.g. TG1 (30); *supE hsdΔ thiΔ(lac-proAB)* F' [*traD36 proAB + laqIq laqZΔM15*]
- TYE agar (Bacto-trzyptone, 16.0 g/L; Bacto-yeast extract, 10.0 g/L; NaCl, 5.0 g/L; Agar, 15.0 g/L)
- Tetracycline
- 2 × TY (Bacto tryptone, 16.0 g/L; Bactoyeast extract, 10.0 g/L; NaCl, 5.0 g/L)
- Plasmid purification system, for example, Wizard Maxiprep kit (Promega)
- TE buffer (Tris pH 8.0, 10 mM; EDTA, 1 mM)
- Caesium chloride (CsCl) (Sigma)
- Ethidium bromide, 10 mg/ml
- 5 ml ultracentrifuge tubes (Beckman)

- 2 L sterile conical flasks
- Ultracentrifuge (e.g. Beckman Vti 65.2 rotor)
- Water-saturated butan-2-ol
- *Sfi*I 20,000 U/ml, NEBuffer 2 (New England Biolabs)
- *Not*I 10,000 U/ml, NEBuffer 3 (New England Biolabs)
- Mineral oil (Sigma)
- TAE buffer (see *Protocol 1*)
- *Agarase I* and manufacturer's recommended buffer (Sigma)
- T4 DNA ligase (400 U/μl) and buffer (New England Biolabs)
- Equipment and reagents for electroporation of DNA into *E. coli* strain TG1, as described in Chapter 13, *Protocol 10*, except note that the 150 mm and 24.5 cm TYE plates and 2 × TY medium should be supplemented only with 15 μg/ml tetracycline.

## Method

1  Transform a suitable strain of *E. coli* (e.g. TG1) with a phage vector. In the case of Fd–Tet–SN, plate the tranformant colonies on TYE agar plates supplemented with 15 µg/ml tetracycline.

2  Using a single transformant colony, innoculate a 2 ml "starter" culture in $2 \times TY$ containing 15 µg/ml tetracycline. Incubate with shaking for 4 h at 37 °C, and then transfer the culture into a 2-L flask containing 1 L of the same medium. Incubate, with shaking, for up to 15 h at 37 °C.

3  Purify the double-stranded replicative form phage DNA from the 1 L bacterial culture using a standard plasmid purification kit such as the Wizard Maxiprep kit.

4  Dissolve 10–100 µg of the recovered phage DNA in 4 ml TE buffer containing 4 g of CsCl and 0.1 ml of ethidium bromide solution (10 mg/ml). Mix the samples to dissolve, and dispense the solution into 5 ml ultracentrifuge tubes. Carry out ultracentrifugations of 370,000g, for 20 h at 20 °C, using a suitable ultracentrifuge. Two bands of DNA should be visible under UV light. Collect the lower (supercoiled plasmid) band with a syringe and extract 4 times with water-saturated butanol-2-ol to remove all traces of ethidium bromide (add an equal volume of water-saturated butan-2-ol, mix by vortexing, centrifuge briefly to separate the phases, carefully remove and discard the upper (organic) phase, and repeat three times or until the pink colouration disappears from the upper phase). Dilute the resulting solution three-fold in water and ethanol precipitate the DNA using standard methods to recover pure phage vector. Wash the DNA pellet in 70% ethanol to remove traces of CsCl.[a]

5  Digest 20 µg of the CsCl-purified phage vector with the appropriate restriction enzymes. For cloning into Fd–Tet–SN these enzymes are *Sfi*I and *Not*I. The procedure is the same as that for preparing the randomized gene cassette (see *Protocol 1*, steps 9–10).

6  Remove the resulting small restriction fragments and prepare the vector for cloning using agarose gel purification: run the digested vector on a 1% low melting point agarose gel in $1 \times TAE$ buffer containing 0.5 µg/ml ethidium bromide. Excise the band of interest from the gel under long-wave UV light and extract the DNA by any suitable method, such as digestion with *Agarase*I (see *Protocol 1*, step 11).

7  Perform ligations with a molar ratio of insert to vector of between 3 : 1 and 5 : 1. It is often worthwhile to carry out small-scale trial ligations to identify those giving optimum ligation efficiency. For library construction, ligate 0.2–0.5 µg of digested vector DNA (from step 6) and the appropriate amount of digested randomized gene cassette (from *Protocol 1*) in a reaction mix containing $1 \times T4$ DNA ligase buffer and 800 U of T4 DNA ligase, in a volume of 10–30 µl. Incubate the reaction mix at 16 °C for 15 h.

**8** Ethanol precipitate the ligated DNA. Carefully wash the pellet in 70% ethanol DNA to remove excess salt which could cause arcing during electroporation. Resuspend the dried pellet in MilliQ $H_2O$ to a final concentration of 0.2–0.5 μg vector per μl.

**9** Electroporate the purified ligated DNA in 1 μl aliquots into electrocompetent *E. coli* strain TG1, as described in Chapter 13, *Protocol 10*,[b] except use 24.5 cm TYE plates and medium supplemented only with 15 μg/ml tetracycline.[c]

[a] Use of CsCl-purified DNA is very important to give the highest transformation efficiencies of phage (as opposed to phagemid) vector libraries.

[b] An alternative optimized bacterial strain and set of protocols for high-efficiency electroporation of bacterial cells is described in Chapter 2.

[c] Because of the large size of phage vectors (~8 kb), electroporation efficiency is reduced to ~$10^6$–$10^7$ transformants per μg of library constructs, even for preparations of electro-competent cells that yield more than $10^9$ transformants per μg of pUC19 supercoiled plasmid control DNA.

## 5  Phage selections

Having constructed a phage display library, peptide variants with novel nucleic acid-binding properties can be rapidly isolated through sequential rounds of affinity selection. This is conveniently done by screening the library against synthetic, biotinylated DNA or RNA targets, which are captured on a streptavidin-coated matrix (e.g. streptavidin-coated tubes, as described in *Protocol 3* or streptavidin-coated paramagnetic beads such as those manufactured by Dynal). Unbound clones are washed away while selected clones are retained, eluted, and amplified by passaging through a bacterial host. By altering binding and washing conditions, a range of peptides with varying affinity can be generated. In addition, the inclusion of soluble competitor DNA (or RNA) allows selection for variants with predetermined binding specificity for one sequence over another. A general method, developed for use with zinc finger domains and applicable both to DNA and RNA binding selections, is described in *Protocol 3*.

## 6  Analysis of selected phage clones

The DNA sequences of phage clones recovered from selections for RNA or DNA binding activity can be identified by sequencing the appropriate part of the phage genome using standard methods. For functional analysis of their nucleic acid binding activity, phage clones can also be characterized expeditiously using a phage ELISA procedure. This method, described in *Protocol 4*, is most suited to a qualitative determination of binding. However, it can be adapted to measure both the apparent binding affinities and specificities of phage expressing novel nucleic acid-binding domains. For example, by testing phage against a range of target-site concentrations, a binding curve may be constructed from which the apparent dissociation constant ($K_d$) of interaction can be estimated. Furthermore, by assaying binding to closely related variants of the binding site, a profile of the sequence discrimination can be determined. The "binding site signature" of a zinc finger (31) is a specific example of a more sophisticated phage ELISA, which allows the preferred binding site(s) of individual fingers to be deduced using positionally randomized DNA libraries.

## Protocol 3

## Standard procedure for phage DNA/RNA binding selections

### Equipment and reagents

- Biotinylated DNA or RNA target oligonucleotides
- DNA annealing buffer ($2 \times$ stock solution: Tris-HCl pH 8.0, 40 mM; NaCl, 200 mM)
- RNA folding buffer ($1 \times$ CM: ammonium cacodylate, 50 mM, pH 6.5; MgCl$_2$, 2 mM)
- $2 \times$ TY (see *Protocol 2*)
- Tetracycline
- Streptavidin-coated immunotubes (or microtitre well plates) (Roche Molecular Biochemicals)
- Phosphate buffered saline (PBS) ($10 \times$ stock solution: NaCl, 80 g/L; KCl, 2 g/L; Na$_2$HPO$_4$.7H$_2$O, 11.5 g/l; KH$_2$PO$_4$, 2 g/L)
- Zinc chloride, 1 M (for selection of zinc finger domain libraries)

- Fat-free freeze-dried milk (Marvel; Premier Brands UK Ltd)
- Tween-20
- Sonicated salmon sperm DNA (10 mg/ml)
- Triethanolamine, 0.1 M
- Tris-HCl, 1 M, pH 7.4
- F$'$ strain of *E. coli*, for example, TG1 (30)
- M9 minimal agar (Na$_2$HPO$_4$, 6.0 g/L; KH$_2$PO$_4$, 3.0 g/L; NH$_4$Cl, 1.0 g/L; NaCl, 0.5 g/L. After autoclaving: 1 M MgSO$_4$, 1.0 ml/L; glucose, 2.0 g/L; 1 M CaCl$_2$, 0.1 ml/L; 1 M thiamine-HCl, 1.0 ml/L; Agar, 15.0 g/L).
- TYE agar (See *Protocol 2*)

### Method

### A. Preparation of phage

1  Transfer cells from fresh or frozen bacterial library stock (*Protocol 2*) to 200 ml of $2 \times$ TY, containing 15 µg/ml tetracycline.[a] We typically transfer a number of viable bacterial cells approximately 10-fold in excess of the library size, to ensure the library diversity is maintained. Viable cell concentration in the frozen stock can be determined by plating out serial dilutions on to appropriate plates.

2  Incubate the culture with orbital mixing at 250 rpm for 16 h at 30 °C.

3  Prepare the phage supernatant by centrifuging the overnight culture at 3700g for 15 min. Remove and keep the supernatant and store at 4 °C. The supernatant can be titered using the procedure given in Chapter 1, *Protocol 1*.

## B. Selections

1 Synthesize the desired target DNA or RNA oligonucleotides, and include a 5′-biotin group. For dsDNA sites, synthesize a complementary, non-biotinylated second strand.

2(a) *For DNA site selections*, anneal the two complementary oligonucleotides by mixing 10 μl of each oligonucleotide strand, from 10 pmol/μl stocks, with 20 μl of 2 × nucleic acid annealing buffer. Heat the samples for 3 min at 94 °C, and then cool them to 4 °C, at a rate of 2 °C per min, in a thermocycler block. When cool, dilute the solution to 1 pmol/μl final concentration with water and store it at − 20 °C.

(b) *For RNA site selections*, make up a 1 pmol/μl solution of the oligonucleotide in RNA folding buffer (1 × CM). Heat the solution to 95 °C and then cool it slowly to room temperature, to allow folding of the RNA. For some sites, "snap" cooling to 4 °C may result in better folding. Store the RNA solution at 4 °C (although some sites may be successfully freeze-thawed).

3 Add biotinylated target sites (typically 1 pmol) in 50 μl PBS to streptavidin-coated tubes. Incubate the tubes for 15 min at 20 °C.

4 Add 1 ml of PBS containing 4% (w/v) skimmed milk to the tube to block nonspecific binding sites. Incubate the tubes for 1 h at 20 °C.

5 Dilute the phage supernatant 1 : 10 in 1 ml PBS containing 2% (w/v) skimmed milk, 1% (v/v) Tween-20 and 20 μg/ml sonicated salmon sperm DNA. To improve the specificity of the selection, add target-related soluble competitor oligonucleotides. Add up to 100-fold excess of competitor, relative to the amount of biotinylated target site.

6 Discard the blocking solution from the nucleic acid-coated tube and apply 1 ml of the diluted phage supernatant solution. Incubate the tube for 1 h at 20 °C.

7 Discard the binding mixture and wash the tube 20 times with 1 ml PBS containing 2% (w/v) skimmed milk, 1% (v/v) Tween-20, and once with 1 ml PBS alone.

8 Elute the retained phage by adding 100 μl of 0.1 M triethanolamine and incubating for 15 s. Transfer the solution to a new tube and immediately neutralize it with an equal volume of 1 M Tris (pH 7.4).

9 Prepare a logarithmic-phase culture of *E. coli* TG1 by inoculating a sample of 2 × TY with 1/100 volume of an overnight culture and incubating at 37 °C for approximately 2 h. The innoculum must be derived from colonies grown on a plate of M9 minimal agar to ensure expression of the F′ pilus, as this is required for phage infection.

10 Infect 0.3 ml logarithmic-phase *E. coli* TG1 with 50 μl of eluted phage by mixing and incubating the mixture for 1 h at 37 °C without shaking.

**Protocol 3** continued

11  Transfer the infected bacterial cultures to 2–5 ml of 2 × TY, containing 15 µg/ml tetracycline. Incubate the cultures with orbital mixing at 250 rpm, for 16 h at 30 °C, in order to produce phage for subsequent rounds of selection.

12  Repeat the selection procedure, from steps 2 or 3 to 12, for 3–5 rounds. After the final round, spread the infected bacteria on plates containing TYE agar with 15 µg/ml tetracycline. The selected clones are now ready for further characterization by sequencing and by ELISA (*Protocol 4*).

[a] Certain nucleic acid-binding domains may require supplements to the growth medium. Zinc fingers, for example, are stabilized by 50 µM ZnCl₂ in all media and selection buffers.

## Protocol 4

# Analysis of DNA/RNA binding activity by phage ELISA

### Equipment and reagents

- Sterile, round-bottom, 200 µl, 96-well plates for tissue culture (Costar; Corning USA)
- 2 × TY (see *Protocol 2*)
- Tetracycline
- Zinc chloride, 1 M (for selection of zinc finger domain libraries)
- Streptavidin-coated microtiter well plates (Roche Molecular Biochemicals).
- PBS (see *Protocol 3*)
- Fat-free freeze-dried milk (Marvel; Premier Brands UK Ltd)
- Tween-20
- Sonicated salmon sperm DNA (10 mg/ml)

- Horse radish peroxidase (HRP)-conjugated anti-M13 IgG (Amersham Biosciences)
- ELISA developer solution [0.1 M Na(CH₃.-COO), pH 5.5; 3′, 3′, 5′, 5′-tetramethylbenzidine (TMB; Sigma), 0.5 mg/ml; dimethyl sulphoxide (DMSO), 1% (v/v); H₂O₂, 0.05% (v/v)]
- Sulphuric acid, 1 M
- ELISA plate reader
- Rotating incubator that can accommodate 96-well plates
- Swinging bucket centrifuge that can accommodate 96-well plates

### Method

### A. Phage preparation

1  Pick single bacterial colonies containing phage clones derived from the library selections (*Protocol 3*, step 12). Use a sterile toothpick to transfer colonies to the wells of sterile round-bottom plates containing 150 µl of 2 × TY and 15 µg/ml tetracycline. As a positive control, use one well to grow phage displaying the wild-type DNA/RNA-binding domain.[a] Incubate the plates with careful orbital mixing at 250 rpm, for 16 h at 30 °C.

**Protocol 4** continued

2 Prepare phage supernatants by centrifuging the 96-well culture plates at 3700g for 15 min, in an appropriate swinging-bucket centrifuge. Remove the supernatants to a fresh plate and store at 4°C.

## B. Phage ELISA

1 Add biotinylated nucleic acid target sites (typically between 0–5 pmol, diluted in 50 μl PBS to the wells of a streptavidin-coated microtitre plate. For the positive control, add an appropriate amount of the wild-type binding site to one well. Use a negative control well, containing PBS only, to measure the ELISA background. Incubate the plate for 15 min at 20°C.

2 To each well, add 150 μl of PBS containing 4% (w/v) skimmed milk as a blocking agent. Incubate the plate for 1 h at 20°C.

3 Dilute the phage supernatants (from Step A2) 1:10 in 1 ml PBS containing 2% (w/v) Marvel, 1% (v/v) Tween-20 and 20 μg/ml sonicated salmon sperm DNA.

4 Discard the blocking solution from the nucleic acid-coated wells and add 50 μl of the diluted phage supernatant solutions to each well. Incubate the plate for 1 h at 20°C.

5 Discard the binding mixture and wash the wells 7 times with 200 μl PBS containing 1% (v/v) Tween-20. Wash a further 3 times with 200 μl PBS alone.

6 To each well, add 50 μl of PBS containing 2% (w/v) Marvel and a 1:5000 dilution of HRP-conjugated anti-M13 IgG antibody. Incubate at 20°C for 1 h.

7 Discard the antibody binding mixture and wash the wells 3 times with 200 μl PBS containing 0.05% (v/v) Tween-20. Wash a further 3 times with 200 μl PBS alone.

8 Develop the ELISA by adding to each well 100 μl of HRP substate, such as the TMB-based ELISA developer solution described above. Stop the colorimetric reaction after approximately 5 min; for TMB add 100 μl of 1 M $H_2SO_4$. Quantitate the ELISA signals immediately using a spectrophotometer fitted with a microtiter plate reader.

9 To assess sequence-specific binding of the selected binding domains, compare their ELISA signals against those of the positive and negative control wells.

[a] Certain nucleic acid-binding domains may require supplements to the growth medium. Zinc fingers, for example, are stabilized by 50 μM $ZnCl_2$ in all media and selection buffers.

## Acknowledgments

We would like to thank Dr Monika Papworth for critical reading of the manuscript and Dr Aaron Klug, in whose lab these techniques were developed, for support and advice. M.I. is supported by the Wellcome Trust, UK.

## References

1. Rebar, E. J. and Pabo, C. O. (1994). *Science*, **263**, 671.
2. Jamieson, A. C., Kim, S.-H., and Wells, J. A. (1994). *Biochemistry*, **33**, 5689.

3. Choo, Y. and Klug, A. (1994). *Proc. Natl. Acad. Sci. USA*, **91**, 11163.

4. Choo, Y., Sanchez-Garcia, I., and Klug, A. (1994). *Nature*, **372**, 642.

5. Wu, H., Yang, W.-P., and Barbas C. F., III, (1995). *Proc. Natl. Acad. Sci. USA*, **92**, 344.

6. Greisman, H. A. and Pabo, C. O. (1997). *Science*, **275**, 657.

7. Isalan, M., Klug, A., and Choo, Y. (1998). *Biochemistry*, **37**, 12026.

8. Choo, Y. (1998). *Nature Struct. Biol.*, **5**, 264.

9. Segal, D. J., Dreier, B., Beerli, R. R., and Barbas, C. F. (1999). *Proc. Natl. Acad. Sci. USA*, **96**, 2758.

10. Isalan, M. and Choo, Y. (2000). *J. Mol. Biol.*, **295**, 471.

11. Beerli, R. R., Dreier, B., and Barbas, C. F. (2000). *Proc. Natl. Acad. Sci. USA*, **97**, 1495.

12. Isalan, M. D. and Choo, Y. (2000). *Curr. Opin. Struct. Biol.*, **10**, 411.

13. Smith, G. P. (1985). *Science*, **228**, 1315.

14. Joung, J. K., Ramm, E. I., and Pabo, C. O. (2000). *Proc. Natl. Acad. Sci. USA*, **97**, 7382.

15. Pavletich, N. P. and Pabo, C. O. (1991). *Science*, **252**, 809.

16. Elrod-Erickson, M., Rould, M. A., Nekludova, L., and Pabo, C. O. (1996). *Structure*, **4**, 1171.

17. Kim, C. and Berg, J. M. (1996). *Nature Struct. Biol.*, **3**, 940.

18. Elrod-Erickson, M., Benson, T. E., and Pabo, C. O. (1998). *Structure*, **6**, 451.

19. Choo, Y. and Klug, A. (1997). *Curr. Opin. Struct. Biol.*, **7**, 117.

20. Isalan, M., Choo, Y., and Klug, A. (1997). *Proc. Natl. Acad. Sci. USA*, **94**, 5617.

21. Pomerantz, J. L., Sharp, P. L., and Pabo, C. O. (1995). *Science*, **267**, 93.

22. Kim, J.-S., Kim, J., Cepek, K. L., Sharp, P. A., and Pabo, C. O. (1997). *Proc. Natl. Acad. Sci. USA*, **94**, 3616.

23. Kim, J.-S. and Pabo, C. O. (1997). *J. Biol. Chem.*, **272**, 29795.

24. Xu, G.-L. and Bestor, T. H. (1997). *Nat. Genet.*, **17**, 376.

25. Beerli, R. R., Segal, D. J., Dreier, B., and Barbas, C. F. (1998). *Proc. Natl. Acad. Sci. USA*, **95**, 14628.

26. Smith, J., Berg, J. M., and Chandrasegaran, S. (1999). *Nucl. Acids Res.*, **27**, 674.

27. Liu, Q., Segal, D. J., Ghiara, J. B., and Barbas, C. F., III (1997). *Proc. Natl. Acad. Sci. USA*, **94**, 5525.

28. Kim, J.-S. and Pabo, C. O. (1998). *Proc. Natl. Acad. Sci. USA*, **95**, 2812.

29. Isalan, M. D., Klug, A. and Choo, Y. (2001). *Nature Biotech.* **19**, 656.

30. Gibson, T. J. (1984). Studies on the Epstein–Barr virus genome. Ph.D. Thesis, Cambridge University, UK.

31. Choo, Y. and Klug, A. (1994). *Proc. Natl. Acad. Sci. USA*, **91**, 11168.

# Chapter 10

# *In vivo* and *ex vivo* selections using phage-displayed libraries

Jason A. Hoffman, Pirjo Laakkonen, Kimmo Porkka, Michele Bernasconi and Erkki Ruoslahti

The Burnham Institute, and Program in Molecular Pathology, Department of Pathology, University of California San Diego School of Medicine, 10901 N. Torrey Pines Road, La Jolla, CA 92037, USA.

## 1 Introduction

Phage display has emerged as a useful method to discover peptide ligands, and to minimize and optimize the structure and function of proteins (1–3). Phage display screens have two basic components: the target macromolecule and the phage. The phage is used as a scaffold to display recombinant libraries of peptides, and provides the means to rescue and amplify peptides that bind a target macromolecule. Target macromolecules are commonly proteins but can be nucleic acids, carbohydrates, and phospholipids, and can be broadly classified as simple or complex depending on their purity (4–6). A pure protein, such as an antibody, immobilized on a solid support is a simple target, while a complex target would be the surface of a cell.

Our laboratory developed *in vivo* selection schemes using phage-displayed libraries to discover peptide and protein ligands for macromolecules that are expressed in an organ- and tissue-specific manner (7). These ligands can distinguish between the vasculatures of different organs and tissues (8), and can be used as homing devices for therapeutic and imaging agents (9, 10), they can be therapeutics themselves (11), and they can provide tools for studying the underlying biology of processes such as lymphocyte homing, cancer metastasis, and the tropism of pathogenic microorganisms (12, 13).

## 2 *In vivo* and *ex vivo* phage display

Our laboratory first reported the successful use of phage display *in vivo* in 1996 and described several peptides that specifically bound mouse brain or kidney vasculature (7). *In vivo* phage display involves intravenously injecting phage libraries, rescuing phage bound to the vasculature of different organs and tissues, amplifying, purifying, and then reinjecting a phage pool enriched for clones that home to specific organs and tissues. One feature of *in vivo* selections is that they are simultaneously positive and subtractive screens because the vasculatures of organs and tissues are different from each other and are spatially separate. Phage that specifically bind the vasculature of organs and tissues other than the target are removed while specific phage become enriched in the target organ.

We use the term *ex vivo* phage display to describe the addition of phage solutions to primary cell suspensions of organs and tissues. A notable difference between *in vivo* and *ex vivo* phage display is that phage are primarily exposed to the vascular endothelial cells in the *in vivo* method but are exposed to all cell types in the *ex vivo* variation.

Finally, we refer to selection on cultured cells, and pure and crude preparations of macromolecules as *in vitro* phage display.

We use *ex vivo* selection as a preselection and selection method, but our primary use is to test individual phage discovered from *in vivo* screens. *Ex vivo* phage display is higher throughput, making it possible to screen many different phage using material from one mouse. We have found that *ex vivo* phage display has a lower background than *in vivo* phage display and that phage which bind to organ and tissue vasculature *in vivo* often show greater binding *ex vivo*, in other words *ex vivo* binding can be more discriminatory than *in vivo* homing. Greater *ex vivo* binding can result because the organ and its vasculature both express the receptor, this is the case for the lung-homing GFE-1 peptide and its receptor, membrane dipeptidase (14); however, *ex vivo* phage display also detects peptides that bind to a minority cell population (P. Laakkonen, K. Porkka, and E. Ruoslahti, unpublished results).

## 3  Phage vectors

We use George Smith's fd-tet-derived fUSE5 phage vector (15) and Novagen's T7Select® vectors. The two phages differ in their size, shape, and life cycle; fd is a filamentous phage that has a temperate life cycle, while T7 has an icosahedral capsid and is a lytic phage. Screens with the lytic system are more rapid and ligation-competent vector preparations are commercially available. fUSE5 displays peptides in five copies (i.e. polyvalently) as an N-terminal fusion with fd's gene III protein, pIII. Novagen's T7 vectors display 1–415 copies of inserts up to ~1200 amino acids as C-terminal fusions with the gene 10 protein. We utilize the T7Select 415-1b, T7Select10-3b, and T7Select1-1b vectors for our peptide libraries, and the T7Select1-1b and T7Select10-3b vectors for displaying larger proteins and cDNA libraries.

## 4  Control comparisons for *in vivo* and *ex vivo* phage display

We monitor our screens using an enrichment function, comparing the ratio of phage output to input in different organs and tissues. The basic comparisons for *in vivo* and *ex vivo* phage display are target–target and phage–phage. A strength of *in vivo* phage display is the presence of built-in target–target controls; a single injection of phage circulates through all of the organs and tissues, and provides the background binding of the phage pool or individual phage in the organ or tissue of interest and in one or more control organs or tissues. Because organs and tissues differ in their mass and amount of vasculature we normalize our *in vivo* selection data using organ and tissue weight. For our *ex vivo* protocol we physically normalize the cell number at $\sim 1 \times 10^7$ or some other constant number.

The phage–phage controls compare individual and pools of phage against insertless phage within the same target organ or tissue. Insertless phage for our two systems are fd-tet, fd-amp, and T7 negative control.

We graph our *in vivo* data with "Phage Output/(Phage Input $\times$ Tissue Weight)" as the *y*-axis (dependent variable) and "target" or "phage" as the *x*-axis (independent variable), and our *ex vivo* data with "Phage Output/Phage Input" as the *y*-axis and "target" or "phage" as the *x*-axis. We physically normalize the number of cells in *ex vivo* selection. We also find it useful to graph our *in vivo* and *ex vivo* data with "Fold relative to nonrecombinant phage" as the *y*-axis and "target" or "phage" as the *x*-axis.

## 5  The selection process

The first step in an *in vivo* selection strategy is to determine the baseline background of nonrecombinant phage. This is a good way to familiarize oneself with the technique and provides a critical data set for subsequent selections. The typical background is $10^{-3}$–$10^{-4}\%$ for *in vivo* phage display and $10^{-4}$–$10^{-5}\%$ for *ex vivo* phage display. The exact numbers depend on the organ, tissue, or cell type, and the perfusion and wash steps in *Protocols 1* and *2*.

We inject or add a sufficient number of phage so as to have 100–1000 copies of each individual phage in the initial pool. The diversity of our libraries is typically $10^7$–$10^9$ so we inject or add a minimum of $10^9$–$10^{11}$ transducing units (TU) or plaque forming units (pfu) for the first round. We inject $10^9$–$10^{10}$ TU or pfu in all subsequent rounds. The maximum volume that can be injected into a single mouse is $\sim 500$ μl while the *ex vivo* procedure can be performed in volumes up to several milliliters. Blood and background phage are partially removed by a perfusion step before the organs and tissues are removed and processed. Background phage are further removed by washing pelleted cells *ex vivo* several times. We lyse the washed pellet of cells with 1% Nonidet P40 (NP-40), rescue bound phage with bacteria and then amplify the phage in liquid culture. When the goal is to isolate vasculature-homing phage, we typically perform up to six rounds *in vivo*, with or without 1–3 rounds of an *ex vivo* preselection. *Protocols 1* and *2* give detailed instructions for these procedures.

## Protocol 1

### *In vivo* selection

#### Equipment and reagents

- A hemostat, a pair of surgical scissors and a pair of tweezers

- Several mice of a specific strain, for example, BALB/c

- One 10 ml plastic syringe filled with 10 ml of Dulbecco's Modified Eagle's Medium (DMEM) or 1 × Phosphate-buffered Saline (PBS) per mouse. Attach a 1-in., 25 gauge needle to each syringe. Prepare common solutions like PBS according to the method by Sambrook (16)

- Two insulin syringes per mouse

- One tared, sterile, 14 ml plastic tube per organ or tissue

- One tared, sterile, 2.2 ml plastic tube per organ or tissue

- ~8 ml of 1% (w/v) bovine serum albumin (BSA) in DMEM or 1 × PBS per organ or tissue

- A hand-held homogenizer with a 1 cm (diameter) × 15 cm (length) shaft or (DAKO's) MediMachine® with a 50 μm Medicon and a 70 μm Filcon. You will need one Medicon, one Filcon, one 1-ml luer slip syringe and one 14-ml conical vial per organ or tissue

- 1.25% (w/v) avertin in water. Prepare the anesthetic, avertin, by adding 2.5 g of 2,2,2-tribromoethanol (avertin, Aldrich-Sigma Chemical Co.) to 5 ml of 2-methyl-2-butanol (tert-amyl alcohol). Place this solution in a hot water bath until the avertin is completely dissolved and then add 200 ml of MilliQ water. The resulting solution is 1.25% (w/v) avertin and is sensitive to light and heat. Avertin is best stored at 4 °C in a red top vacutainer tube wrapped with aluminum foil and may be kept for up to 6 months. Discard any avertin stock that has a pH < 5

- A 200–500 μl phage solution of $10^9$–$10^{11}$ pfu or TU (for fUSE5 or T7, respectively). If necessary use DMEM or PBS as a diluent.

A protocol for propagation and titration of Ff phage vectors, such as fUSE5, is given in Chapter 1, *Protocol 1*

- A microcentrifuge, for example, Eppendorf® centrifuge 5415D

- A bench-top clinical centrifuge with a swinging bucket rotor, for example, Beckman's Allegra 6R centrifuge with a GH-3.8 centrifuge

- A centrifuge, and a Sorvall® SLA-3000, and SS-34 or equivalent rotors

- 400 ml centrifuge tubes

- 30 ml Oak Ridge centrifuge tubes

- 1% solution of NP-40 (prepare from 10% stock solution, Pierce Chemicals)

#### For fUSE5

- K91Kan culture. Prepare a culture by inoculating a volume of Terrific broth with one colony of K91Kan. The protocol uses 1 ml of culture per organ or tissue homogenate. Incubate and shake at 37 °C until the optical density (OD) at 600 nm is 1. It takes about 4 h for 10 ml to reach this OD

- Tetracycline solution (10 mg/ml in 50% (v/v) ethanol in water; store at − 20 °C)

- NZY media (15)

- 0.2 μg/ml tetracycline in NZY media. To prepare add 100 μl of stock tetracycline to 500 ml of NZY media. Store at 4 °C

- 40 μg/ml tetracycline in NZY media. To prepare add 2 ml of stock tetracycline to 500 ml of NZY media. Store at 4 °C

- 40 μg/ml tetracycline Luria-Bertani (LB) agar plates

- One sterile, 50 ml conical vial containing 9 ml of 0.2 μg/ml tetracycline in NZY media per organ or tissue

- Sterile, 2 L Erlenmeyer flasks

- Vacuum-driven bottle-top 0.2 μm filters (e.g. Millipore's SteriTop® filters)

- Sterile, 500 ml glass bottles
- 16.6% polyethylene glycol (PEG), 3.3 M NaCl solution. Prepare by adding 100 g PEG 8000 (Sigma 2139), 116.9 g NaCl, and 475 ml water in a 1 L bottle. Autoclave to sterilize and to dissolve all solids, you will have to periodically shake the bottle as it cools to prevent the formation of several phases
- 1 × Tris-Buffered Saline (TBS)
- 0.1% (w/v) gelatin in 1 × TBS

## For T7

- M9LB media
- An overnight culture of the BL21 or BLT5615 strain of *E. coli* (Novagen) in M9LB. Use BL21 for the T7Select 415-1 vector and BLT5615 for the T7Select 1-1, and T7Select10-3 vectors. Prepare by inoculating a volume of M9LB with one colony of BL21 or BLT5615 and incubating at 37 °C for 16 h (overnight)
- 2 M stocks of isopropyl thiogalactoside (IPTG) (Sigma) in water
- LB top agar
- LB agar plates when using BL21 and T7 Select415-1 phage, or 50 µg/ml carbenicillin LB agar plates when using BLT5615 and T7Select1-1 or T7Select10-3 phage
- 50 ml conical vials
- 0.2 µm filter (e.g. Gelman Sciences Supor® Acrodisc® 25, product no. 4612) tipped 30 ml syringes

## A. Common *in vivo* selection steps

1   Anesthetize the mouse by intraperitoneally injecting 1.25% (w/v) avertin; the dosage for avertin is 15 µl/g body weight. Start by injecting 300 µl using an insulin syringe. To pick up the mouse, grasp the tip of its tail with your dominant hand, and lift its hind legs from the ground. Then use your other hand to grasp the mouse behind the head and neck. Lift and turn the mouse over to expose its underside. If properly held the mouse should not be able to twist or turn. Use your dominant hand to inject the initial 300 µl of the anesthetic, which should take effect within 1 min. If the mouse is still not anesthetized then inject additional anesthetic in increments of 100 µl until it shows signs of being under deep anesthesia. The anesthetic should last 30–60 min.

2   Place the mouse's tail in warm water to dilate the tail veins.

3   Use an insulin syringe to inject the 200–500 µl phage solution into one of the tail veins. The tail veins are on the sides of the tail and are very close to the surface of the skin. Insert the needle into the tail vein bevel side up and apply brief positive pressure to the syringe's plunger, thereby injecting ∼20 µl of the phage solution. If the vein briefly empties of red blood and refills, you are correctly positioned in the vein. Slowly inject the remaining phage by injecting 20 µl at a time and allowing the vein to refill after each 20 µl.

4   Allow the phage to circulate for a minimum of 5 min.

5   Place the mouse on its back and use your scissors to make a cut in the abdomen directly under the rib cage. You will be looking at the inferior side of the liver. Use the tip of your scissors to move the liver aside, and expose the diaphragm. Puncture the diaphragm and cut up one side of the rib cage; then cut transversely across the diaphragm and up the opposite side of the rib cage. Clasp the xyphoid process with

your hemostat and roll the rib cage back to expose the entire heart. Use your scissors to cut any connective tissue between the heart and the back of the sternum. Make a final ventral to dorsal cut on one side of the chest cavity to allow for drainage. When done correctly, the entire procedure is relatively bloodless. The skin will bleed slightly but there should be no blood accumulated in the chest cavity or any active bleeding from the heart or any major vessels.

6   Perfuse the mouse with 10 ml of DMEM or $1 \times$ PBS. First insert the syringe's needle 3–5 mm into the heart's left ventricle just to the side of the apex, then apply brief positive pressure to the plunger; if you have correctly positioned the needle you will see the left side of the heart pale and slightly bulge. Now use the scissors in your left hand to cut the right atrium of the heart. Begin to perfuse with the DMEM or PBS at a rate that keeps the heart inflated but still beating, approximately 2 ml/min.

7   Once the perfusion is complete, remove the organs and tissues of interest along with one or more control organs.

8   Put each organ or tissue into a separate, tared, sterile, labeled, 14-ml plastic tube. Weigh the tubes and subtract the tare to get the organs' and tissues' weights. Note these weights because you will be using them to normalize your data.

9   Add 2 ml of 1% BSA in DMEM to each organ or tissue.

10  Gently disrupt each organ or tissue with the hand-held homogenizer while on ice. Between different organs and tissues wash the shaft with 10% bleach, and thoroughly rinse the bleach away with water. Alternatively, use DAKO's MediMachine® to prepare single-cell suspensions from each organ or tissue. Place an organ or tissue into a separate 50 μm Medicon, and disrupt in the MediMachine for 2 min. Evacuate the Medicon's chamber with a 1-ml luer slip syringe and filter the single cell suspension through a 70 μm Filcon into a 14-ml conical vial. Add 1 ml of 1% BSA in DMEM to the Medicon's chamber, and repeat the disruption, evacuation, and filtration. Repeat for a total of three cycles, and repeat the overall process for each organ or tissue. Centrifuge the cell suspensions in a bench top clinical centrifuge at 1500 rpm for 10 min. Discard the supernatants and resuspend the cell pellets in 2 ml of 1% BSA in DMEM.

11  Pour the homogenates (cell suspensions) into the 2.2 ml plastic tubes.

12  Microcentrifuge the homogenates at 5000 rpm (2660g in an Eppendorf® 5415D centrifuge) for 5 min. You can use lower speeds (300–800g) if nondisrupted cells are desired or required. Pour off and discard the supernatant.

13  Resuspend the pellets in 1 ml of 1% BSA in DMEM by gently pipetting up and down. Microcentrifuge the homogenates at 5000 rpm for 5 min. Pour off and discard the supernatant. Repeat three more times, for a total of four washes. For the final fifth wash resuspend the pellets in 1 ml of 1% BSA in DMEM and transfer to a new, sterile, 2.2 ml plastic tube. Microcentrifuge the homogenates at 5000 rpm (2660g) for 5 min. Pour off and discard the supernatant. You can weigh these pellets if you want to monitor any losses during disruption and washing.

14  Add 100 μl of 1% NP-40 to each pellet and allow the cells to lyse on ice for 5 min.

## B. fUSE5 specific steps

1   For bacterial elution, add 900 μl of a K91Kan Terrific broth culture to the lysed cells and incubate at room temperature for 20 min.

2   Decant the entire contents of the tubes from step 1 into separate, 50 ml conical vials containing 9 ml of 0.2 μg/ml tetracycline in NZY media. Incubate at room temperature for 30 min.

3   For titering fd phage output, triplicate plate 1, 10, and 100 μl from step 2 onto 40 μg/ml tetracycline LB agar plates. You can spread the bacteria by using a glass or metal spreader, or sterile glass beads. Incubate the plates for 16 h (overnight) at 37 °C.

4   To determine the total phage output per organ or tissue, divide the number of colonies (TU) on each plate by the volume plated (1, 10, or 100 μl) and multiple by $10^4$ μl. This factor accounts for a 10-fold dilution in step 2 and the 1000 μl of K91Kan culture added to the pellets. For example, if you have a mean of 200 TU on the brain's 10 μl plates, then the total phage output for the brain is $2.0 \times 10^5$ TU per gram weight of brain. Average the counts from the triplicate plates for each organ and tissue, and graph your data according to Section 4.0.

5   For amplifying and purifying recovered fd phage, after you have plated your samples, add the remaining 9.67 ml from step 2 above to 200 ml of 40 μg/ml tetracycline NZY media in a 2L flask. You will need 200 ml of 40 μg/ml tetracycline NZY media per organ or tissue that you are selecting phage for. Incubate and shake at 37 °C for 16 h (overnight).

6   Decant the overnight cultures into 300 ml centrifuge tubes and centrifuge in a Sorvall® SLA-3000 rotor at 8000 rpm for 15 min. Filter the supernatants through vacuum driven bottle top 0.2 μm filters into sterile, 500 ml bottles.

7   Precipitate the phage from the supernatant by adding 40 ml of the PEG/NaCl solution to each 200 ml of phage and incubate at 4 °C for 4–16 h.

8   Centrifuge the precipitated phage in a Sorvall® SLA-3000 or equivalent rotor at 8000 rpm for 40 min. Pour off and discard the supernatant, then invert the centrifuge tubes onto some paper towels and thoroughly drain each white phage pellet.

9   Dissolve each phage pellet in 20 ml of 1× TBS, transfer to a 30 ml Oak Ridge centrifuge tube, and vortex. Re-precipitate the phage by adding 4 ml of the PEG/NaCl solution and placing on ice for 1 h.

10  Centrifuge the precipitated phage in a Sorvall® SS-34 or equivalent rotor at 10,000 rpm (11,984g) for 10 min. Pour off and discard the supernatant, then invert the centrifuge tubes onto some paper towels and thoroughly drain each white, phage pellet.

11  Dissolve each phage pellet in 500 μl to 1 ml of 0.1% gelatin in 1× TBS. Microcentrifuge at 10,000 rpm (10,640g) for 5 min to remove an insoluble material. Transfer the supernatant to a new, 1.7 ml Eppendorf tube, label and store at 4 °C. A typical titer range is $10^8$–$10^9$ TU/μl.

**Protocol 1** continued

## C. T7 specific steps

1  For bacterial elution, dilute the overnight BL21 or BLT5615 cultures to an OD at 600 nm of 1 using M9LB. If you are using the BLT5615 strain along with T7Select1-1 or T7Select10-3 vectors you must add IPTG to the culture to a final concentration of 10 μM.

2  Add 900 μl of this culture to the lysed cells and incubate at room temperature for 10 min.

3  For titering T7 phage output, triplicate plate 1 μl of a 1 : 10 dilution, 1 and 10 μl from step 2 (above) tubes. Add 1 μl of a 1 : 10 dilution, 1 or 10 μl into a 14-ml polystyrene tube along with 300 μl of a BL21 or BLT5615 (with IPTG) culture and mix with 4 ml of melted (45–50 °C) top agar. Pour the top agar containing the phage and bacteria onto LB agar (BL21) or 50 μg/ml carbenicillin LB agar (BLT5615) plates. Incubate the plates 1.5–3 h at 37 °C or overnight at room temperature.

4  To determine the total phage output per organ or tissue, divide the number of plaques (pfu) on each plate by the volume plated (0.1, 1, or 10 μl) and multiple by $10^3$ μl. This factor accounts for the 1000 μl of bacteria culture added to the cell pellets. For example, if you have a mean of 200 pfu on the brain's 1 μl plates, then the total phage output for the brain is $2.0 \times 10^5$ pfu/g weight of brain. Average the counts from the triplicate plates for each organ and tissue, and graph your data.

5  For amplifying and purifying T7 phage, add the remaining phage from step 2 to 10 ml of an OD at 600 nm of 0.5 culture of BL21 or BLT5615 (plus 10 mM IPTG). Incubate and shake at 37 °C for 2 h until the culture is lysed and clarified.

6  Decant the cultures into 30 ml Oak Ridge centrifuge tubes and centrifuge in an SS-34 or equivalent rotor at 8000 rpm (7670g) for 10 min. Filter each supernatant through a 0.2 μm filter tipped syringe into a sterile, 50 ml conical vial and store at 4 °C. Typical titers for T7 culture supernatant is $10^7$ pfu/μl. We use T7 lysates for subsequent rounds without further purification by precipitation.

## Protocol 2

### *Ex vivo* selection

#### Equipment and reagents

- Protocol 1's equipment and reagents, and the fUSE5 or T7 equipment and reagents

- One sterile, 2.2 ml plastic tube per organ or tissue

- Sterile razor blades

- One sterile, 3 cm cell culture dish per organ

- One sterile 50 ml polystyrene or polypropylene tube per organ or tissue

- 0.5 mg/ml cell culture-grade collagenase or DAKO's MediMachine with a 50 μm Medicon. You will need one Medicon, one 1-ml luer slip syringe, one 70 μm Filcon and one 14-ml conical vial per organ or tissue

- One 70 μm nylon cell strainer (Falcon #2350) per organ or tissue

- One sterile, 1.7 ml plastic tube per organ or tissue

## A. Common *ex vivo* selection steps

1  Anesthetize the mouse by intraperitoneally injecting 1.25% (w/v) avertin.[a]

2  Open the chest and expose the heart.[a]

3  Perfuse the mouse with 10 ml of DMEM or PBS.[a]

4  Remove organs and tissues of interest along with one or more control organs, and place each on a 3 cm cell culture dish.

5  Mince each organ or tissue with a sterile razor blade.

6  Put the minced organs or tissues, place into sterile, 2.2 ml plastic tubes and add 1.8 ml of 0.5 mg/ml collagenase. Vortex and incubate at 37 °C for 30 min. Continue to shake the samples while incubating at 37 °C.

7  Microcentrifuge the cell suspensions at 1500 rpm for 5 min and discard the supernatants.

8  Add each collagenase-treated organ to the top of a 70 μm nylon cell strainer and let drain into a 50 ml conical vial. Twice add 10 ml of 1% BSA in DMEM to the top of the cell strainer and let drain into the 50-ml conical vial. This brings the total volume of the cell suspension to 20 ml. Centrifuge in a bench top clinical centrifuge at 1500 rpm for 5 min to pellet the cells. Discard the supernatant and wash each pellet twice with 20 ml of 1% BSA in DMEM.

9  Alternative to steps 6–8 is to use DAKO's MediMachine to prepare single cell suspensions from each organ or tissue. Place each organ or tissue into a separate 50 μm Medicon, and disrupt in the MediMachine for 2 min. Evacuate the Medicon's chamber with a 1 ml luer slip syringe and filter the single cell suspension through a 70 μm Filcon into a 14-ml conical vial. Add 1 ml of 1% BSA in DMEM to the Medicon's chamber, and repeat the disruption, evacuation, and filtration. Repeat for a total of three cycles. Centrifuge the cell suspensions in a bench top clinical centrifuge at 1500 rpm for 10 min. Discard the supernatants, and retain the pellets of cells.

10  Resuspend the cells from either steps 8 or 9 in 1 ml of 1% BSA in DMEM and count them using an inverted, light microscope and a Neubauer-ruled hemocytometer (17).

11  Remove an aliquot of each organ's or tissue's cell suspension so that you have $\sim 1 \times 10^7$ cells total and place in a 1.7-ml plastic tube.

12  Chill the cells on ice for 5 min.

13  Add $10^9$–$10^{10}$ TU or pfu of phage to cells and bring total volume to 0.5–1 ml by using 1% BSA in DMEM. You can use larger volumes of phage if necessary or desired.

14  Incubate the phage with the cells for 30 min to overnight at 4 °C. The low temperature is to prevent endocytosis of any bound phage. We use shorter times when testing individual clones, and longer times when screening with diverse, primary libraries.

**15**  Microcentrifuge the cell suspensions at 5000 rpm for 2–3 min. Pour off and discard the supernatants.

**16**  Resuspend the cells in 1 ml of 1% BSA in DMEM by gently pipetting up and down. Microcentrifuge the cells at 5000 rpm for 2–3 min. Pour off and discard the supernatant. Repeat three more times, for a total of four washes. For the final, fifth wash resuspend the cell pellets in 1 ml of 1% BSA in DMEM and transfer to a new, sterile, 2.2-ml plastic tube. Microcentrifuge the cells at 5000 rpm for 2–3 min. Pour off and discard the supernatant.

## B. fUSE5 specific steps

**1**  For bacterial elution, resuspend the cell pellets in 1 ml of a Terrific broth culture of K91Kan and incubate at room temperature for 20 min. Optionally, you can lyse the cells by adding 100 μl of 1% NP-40 to each pellet and allowing the cells to lyse on ice for 5 min before adding 900 μl of a K91Kan Terrific broth culture to the lysed cells and incubating at room temperature for 20 min. The lysis step is optional because the phage binding is performed at 4 °C.

**2**  For titering fd phage output, triplicate plate 1 μl, 10 μl, and 100 μl from step 1 onto 40 μg/ml tetracycline LB agar plates. You can spread the bacteria by using a glass or metal spreader, or sterile glass beads. Incubate the plates for 16 h (overnight) at 37 °C.

**3**  To determine the total phage output per organ or tissue, divide the number of colonies (TU) on each plate by the volume plated (1 μl, 10 μl, or 100 μl) and multiply by $10^3$ μl. Average the counts from the triplicate plates for each organ and tissue, and graph according to Section 4.0.

**4**  Amplify and purify recovered fd phage (see *Protocol 1*).

## C. T7 specific steps

**1**  For bacterial elution, dilute the overnight BL21 or BLT5615 cultures to an OD at 600 nm of 1 using M9LB. If you are using the T7Select1-1 or T7Select10-3 vectors you must add IPTG to the bacteria culture to a final concentration of 10 μM. Resuspend the cell pellets in 1 ml of this culture and incubate at room temperature for 10 min. Optionally, you can lyse the cells by adding 100 μl of 1% NP-40 to each pellet and allowing the cells to lyse on ice for 5 min before adding 900 μl of BL21 or BLT5615 culture to the lysed cells and incubating at room temperature for 10 min. The lysis step is optional because the phage binding is performed at 4 °C.

**2**  For titering the phage output, triplicate plate 1 μl of a 1 : 10 dilution, 1 μl and 10 μl from step 1. Add 1 μl of a 1 : 10 dilution, 1 or 10 μl into a 14-ml Falcon tube along with 300 μl of an OD 600 nm of 1 culture of BL21 or BLT5615 (with IPTG) and mix with 4 ml of melted (45–50 °C) top agar. Pour the top agar containing the phage and bacteria onto LB agar (BL21) or 50 μg/ml carbenicillin LB agar (BLT5615) plates. Incubate the plates 1.5–3 h at 37 °C or overnight at room temperature.

**3** To determine the total phage output per organ or tissue, divide the number of plaques (pfu) on each plate by the volume plated (0.1, 1 or 10 μl) and multiple by $10^3$ μl. This factor accounts for the 1000 μl of bacteria culture added to the cell pellets. Average the counts from the triplicate plates for each organ and tissue, and graph according to Section 4.0.

**4** Amplify and purify recovered T7 phage (see *Protocol 1*).

[a] see *Protocol 1*, Common *in vivo* selection steps for more detailed instructions.

# 6 Identification of individual phage

The selection process is successful when an enrichment is seen. The first criterion a phage pool must meet is to show an enrichment relative to insertless phage, and this is why we suggested in Section 4 to graphically represent data using "Fold relative to nonrecombinant phage" as the dependent variable. This phage–phage comparison on the same target informs us that the recombinant peptides have bestowed a cell-binding activity greater than the nonspecific binding of the phage itself. This control is not exclusive from the target–target comparison but precedes it in importance because it is a more inclusive benchmark. A phage pool must first show an enrichment relative to nonrecombinant phage before it can show an enrichment relative to other organs and tissues.

We consider two-fold differences to be the normal variation in our data and have routine *in vivo* enrichments in the range of 5–200-fold, and *ex vivo* enrichments in the range of $10–10^4$-fold.

Once we have a phage pool that minimally shows an enrichment relative to nonrecombinant phage, we sequence individual phage and use this sequence data to choose candidates for further testing. We then determine the *ex vivo* binding (*Protocol 2*) and *in vivo* homing (*Protocol 1*) of these chosen clones relative to nonrecombinant phage. Positive clones that bind *ex vivo* or home *in vivo* by phage counts are further analyzed by immunohistochemistry (*Protocol 5*) and immunofluorescence (*Protocol 6*). And finally, we analyze the specificity by inhibition and bio-distribution studies (*Protocol 7*) using cognate, fluorescently labeled, synthetic peptides.

## 6.1 PCR and sequencing of the insert coding region

We use PCR to amplify the recombinant phage's insert coding region directly from colonies or plaques. This protocol is faster and of higher throughput than growing individual phage in liquid culture, purifying the phage and then purifying the phage DNA. A single PCR reaction using 5% of a single colony or plaque makes enough insert DNA for 15 sequencing reactions.

We sequence 48–96 clones in a 96-well format from each round that shows a significant enrichment, generally the second *ex vivo* round, the second or third *in vivo* rounds that are post-*ex vivo*, or the fifth and sixth *in vivo* rounds of *in vivo*-only screens.

## Protocol 3

## PCR amplification of the insert region

### Equipment and reagents

- A thermocycler
- Platinum® PCR Supermix (Invitrogen)
- MilliQ water
- 96-well reaction tube plates (Perkin-Elmer)

### For fUSE5 PCR

- FUSE5 "Forward" primer, a 43-mer with the sequence 5′-TAA TAC GAC TCA CTA TAG GGC AAG CTG ATA AAC CGA TAC AAT T-3′(10 pmol per reaction)

- fUSE5 "Super Reverse" primer, a 20-mer with the sequence 5′-CCG TAA CAC TGA GTT TCG TC-3′(10 pmol per reaction)

### For T7 PCR

- T7 "Super Up" primer, a 20-mer with the sequence AGC GGA CCA GAT TAT CGC TA (10 pmol per reaction)
- T7 "Down" primer, a 20-mer with the sequence AAC CCC TCA AGA CCC GTT TA (10 pmol per reaction)

### Method

1  Prepare a primer pair solution containing 5 pmol/μl (5 μM) of each primer in water.

2  Add 200 μl of the 5 pmol/μl primer pair solution to 2.25 ml (two tubes) of Platinum® PCR Supermix to make the PCR reaction mix. Aliquot 24.5 μl of the PCR reaction mix into each well of a 96-well reaction tube plate.

3  Use a sterile pipette tip to pick a colony (FUSE5) or a plaque (T7) and suspend it in 10 μl of 1× TBS. You will need to retain these 10 μl suspensions in their own 96-well reaction plates. These plates will be the source for each individual phage when it becomes necessary to amplify and test individual clones. We recommend that you have a naming scheme that allows you to backtrack from the sequence data to the PCR reaction plate and finally to the corresponding phage plates.

4  Add 0.5 μl of phage or bacteria suspension to each 24.5 μl PCR reaction mix.

5  Run a PCR reaction using the following conditions: hold at 72 °C for 10 min; hold at 95 °C for 5 min; run three-temperature PCR, 35 cycles of 94 °C for 50 s, 50 °C for 1 min and 72 °C for 1 min; hold at 72 °C for 6 min; hold at 4 °C until ready to sequence.

6  Check PCR reactions by running products on a 2–4% agarose gel.

## Protocol 4

## Sequencing the PCR products

### Equipment and reagents

- ABI 377 sequencer
- ABI's Big Dye
- ABI 9700 thermocycler

- 0.4 pmol/μl of fUSE5 "Super Reverse" primer or the T7 "Super Up" primer. (see *Protocol 3*)

**Protocol 4** continued

- ABI Blue Dextran/EDTA loading buffer
- 3 M sodium acetate
- 95 and 70% ethanol
- A Qiagen 4K15C centrifuge with a Qiagen Plate Rotor 2 × 96 or equivalent

## Method

1 Mix the sequencing reaction components in a 10 μl reaction volume: 1.5 μl of the PCR products from *Protocol 3*, 2.5 μl MilliQ water, 2.0 μl of 0.4 pmol/μl of FUSE5 "Super Reverse" primer or the T7 "Super Up" (total of 0.8 pmol), 4.0 μl Big Dye.

2 Perform thermal cycling with the following conditions: 94 °C for 2 min; three-temperature PCR, 30 cycles of 96 °C for 45 s, 50 °C for 20 s and 60 °C for 3 min; hold at 4 °C until ready to purify.

3 Add 1 μl of 3 M sodium acetate and 25 μl of 95% ethanol to the 10-μl sequencing reactions.

4 Precipitate the products by incubation at − 20 °C for 60 min.

5 Centrifuge at 6000 rpm (5788g) in a Qiagen 4K15C centrifuge with a Qiagen Plate Rotor 2 × 96 or equivalent for 20 min.

6 Rapidly invert the 96-well plate and gently shake out the ethanol.

7 Wash with 125 μl of 70% ethanol.

8 Centrifuge at 6000 rpm in a Qiagen 4K15C centrifuge with a Qiagen Plate Rotor 2 × 96 or equivalent for 5 min.

9 Rapidly invert the 96-well plate and gently shake out the ethanol.

10 Dry on a 96-well heat block at 90 °C for 15 min.

11 Resuspend the samples in 1.5 μl of ABI Blue Dextran/EDTA loading buffer.

12 Load 0.8 μl of each samples into a precast, polyacrylamide sequencing gel and run for 3.5 h in an ABI 377 Sequencer.

13 Analyze the data using sequence analysis software such as Sequencher™ (Gene Codes Corp.).

The sequence generated by sequence analysis software must be translated in order to obtain the encoded peptide sequence. An example of raw fUSE5 DNA sequence data is shown here with the flanking regions in bold: 5′-TCTGGCTCTTTCAGATTATAGTAACTCCATTTTCTGTATGATGTTTTGCTAAACA-ACTTTCAACAGTTTCGG**CCCCAGCGGCCCC**ACGAGGAGCCTTAATCCCACAACA-CGCCAA**AGCCCCG**TCGGCCGAGTGAGAATAGAAAGGAACAACTAAAGGAATTGC-GAATAATAATTTTTTCACGTTGAAAATCTCCAAAAAAAA-3′. From this sequence, a reverse, complementary sequence of the underlined insert sequence yields the following: TTGGCGTGTTGTGGGATTAAGGCTCCTCGT. Translating the reversed/complemented insert yields the X$_{10}$ peptide sequence: LACCGIKAPR. Note that the first flanking region of CGGCCCCAGCGGCCCC has the same sequence appearing twice. A common mistake is to read the insert after the first CGGCCCC and thereby tag AGCGGCCCC onto the insert sequence. The insert will be three amino acids longer than the peptide encoded by the library and have the additional sequence of Gly-Ala-Ala.

An example of raw T7 DNA sequence data is shown here with the flanking regions in bold: ATGCGGCTAGGATNGGCCNTGCAAGNCTTGGGAAAGGTCCGAG-ACCAGCAGCTGCAGGAGCTGTCGTATTCCAGTCAGGTGTGATGCTCGGG**GATCC-GAATTCA**TGTAAGCGTATGGGTCGGGCGACGTGT**AGCTTGCGG**CCGCACTCGA-GTAACTAGTTAACCCCTTGGGGCCTCTAA ACGGGTCTTGAGGG. In this case, the sequencing product is the same as the coding strand in the T7 sequences, so the insert sequence is TGTAAGCGTATGGGTCGGGCGACGTGT. Translating the inserted sequence yields the $CX_7C$ peptide sequence: CKRMGRATC. One can verify that the reading frame is correct by using the 5′ flanking region, which should translate to DPNS.

We align and cluster all of the sequences from a screen using the multiple alignment programs within GCG's software suite. We look for sequences that appear multiple times in a given round but also look for their appearance in multiple rounds. We often observe enrichments on the sequence level by comparing the *in vivo* sequences between two *in vivo* rounds and with the *ex vivo* preselection sequences. The enrichments can be an increase in the number of times a sequence appears or it can be the victory of one sequence over several closely variant sequences that differ in one, two, or three amino acid residues. The later type can yield important clues about the plasticity of that peptide's primary sequence.

Sequence enrichment is not an absolute requirement; because of the target complexity in our screens, it is entirely possible to have a successful pool that is composed of many peptides against many targets. Some of these successful peptides may appear multiple times in the sequence data and others may only appear once or twice in the sample.

## 6.2 Establishing that individual clones specifically bind organ or tissue vasculature

The next step after selection, and sequencing is to show binding of individual clones in three ways: first, an increased number of phage recovered from the target organ relative to insertless phage and relative to other organs and tissues; second, positive immunohistochemical and immunofluorescent staining of the vasculature in an organ or tissue; third, we synthesize, inject and analyze the biodistributions of cognate fluorescently labeled peptides to establish the peptide's activity outside of the phage context.

### 6.2.1 *In vivo* and *ex vivo* binding

One can quickly test several individual clones for *ex vivo* binding according to *Protocol 2*, followed by testing individual clones for *in vivo* binding according to *Protocol 1*. Positive clones have an increased number of phage recovered from the target organ relative to insertless phage and to other organs, tissues, and cells; perhaps a lower background in control organs and tissues than insertless phage; and often perform better than the source phage pool.

A comparison between *ex vivo* binding and *in vivo* homing for an individual clone can also provide some information about its receptor distribution. For example, the lung-homing GFE-1 peptide was discovered in an *in vivo* screen and found to only home to lung vasculature *in vivo* when compared with other organs and tissues. However, when tested *ex vivo*, it is bound to both lung and kidney primary cell preparations. The reason became clear after its receptor, membrane dipeptidase, was discovered. Membrane dipeptidase is found only on lung vasculature but is present on both lung and kidney parenchymal cells (14); a distribution that correlates exactly with the *in vivo* and *ex vivo* data.

### 6.2.2 Immunohistochemistry and Immunofluorescence

Immunohistochemistry and immunofluoresecence are less sensitive than the infectivity assays we have performed in the previous protocols, so in our protocols it is possible to have staining of positive vasculature while the background in negative control organs do not stain.

We make the same control comparisons in the immunohistochemistry and immunofluorescence that we make in the selections: phage–phage and target–target. You will generate a set of data looking at the presence of insertless phage in multiple organs and tissues. The liver is both a positive and negative control for phage immunohistochemistry because the Kupffer cells (macrophages) stain strongly positive; however, the vasculature is negative. The reason for the Kupffer cell staining is that phage is eliminated from circulation by these cells.

Negative controls for the histological procedures for both the fd and T7 polyclonal sera should include purified rabbit serum or IgG from an animal that was similarly immunized with an unrelated antigen. Use these at the same concentration as the primary antibody.

## Protocol 5

## Immunohistochemistry on paraffin embedded tissues

### Equipment and reagents

- A 10-ml plastic syringe filled with 10 ml of cold freshly made 4% paraformaldehyde (PFA), and with a 1-in. 25 gauge needle attached (one per mouse)
- Two insulin syringes per mouse
- One 6 cm tissue culture dish per organ or tissue
- 1.25% (w/v) avertin in water
- Phage solution of $10^9$–$10^{10}$ TU or pfu

- 100% absolute ethanol
- Paraffin and an automated tissue processor (e.g. Tissue-Tek VIP; Miles Scientific)
- 3% (v/v) hydrogen peroxide ($H_2O_2$) in $1 \times$ PBS
- $1 \times$ PBS
- MilliQ water
- Trypsin (Zymed)

**Protocol 1** continued

- Primary fd antibody: polyclonal rabbit anti-fd IgG (Sigma)
- Primary T7 antibody: polyclonal rabbit anti-T7 IgG (Ruoslahti Lab)
- Secondary fd and T7 antibody: goat anti-rabbit IgG-HRP (DAKO Corporation)
- DAKO antibody diluent (Tris-HCl buffer with carrier protein and 0.015 M $NaN_3$)

- DAB (3,3'-diaminobenzidine tetrahydro-chloride (DAKO)) substrate solution. Prepare by first dissolving 6 mg of DAB in 10 ml of 50 mM Tris HCl pH 7.5. Then mix in 0.1 ml of 3% $H_2O_2$, and filter if any precipitate forms. You must use the solution within 2 h after preparing it

## A. Phage injection and tissue preparation

1 Anesthetize the mouse by intraperitoneally injecting 300–500 µl of 1.25% (w/v) avertin.[a]

2 Place the mouse's tail in warm water to dilate the tail veins.

3 Using the other insulin syringe inject the 200 µl phage solution.[a]

4 Allow the phage to circulate for a minimum of 5 min.

5 Open the chest and expose the heart.[a]

6 Perfuse animal with 10–20 ml cold freshly made 4% PFA.

7 Remove and cut bulky organs or tumors into smaller pieces and immerse in 4% PFA. Do not fix in PFA for more than 24 h.

8 Process tissues in an automated tissue processor (Tissue-Tek VIP; Miles Scientific).

9 Embed in paraffin and cut in 5 µm sections.

## B. Antigen retrieval and blockings

1 Block endogenous peroxidase activity by incubating slides in 3% (v/v) $H_2O_2$ in PBS for 15 min at room temperature. Wash twice with $1 \times$ PBS.

2 Perform enzymatic digestion with trypsin (Zymed Laboratories, Inc.) 10 min at 37 °C to reveal antigenic sites. Wash 3 times with $1 \times$ PBS.

3 Block endogenous avidin and biotin activity with a avidin/biotin block kit (Vector Laboratories).

4 No protein block is necessary if DAKO antibody diluent is used (see below).

## C. Antibodies and counterstaining

1 Dilute the primary antibody with DAKO antibody diluent (1 : 1000 for polyclonal), and add approximately 100 µl to the section (a sufficient volume to completely cover the section). Incubate for 1 h at room temperature. Wash the section twice with $1 \times$ PBS.

2 Add approximately 100 µl (or a sufficient volume to completely cover the section) of the prediluted polyclonal secondary antibody to the section. Incubate for 30 min at room temperature. Wash twice with $1 \times$ PBS.

**3**  Add 100 μl of DAB substrate solution to a section and incubate for 4–6 min at room temperature.

**4**  Counterstain with Harris Hematoxylin (Surgipath) for 3 min at room temperature, wash once with tap water, dehydrate, and coverslip with an acrylic resin mounting media (Cytoseal 60; EMS).

[a] See *Protocol 1,* Common *in vivo* selection steps for more detailed instructions.

# Protocol 6

# Immunofluorescence on fresh frozen tissues

## Equipment and reagents

- A 10 ml plastic syringe filled with 10 ml of cold freshly made 4% paraformaldehyde (PFA), and with a 1-inch 25 gauge needle attached (one per mouse)
- 2 insulin syringes per mouse
- One 6 cm tissue culture dish per organ or tissue
- 1.25% (w/v) Avertin in water
- Phage solution of $10^9$–$10^{10}$ TU or pfu
- Tissue-Tek O.C.T.® (Optimal Cutting Temperature) embedding media (Sakura Finetek)
- Cryomold® Standard (Miles)
- Dry ice

- 1x PBS
- MilliQ water
- Primary fd antibody: polyclonal rabbit anti-fd IgG (Sigma)
- Primary T7 antibody: polyclonal rabbit anti-T7 IgG (Ruoslahti Lab)
- Secondary fd and T7 antibody: goat anti-rabbit IgG-Alexa conjugates (Molecular Probes)
- DAKO antibody diluent (Tris-HCl buffer with carrier protein and 0.015M $NaN_3$)
- 28% sucrose in 1x PBS
- VECTASHIELD® Mounting Media with DAPI (Vector Laboratories)

## A. Phage injection and tissue preparation

**1**  Anesthetize the mouse by intraperitoneally injecting 300–500 μl of 1.25% (w/v) Avertin.[a]

**2**  Place the mouse's tail in warm water to dilate the tail veins.

**3**  Using the other insulin syringe, inject the 200 μl phage solution.[a]

**4**  Allow the phage to circulate for a minimum of 5 min.[a]

**5**  Open the chest and expose the heart.[a]

**6**  Perfuse animal with 10–20 ml cold freshly made 4% paraformaldehyde (PFA).[a]

**7**  Remove and cut bulky organs or tumors into smaller pieces and immerse in 4% PFA. Do not fix in paraformaldehyde for more than 2 h at 4 °C (NOTE: this is an important time difference from the paraffin fixation protocol).

**8** Incubate fixed tissues overnight at 4 °C in 28% sucrose in 1x PBS for cryoprotection.

**9** Embed in Cryomold® Standard molds with Tissue-Tek O.C.T. media by filling a mold with Tissue-Tek O.C.T. media and then placing 1–2 pieces of tissue into the center of the mold. Push the tissue pieces to the bottom of the O.C.T. media and then completely fill the mold with O.C.T. media. Freeze the sections by placing on dry ice. The blocks can be stored at − 80 °C until cut in 10 μm sections.

### B. Antigen blockings

**1** No protein block is necessary if DAKO Antibody diluent is used.

### C. Antibodies and counterstaining

**1** Wash the slides for 10 min in 1x PBS at room temperature to remove the Tissue-Tek O.C.T. embedding media.

**2** Dilute the primary antibody with DAKO antibody diluent (1:1000 for polyclonal), and add approximately 100 μl to the section (a sufficient volume to completely cover the section). Incubate for 1 h at room temperature. Wash the section three times with 1x PBS.

**3** Add approximately 100 μl (or a sufficient volume to completely cover the section) of the polyclonal secondary antibody (1:200 dilution in DAKO antibody diluent) to the section. Incubate for 30 min at room temperature. Wash three times with 1x PBS.

**4** Mount the slides with VECTASHIELD® Mounting Media with DAPI for nuclear counterstaining.

[a] see *Protocol 1*, Common *in vivo* selection steps for more detailed instructions.

### 6.2.3 Binding specificity

We establish specificity by showing that the synthetic, cognate peptide can diminish or abolish targeting as detected by phage counts (*Protocols 1 and 2*) and immunohistochemistry (*Protocol 5*) and immunofluorescence (*Protocol 6*). We generally transfer a binding or homing peptide into the monovalent T7Select 1-1 vector or use the isolated T7Select 415-1 clone, and start an inhibition curve at a 1 mg/ml solution of peptide (~1 mM), then step down in half-log increments +0.01 μM. If the monovalently displayed peptide is used, then the IC50 determined from the inhibition curve is the apparent $K_d$ for the peptide.

### 6.2.4 Bio-distribution of fluorescently labeled peptides

The final step in our process is to synthesize each binding or homing peptide as a fluorescently labeled (typically fluorescein or rhodamine) peptide and then determining its bio-distribution after intravenous injection. The fluorescent label also allows the peptide to be used in subsequent receptor isolations without any modifications: fluorescent peptides can be used to sort receptor-positive cells in cell-based expression cloning systems and fluorescein-peptides can also be directionally immobilized to solid supports using a high-affinity mouse anti-fluorescein monoclonal antibody.

We synthesize selected peptides as N-terminally labeled fluorescein isothiocyanate (FITC) or Rhodamine conjugates by adding fluorescein isothiocyanate isomer I (Sigma) or Rhodamine B (Sigma) in the presence of the coupling reagent HATU (Sigma) and DMF (Sigma) as the last step of a peptide's solid-phase synthesis.

We typically inject mice with 100 µl of a 1 mM (100 nmol injected) peptide solution in 1x PBS; however, the optimal concentration has to be empirically determined. For example, a 1 mM solution of a 1000 Da peptide is 1 mg/ml, and injecting 100 µl of this solution is a total of 100 µg or 100 nmol of the peptide. A mouse has approximately 2 ml of circulating blood, so this gives a peptide concentration of 100 nmol/2ml or 50 µM, which is in range of expected affinity constants for these peptides.

The best peptide control for these experiments is a fluorescently labeled point mutant or scrambled version of a peptide. Alternatively, an appropriate control is a FITC- or Rhodamine-labeled NSSSVDK peptide: the sequence that is displayed by the nonrecombinant T7Select 415-1 phage.

The fluorescent labels can make the peptides more hydrophobic and difficult to directly solubilize in aqueous buffer. If this is the case, bring the peptide into solution in a small volume of 100% DMSO (e.g. 100 mg/ml) and then slowly add 1x PBS or MilliQ water until peptide is 1 mg/ml; the solubility can also be improved by using a lower peptide concentration, mild heating, sonication and agitation. Prepare the control peptide under the same conditions. To visualize blood vessels without the need for additional antibody staining, we inject tomato lectin, either FITC-labeled (for Rhodamine peptides) or biotinylated with Streptavidin-Alexa Fluor 594 as the secondary reagent (for FITC-labeled peptides).

## Protocol 7

## Injection of fluorescently labeled peptides and visualization of the blood vessels by lectin staining *in vivo*

### Equipment and reagents

- A 10 ml plastic syringe filled with 10 ml of cold freshly made 4% paraformaldehyde (PFA), and with a 1-inch 25 gauge needle attached (one per mouse)
- 3 insulin syringes per mouse
- One 6 cm tissue culture dish per organ or tissue

- 1.25% (w/v) Avertin in water
- FITC- or Rhodamine-peptide (1 mM in 1x PBS)
- Tissue-Tek O.C.T.® (Optimal Cutting Temperature) embedding media (Sakura Finetek)

- Cryomold® Standard (Miles)
- Dry ice
- 1x PBS
- MilliQ water
- *Lycopersicum esculentum* [tomato] lectin (LEA), biotinylated or FITC labeled (Vector Laboratories), reconstituted at 0.5 mg/ml in 1x PBS

- Streptavidin-Alexa Fluor 594 (Molecular Probes) if injecting FITC-peptide and biotinylated LEA.
- DAKO antibody diluent (Tris-HCl buffer with carrier protein and 0.015M $NaN_3$)
- 28% sucrose in 1x PBS
- VECTASHIELD® Mounting Media with DAPI (Vector Laboratories)

## A. Peptide and lectin injections and tissue preparation

1  Anesthetize the mouse by intraperitoneally injecting 300–500 µl of 1.25% (w/v) Avertin.[a]

2  Place the mouse's tail in warm water to dilate the tail veins.

3  Using the second insulin syringe, inject 100 µl peptide solution.[a]

4  Allow the peptide to circulate for a minimum of 5 min.

5  Using the third insulin syringe slowly inject 200 µl of 0.5 mg/ml LEA in 1x PBS.[a]

6  Let the LEA circulate for a maximum of 5 min.

7  Open the chest and expose the heart.[a]

8  Perfuse animal with 10–20 ml cold freshly made 4% paraformaldehyde (PFA).[a]

9  Remove and cut bulky organs or tumors into smaller pieces and immerse in 4% PFA. Do not fix in paraformaldehyde for more than 2 h at 4 °C.

10  Incubate fixed tissues overnight at 4 °C in 28% sucrose in 1x PBS for cryoprotection.

11  Embed in Cryomold® Standard molds with Tissue-Tek O.C.T. media by filling a mold with Tissue-Tek O.C.T. media and then placing 1–2 pieces of tissue into the center of the mold. Push the tissue pieces to the bottom of the O.C.T. media and then completely fill the mold with O.C.T. media. Freeze the sections by placing on dry ice. The blocks can be stored at − 80 °C until cut in 10 µm sections.

## B. Peptide and lectin detection for mice injected with rhodamine-peptide and FITC-LEA

1  Wash the slides for 10 minutes in 1x PBS at room temperature to remove the Tissue-Tek O.C.T. embedding media.

2  Mount the slides with VECTASHIELD® Mounting Media with DAPI for nuclear counterstaining.

## C. Peptide and lectin detection for mice injected with FITC-peptide and Biotin-LEA

1 Wash the slides for 10 minutes in 1x PBS at room temperature to remove the Tissue-Tek O.C.T. embedding media.

2 Dilute the streptavidin-Alexa Fluor 594 1:500 in DAKO antibody diluent, and add approximately 100 µl to the section (a sufficient volume to completely cover the section). Incubate for 30 min at room temperature. Wash the section three times with 1x PBS.

3 Mount the slides with VECTASHIELD® Mounting Media with DAPI for nuclear counterstaining.

[a] see *Protocol 1*, Common *in vivo* selection steps for more detailed instructions.

## 7 Concluding remarks

Our ongoing, methodical goals are improvements in the speed, consistency, and success of *in vivo* and *ex vivo* selections. We would also like to understand how our methods relate to a theoretical framework of phage display (18).

## Acknowledgments

We thank all past and present members of the Ruoslahti laboratory's phage group. We also thank the Burnham Institute DNA Chemistry, Histopathology, and Animal core facilities. JAH is a recipient of a National Cancer Institute Training Grant in the Molecular Pathology of Cancer. KP is a recipient of an American Cancer Society Fellowship. PL is supported by grant 42782 from the Academy of Finland, and a grant from the Finnish Cultural Foundation. This work was supported by National Cancer Institute grants CA74238 to ER, and Cancer Center Support grant, CA 30199.

## References

1. Smith, G. P. and Petrenko, V. A. (1997). *Chem. Rev.*, **97**, 391.
2. Zwick, M. B., Shen, J., and Scott, J. K. (1998). *Curr. Opin. Biotechnol.*, **9**, 427.
3. Forrer, P., Jung, S., and Pluckthun, A. (1999). *Curr. Opin. Struct. Biol.*, **9**, 514.
4. Choo, Y. and Klug, A. (1995). *Curr. Opin. Biotechnol.*, **6**, 431.
5. Yamamoto, M., Kominato, Y., and Yamamoto, F. (1999). *Biochem. Biophys. Res. Commun.*, **255**, 194.
6. Vant-Hull, B., Payano-Baez, A., Davis, R. H., and Gold, L. (1998). *J. Mol. Biol.*, **278**, 579.
7. Pasqualini, R. and Ruoslahti, E. (1996). *Nature*, **380**, 364.
8. Rajotte, D., Arap, W., Hagedorn, M., Koivunen, E., Pasqualini, R., and Ruoslahti, E. (1998). *J. Clin. Invest.*, **102**, 430.
9. Arap, W., Pasqualini, R., and Ruoslahti, E. (1998). *Science*, **279**, 377.
10. Ladner, R. C. (1999). *Q. J. Nucl. Med.*, **43**, 119.

11. Koivunen, E., Arap, W., Valtanen, H., Rainisalo, A., Medina, O.P., Heikkila., P. *et al.* (1999). *Nat. Biotechnol.*, **17**, 768.

12. Ruoslahti, E. and Rajotte, D. (2000*). Annu. Rev. Immunol.*, **18**, 813.

13. Ruoslahti, E. (1994). *Princess Takamatsu Symp.*, **24**, 99.

14. Rajotte, D. and Ruoslahti, E. (1999). *J. Biol. Chem.*, **274**, 11593.

15. Smith, G. P. and Scott, J. K. (1993). Libraries of peptides and proteins displayed on filamentous phage, *Methods in enzymology*, Vol. 217 (ed. R. Wu), p. 228, Academic Press, San Diego.

16. Sambrook, I., Fritsch, E. F., and Maniatis, T. (1989). *Molecular cloning: A laboratory manual* 2nd edn., Cold Spring Harbor Laboratory Press, New York.

17. McAteer, J. A. and Davis, J. (1994). In *Basic cell culture: A practical approach*, Vol. 146 (ed. Davis, J. M.), p. 93, IRL Press at Oxford University Press, New York.

18. Levitan, B. (1998). *J. Mol. Biol.*, **277**, 893.

# Chapter 11

# Screening phage libraries with sera

Paolo Monaci and Riccardo Cortese

Istituto di Ricerche di Biologia Molecolare P. Angeletti, Via Pontina km 30,6, 00040 Pomezia (Rome) Italy.

## 1 Introduction

This chapter comprises a description of methods and procedures for screening phage-displayed peptide libraries using complex mixtures containing the target receptor. This is the case for many interesting biological systems such as serum, cerebrospinal and other biological fluids, cell or tissue surfaces, etc. In such cases the goal is to identify peptides that bind to a particular receptor molecule or to a set of receptors with specific properties.

We and others have concentrated on identifying peptides that bind to antibodies specifically associated with a disease (disease-specific antibodies, DS-Ab). Here we describe the methods used to identify ligands for antibodies specifically associated with hepatitis C infection (HCV), a major cause of chronic liver disease worldwide (1). Phage-displayed libraries are screened using sera from HCV-infected patients and non-infected individuals (referred to henceforth as positive and negative sera, respectively). The protocols cover all the experimental steps, from generation of the library to characterization of ligands for HCV-specific antibodies. The majority of the considerations and protocols here reported are equally applicable to other complex receptor systems.

## 2 Construction of phage libraries

### 2.1 Random peptide libraries

Two major phage-display systems exploiting filamentous phage have been developed and widely used over the last 15 years. Foreign peptides can be expressed as N-terminal fusions to either the minor (pIII) or the major coat proteins (pVIII). A feature that distinguishes the two systems is the number of copies displayed per phage particle: pIII-based systems range from less than one up to five copies per phage, whereas pVIII-based systems display from tens to several hundreds of peptides per particle on the phage capsid (see Chapter 1). This latter feature has proved a remarkable advantage when screening with sera, especially for random peptide libraries. In fact, multiple display of peptides on the phage surface augments the avidity of phage–antibody interaction and enables selection of ligands for most, if not all the antibodies present in the serum sample. In this chapter, we describe the use of phagemid vectors for display either on pIII (for monovalent display) or pVIII (for polyvalent display).

Many protocols have been published which describe the construction of random peptide libraries displayed on filamentous phage (for a review, see (2)). Libraries are generated by cloning an *in vitro* synthesized random sequence into a display vector. The ligation products are then transformed into bacteria, and the efficiency of this last step has previously limited the complexity of the library to include not more than $10^8$ different sequences (see (3); however, see Chapter 2). Thus random peptide libraries usually sample only a very small number of all the possible sequences. We have tried to improve this situation by increasing the effective complexity of the library. The random sequences are synthesized using a codon-based method to include at each position only 20 triplets which code for 20 different amino acids, hence eliminating codon degeneracy and excluding stop codons (4).

## 2.2 Natural epitope libraries

In those cases where an infectious agent is known and its genome available, an attractive way to identify ligands for DS-Ab is to generate a phage-displayed complementary DNA (cDNA) library and screen it using sera from patients. This approach shows clear advantages, since natural epitope sequences are preferentially included in the library. It must be remembered, however, that ligands fused to pIII or pVIII move through the *Escherichia coli* secretion apparatus and are transported to the periplasmic space where they fold, before being incorporated into phage particles. Impairment of correct processing and/or folding of fusion-peptides can dramatically influence the effective complexity of these types of libraries.

In natural epitope libraries, especially for conformational epitopes, the size of the foreign peptide may be more important than the number of copies displayed on the phage surface, which is why either pIII- or pVIII-based systems can be used. Here we describe the construction of a library in which sequences from the HCV genome are displayed on the phage surface.

# 3 Selection of phage-displayed peptides binding to serum antibodies

When a phage library is screened using a serum sample, a high concentration of antibodies is desirable during panning. However, very few copies of a given DS-Ab may be present in the serum. Furthermore, the sensitivity threshold imposed on each panning round critically influences the selection of ligands for DS-Ab. It must be also taken into account that when multiple panning rounds are performed, the competition between different ligand–antibody pairs can eliminate the ligands of interest.

## Protocol 1

# Construction of pIII- and pVIII-based natural epitope libraries (HCV genome)

### Equipment and reagents

- *Fse*Ia primer:  5′-GGCCGGCCAAC(N)9-3′ (30 pmol)
- *Not*Ia primer:  5′-GC GGCCGCAAC(N)9–3′ (30 pmol)
- *Fse*Ib primer:  5′-AGGCATCTAGGGCCGGCCAAC-3′ (100 pmol)
- *Not*Ib primer:  5′-GGAGGCTCGAGCGGCCGCAAC-3′ (100 pmol)
- Template DNA (180 ng)
- Universal buffer (10 mM Tris pH 7.5, 5 mM MgCl₂, 7.5 mM DTT)
- Klenow polymerase (New England Biolabs)
- Klenow buffer (New England Biolabs)
- Deoxyribonucleotide triphosphates (dNTPs) (25 mM dTTP, 25 mM dCTP, 25 mM dGTP, 25 mM dATP; Roche Biochemicals)
- Bovine serum albumin (BSA) (Sigma)
- QIA quick column (Qiagen)
- Wizard columns (Promega)
- *Fse*I restriction enzyme (New England Biolabs)
- *Not*I restriction enzyme (New England Biolabs)
- pEL8FN vector (5)
- pEL3FN vector (5)
- XL1-Blue competent *E. coli* cells (Stratagene)
- Gene Pulser apparatus (BioRad)

### Method

1  Mix 180 ng of the template DNA with 30 pmol of the tagged random *Fse* Ia primer in 0.5× universal buffer. Heat the mixture at 95 °C for 5 min and then place it on ice for an additional 5 min.

2  Add 10 µl of Klenow mixture (10 units Klenow polymerase, 33 µg BSA, 5 mM dNTPs in Klenow buffer) and incubate the mixture for 6 h at 37 °C.

3  Purify the products of the first strand elongation reaction through QIA quick column following the manufacturer's instructions (final elution volume 50 µl).

4  Transfer 10 µl into a new tube and add 30 pmol of tagged random *Not* Ia primer in a final volume of 20 µl. Heat the mixture at 95 °C for 5 min and cool on ice for additional 5 min.

5  Add 10 µl of Klenow mixture (10 units Klenow polymerase, 33 µg BSA, 5 mM dNTPs in Klenow buffer) and incubate the mixture for 6 h at 37 °C.

6  Purify the products of the second strand elongation reaction through a QIA quick column following the manufacturer's instructions (with a final elution volume of 50 µl).

7  Amplify 10 µl of the purified second strand products by 30 cycles of polymerase chain reaction (PCR) using 100 pmol of *Fse*Ib and *Not*Ib primers in a total volume of 100 µl. The cycle profile is: 5 min at 94 °C, 1 cycle; 30 s 94 °C, 30 s at 52 °C, 1 min at 72 °C, 30 cycles; one additional elongation step of 5 min at 72 °C is performed at the end.

8  Purify the PCR products through Wizard columns following the manufacturer's instructions.

9   Digest purified PCR products with *Fse* I and *Not* I restriction enzymes.

10  Purify the digested PCR products through Wizard columns following the manufacturer's instructions.

11  Digest 5 µg of pEL3FN and pEL8FN vectors with *Fse* I and *Not* I restriction enzymes, and clean the digested products by phenol/chloroform extraction and ethanol precipitation.

12  Set up 10 ligations, each containing 100 ng of digested vector and 5–20 ng of digested PCR products into 20 µl total volume (tailoring of the insert/vector ratio may be required).

13  Transform the products into electro-competent XL1-Blue cells.

14  Collect bacteria and amplify as previously described (6). A procedure for amplification of phagemid-based display vectors is also given in Chapter 4, *Protocol 1*.

Many different methods can be used for panning a phage library using serum. To sample the highest fraction of serum antibodies we exploited the large coating surface offered by magnetic micro-beads. The serum antibodies are oriented on the beads' surface through the use of the secondary antibodies, thus avoiding the partial denaturation that may occur in direct coating process. A variant to this protocol involves the pre-incubation of serum and library, which are then captured by micro-beads coated with an antihuman antibody.

The panning procedure is described in *Protocols 4* and *5*. *Protocol 2* describes a way to maximize F-pilus display on the bacterial surface and thus optimize the *E. coli* susceptibility to infection by filamentous phage. *Protocol 3* details the preparation of carrier phage particles to eliminate the interference of serum antibodies directed toward the wt phage particles.

## Protocol 2

## Preparation of bacterial cells for phage infection (TB cells)

### Equipment and reagents

- XL1-Blue (Stratagene) or DH5αF' *E. coli* (7)
- Terrific broth (TB; 12 g of bacto-tryptone, 24 g of bacto-yeast extract and 4 ml of glycerol in 900 ml of distilled water. Autoclave on liquid cycle; allow to cool to 60 °C and then add 100 ml of a sterile solution of 0.17 M $KH_2PO_4$, 0.72 M $K_2HPO_4$)

### Method

1   Inoculate bacterial cells from a fresh plate of minimal medium in 20–30 ml of TB containing 20 µg/ml tetracycline.

2    Grow with vigorous shaking at 37 °C until $OD_{600} = 0.15$–0.25 (as measured from a 1:10 dilution of the culture in TB; corresponding to a concentration of viable cells of about $5 \times 10^9$ cells/ml).

3    Incubate the cells for 15 min at 37 °C with gentle agitation to allow pili regeneration and use for infection within 60 min.

# Protocol 3

## Preparation of UV-killed carrier phage

### Equipment and reagents

- XL1-Blue or DH5αF′ TB cells (see *Protocol 2*)
- Phage f1R1 (8)
- TB (see *Protocol 2*)
- TBS, (tris buffered saline; 50 mM Tris-HCl pH 7.6, 150 mM NaCl, 0.05% Thimerosal)
- PEG/NaCl solution (20% PEG, 2.5 M NaCl)
- Petri dishes, 150 mm diameter (Falcon)
- UV Stratalinker 2400 (Stratagene)
- Thimerosal (Sigma)

### Method

1    Infect 2 ml of XL1-Blue or DH5αF′ TB cells with ≈$10^9$ plaque forming units (pfu) of f1R1 phage from an old stock or directly from a plaque. Incubate the cells for 15 min at 37 °C without agitation, and then 2–3 h with agitation.

2    Dilute the infection mixture into 800 ml of TB containing 20 μg/ml tetracycline in a 2-L flask. Incubate for 12–14 h at 37 °C with strong agitation (at least 150 rpm).

3    Centrifuge the bacteria/phage suspension in 1000 ml plastic bottles 60 min, 4700 rpm at 10 °C.

4    Precipitate the phage by adding PEG-8000 (4% final) and NaCl (0.5 M final). Dissolve completely and leave on ice for 4 h or overnight at 4 °C.

5    Centrifuge at 4700 rpm for 60 min. Carefully remove the supernatant. Dissolve the pellet in 1/10 of the original volume with TBS.

6    Precipitate the phage again by adding $\frac{1}{4}$ volume of PEG 8000/NaCl solution to the resulting supernatant, and incubate on ice for 2 h at 4 °C.

7    Centrifuge the phage for 30 min, 10,000 rpm, 4 °C and resuspend the phage pellet in 10 ml of TBS.

8    Centrifuge the phage again for 15 min at 8000 rpm to remove insoluble material. Titer the phage as pfu.

9    Dilute a purified phage preparation to about $10^{13}$ particles/ml in TBS.[a] Measure the number of particles/ml through reading the absorbance at 269 nm ($A_{269}$): particles/ml = $A_{269} \times$ dilution factor $\times 6 \times 10^{16}/6407$.

Protocol 3 continued

10  Distribute 20–25 µl drops of each phage dilution on petri dishes (150 mm diameter). An automatic dispenser (e.g. Eppendorf 4780) can be used, in order to save time and avoid pipette contamination.

11  Perform 2–3 cycles of irradiation with the Stratalinker at 200.000 µJ.

12  Collect the irradiated phage, add 0.05% Thimerosal and store in 1 ml aliquots at 4 °C.

13  Check the phage survival by counting pfu. Phage titers should be less than $10^3$ phage/ml.

[a] To efficiently kill the phage it is very important that the number of particles/ml is not higher than $10^{13}$.

# Protocol 4

## Bead coating with antihuman antibodies

### Equipment and reagents

- Magnetic beads (Dynabeads M-450, Tosyl Activated, Dynal)
- Borate buffer (50 mM sodium borate, pH 9.5)
- Antihuman Fc polyclonal antibody (Immunopure Goat antihuman IgG Fc-specific unconjugated, Pierce)
- Magnetic apparatus (Dynal)
- PBS (phosphate buffered saline: 2 mM $NaH_2PO_4 \cdot H_2O$, 16 mM $Na_2HPO_4 \cdot 2H_2O$, 150 mM NaCl)
- PBBS (PBS, 0.1% BSA)
- Thimerosal (Sigma)

### Method

1  Transfer 500 µl of thoroughly suspended magnetic beads into an Eppendorf tube. Insert the tube into the magnetic apparatus and remove the liquid.

2  Resuspend the beads in 1 ml of borate buffer, mix and remove the supernatant as above.

3  Resuspend the beads in borate buffer containing 75 µg of antihuman Fc-specific polyclonal antibody in a final volume of 1 ml and incubate for 24 h at room temperature on a rotating wheel.

4  Discard the supernatant using the magnetic apparatus and transfer the beads into a Falcon tube in 10 ml of PBBS. Wash the magnetic beads by rotating at 4 °C, once for 10 min, 3 times for 30 min and then overnight. After each washing step, insert the tubes containing the beads into the appropriate magnetic apparatus, removing the supernatant.

5  The coated beads can be stored at 4 °C (adding 0.05% Thimerosal). For long-storage (more than 2 weeks), wash the beads 3 times with 10 ml of PBBS (5 min each time) just before usage.

## Protocol 5

## Library selection

### Equipment and reagents

- PBBST (PBS, 0.1% BSA, 0.1% Tween-20)
- Magnetic apparatus (Dynal).
- f1R1 UV-killed phage (see *Protocol 3*).
- Elution buffer (0.1 N HCl adjusted to pH 2.2 with glycine, 1 mg/ml BSA)
- 2 M Tris-HCl pH 9.0

- L-agar (1% bacto-tryptone, 0.5% yeast extract, 1.5% bacto-agar, 0.17 M NaCl, pH 7.2)
- L-agar + Amp/X-gal/IPTG (L-agar, 50 μg/ml Ampicillin, 35 μg/ml X-gal (5-bromo-4 chloro-3-indolyl-β-D-galactoside), 35 μg/ml IPTG (isopropyl β-D-thiogalactoside))

### Method

1  Transfer coated magnetic beads into Eppendorf tubes adding 50 μl of serum sample in PBBST in a total volume of 1 ml. Incubate on a rotating wheel at 4 °C for 12–16 h.

2  Wash the beads 4 times, 30 min each time at 4 °C with PBBST; use 10 ml for each wash.

3  Resuspend the magnetic beads in PBBST, containing $1 \times 10^{13}$ particles of UV-killed f1R1 in a volume of 1 ml and place on a rotating wheel for 3–4 h at 4 °C.

4  Add $2 \times 10^{11}$ library phage particles to the above mixture, and leave for 12–16 h at 4 °C with gentle agitation.

5  Remove unbound phage by washing the beads 6–7 times with 10 ml of PBBST at 4 °C, 10 min each time, removing all liquid from beads after each wash using the magnetic apparatus.

6  Elute bound phage by incubating the beads with 0.5 ml of elution buffer for 30 min with gentle agitation at room temperature.

7  Transfer the eluate into an Eppendorf tube and neutralize by adding 50 μl of 2 M Tris-HCl pH 9.0.

8  Titrate eluted phage (A) and non-adsorbed phage (N) as transducing units (TU) on L-agar + Amp/X-gal/IPTG plates.

## Protocol 6

## Amplification of phage pool derived from panning[a]

### Reagents

- Phage buffer (see *Protocol 6*)
- XL1-Blue or DH5αF′ TB cells (see *Protocol 2*)
- IPTG (30 mg/ml stock solution in water; store at −20 °C)

- X-gal (30 mg/ml stock solution in N,N-dimethyl formamide; store in the dark −20 °C)
- L-agar +Amp/Glu (L-agar (see *Protocol 5*), 50 μg/ml Amp, 1% glucose)

- L-agar + Amp/X-gal/IPTG (see *Protocol 5*)
- Luria–Bertani (LB) (L-broth (1% bacto-tryptone, 0.5% yeast extract, 0.085 M NaCl, pH 7.2))
- LB + Amp (L-broth, 50 µg/ml Amp)
- Glycerol (Sigma)
- M13K07 helper phage (Amersham Biosciences)
- Thimerosal (Sigma)

## Method

1   Infect approximately 5 ml of TB cells with the neutralized phage eluted from magnetic micro-beads. Incubate for 10 min at 37 °C without agitation.

2   Pellet the bacterial cells by brief centrifugation. Discard the supernatant and resuspend the cells in 1–2 ml of LB. Plate cells onto 15 mm diameter plates of L-agar + Amp/Glu plates and incubate overnight at 37 °C at a maximum density of $5 \times 10^3$ cell/plate.

3   On the following day, scrape cells using approximately 10 ml of LB + 10% glycerol. Store in aliquots at $-20$ °C.

4   Inoculate cells in 20 ml of LB + Amp. Adjust the cell concentration to $OD_{600} \approx 0.05$ and grow with vigorous shaking (at least 300 rpm) to an optical density of $OD_{600} \approx 0.25$.

5   Incubate the cells for 15 min at 37 °C with very gentle agitation.

6   Super-infect the bacterial culture with helper phage M13K07 (multiplicity of infection (moi) = 30). Add IPTG (0.1 mM final concentration) and mix gently for 10 min and then incubate at 37 °C for 5 h with strong agitation.

7   Centrifuge the bacteria/phage suspension for 30 min, 5000 rpm at 4 °C and recover the supernatant. Titrate phage supernatant as TU on L-agar + Amp/X-gal/IPTG plates (usually titers are above $1 \times 10^{10}$ TU/ml). Store an aliquot of phage supernatant at 4 °C following addition of 0.05% Thimerosal.

8   Precipitate phage particles by adding PEG 8000 and NaCl solution (4% and 0.5 M final concentrations, respectively) and incubate on ice for 2–4 h at 4 °C.

9   Recover phage pellet by centrifugation for 40 min, 5000 rpm at 4 °C and resuspend it in 1/10 of the original volume with TBS.

10   Incubate the phage suspension for 30 min at 70 °C in a water bath to denature contaminating bacterial proteins, which are removed by centrifugation for 30 min, 10,000 rpm at 4 °C.

11   Titrate the phage suspension as TU on L-agar + Amp/X-gal/IPTG plates. Use about $10^9$–$10^{11}$ phage for the next round of panning.

[a] Assaying the reactivity of phage-pool supernatant derived from a panning round by ELISA (with sera) is highly informative on the efficiency and specificity of selection.

## 4 Selection strategies

Panning a phage library with serum can select a large pool of peptides that bind to a myriad of different available antibodies. Thus it is essential that the screening strategy includes a way to distinguish ligands for DS-Ab from ligands to irrelevant antibodies present in the sample. Peptides that bind to DS-Ab (DS-peptides) usually constitute a tiny minority of this pool. An effective way to select for this small population is to enrich ligands that are recognized by antibodies from different positive sera, as within this common subset the frequency of DS-peptides is considerably higher.

This positive selection can be implemented through two alternative strategies: serial selection or parallel selection. The basic protocol is the same for both strategies but the order of operations is completely different.

### 4.1 Serial selection

In serial selection, a phage library is panned on a positive serum. The pool of phage thus derived is then panned on a different positive serum, and so on. This process can be performed using different sera in different order. It is a good rule to carry out a limited number of panning rounds (usually no more than three) to preserve the complexity of the pool.

The pools of phage are evaluated by their ELISA reactivity with positive and negative sera (*Protocol 10*). Clones from the pools exhibiting the most interesting reactivity profile are then individually tested for their reactivity with positive and negative sera.

### 4.2 Parallel selection

In this case, the same phage library is simultaneously panned on antibodies from various positive and negative sera. This operation generates a pool of phage for each serum, which bind to the antibodies present in that serum. DS-peptides are then identified by isolating those ligands that are specifically present in the pool of phage derived from positive sera, according to the protocols described in the next section.

## 5 Screening individual clones

The final step of selection is the individual analysis of clones from the pool of selected phage: the more clones screened, the higher the number and quality of the specific ligands that can be identified.

### 5.1 Immuno-screening

The protocol we describe below is a robust and sensitive procedure for screening a large number of clones from phage-secreting colonies using sera. Screening colonies has several advantages: they are easier to generate, their size does not vary greatly from clone to clone and the corresponding signals are highly reproducible among different replicas. Unfortunately, a high level of background

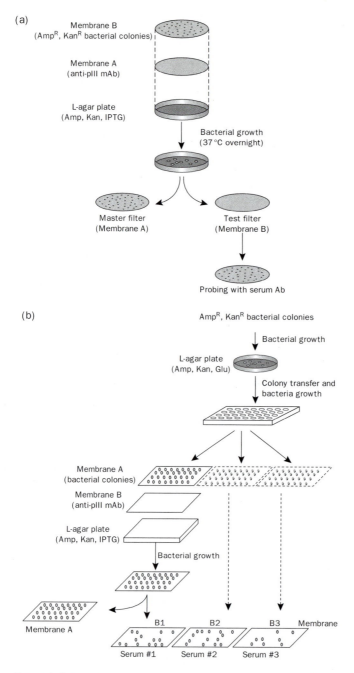

**Figure 1** Schematic representation of the procedure used for the expression and detection of phage-displayed peptides that are secreted from bacterial colonies and bind to serum antibodies. (a) Screening colonies plated on single-copy filter. (b) Screening replicated filters of the same ordered array of colonies using different sera.

generated by the serum antibodies cross-reacting with bacterial surface proteins hampers detection of serum antibodies that bind to phage-displayed ligands secreted from bacterial colonies. We overcome this problem by physically separating the site where the colony grows from that where the secreted phage binds to the antibody. Bacteria are grown on a first nitrocellulose membrane (membrane A) that is placed on top of a second membrane (membrane B), where an anti-phage monoclonal antibody (mAb) has been immobilized, layered in its turn on the surface of an LB-agar plate. Membrane A, which has a defined pore size, acts as a selective barrier; it permits the diffusion of phage and other proteins released from the periplasmic space of *E. coli* to membrane B, but prohibits the transfer of bacteria. Phage that diffuse through membrane A toward the agar surface are captured on membrane B by anti-phage mAb. Following growth of bacterial colonies, the membrane cast is dismantled, and colonies regrown on membrane A, which acts as a master filter. Phage captured on membrane B are probed with serum antibodies, whose binding is detected by using an antihuman enzyme conjugated antibody. The stained spots are readily correlated to the positions of the colonies on membrane A.

The same protocol can be used to quickly determine binding specificity of each individual clone. Clones are individually inoculated in the wells of an ELISA plate that have been filled with LB-agar. Following their growth, the entire set of clones is spotted in an ordered array on filter A using a multi-pin device. Many replicas of the same ordered array of clones can be generated and simultaneously screened with different positive or negative sera.

This protocol is particularly suitable when sufficient amounts of serum samples are available following serial selection.

## Protocol 7

# Colony assay for phage-displayed libraries

### Equipment and reagents

- Nitrocellulose membrane (Protran, 0.45 μ pore size, Schleicher & Schuell)
- Stratalinker UV crosslinker (Stratagene)
- anti-pIII mAb 57D1 (9) solution (2 μg/ml in 50 mM NaHCO$_3$, pH 9.6)
- Blocking solution (5% nonfat milk, 0.05% Tween-20 in PBS)
- PBST (PBS, 0.05% Tween-20)
- L-agar + Amp/Kan + IPTG (L-agar (see *Protocol 5*) + 50 μg/ml Amp, 20 μg/ml Kan, 35 μg/ml IPTG)
- L-agar + Amp/Glu (L-agar (see *Protocol 5*), 50 μg/ml Amp, 1% glucose)
- IPTG (see *Protocol 6*)
- LB (see *Protocol 6*)
- LB + Amp (LB, 50 μg/ml Amp)
- PBS (see *Protocol 4*)
- TB (see *Protocol 2*)
- M13K07 helper phage (Amersham Biosciences)
- DH5αF′ bacterial cells (7)
- ELISA multi-well plate (Immuno plate Maxisorp, Nunc)

- 96-pin gridding device ('Q' rep, Genetix)
- f1R1 (8)
- Antihuman (Fc-specific) alkaline-phosphatase conjugated antibodies (Sigma; dilute 1 : 5000 in blocking solution)

- Developing buffer (100 mM NaCl, 5 mM MgCl$_2$, 100 mM Tris-HCl, pH 9.6)
- Developing solution (330 mg/ml nitro blue tetrazolium, 165 mg/ml 5-bromo-4-chloro-3-indolyl phosphate in developing buffer)

## A. Membrane A preparation

1  Sterilize a 150 mm diameter nitrocellulose membrane by UV-exposure for 4 min for each side in the Stratalinker UV crosslinker and incubate in 10 ml of anti-pIII mAb 57D1 (9) solution overnight at 4 °C.

2  Incubate the membrane with freshly prepared blocking solution for 2 h at room temperature. Discard the blocking solution, wash the membrane with sterile PBST and layer it onto the surface of a 150mm L-agar + Amp/Kan/IPTG plate.[a]

## B. Membrane B preparation (for screening single-copy filters)

1  Infect DH5αF′ bacterial cells with a pool of phage and incubate for 2 min at 37 °C[b] (*Figure 1(a)*).

2  Afterwards, super-infect the same cells with the helper phage M13KO7 at a moi of 20–50 and incubate for 2 min at 37 °C.

3  Remove the non-adsorbed helper phage by centrifugation and carefully plate the infected cells onto a nitrocellulose membrane, previously layered on top of membrane A, placed on the surface of L-agar plate. A sterile glass spreader is used to remove the bubbles between the filters and the plate and to help adsorption of the bacterial suspension. Alternatively, infect DH5αF′ bacterial cells with the pool of phage and plate onto L-agar + Amp/Glu plates. After overnight incubation at 37 °C, scrape the bacterial colonies, resuspend in L-broth containing 100 μg/ml Amp, and grow at 37 °C to an A$_{600}$≈0.2. Infect the cells with the helper phage M13KO7 at a moi of 20–50. After 15 min incubation at 37 °C, remove the non-adsorbed helper phage by centrifugation and carefully plate infected cells on the UV-sterilized nitrocellulose membrane A, as described above.

## C. Membrane B preparation (for screening replicated filters)

1  Using the above protocol, plate Amp[R], Kan[R] bacterial colonies onto L-agar plates containing 100 μg/ml Amp, 20 μg/ml Kan, 1% Glu and incubate overnight at 37 °C. Use single clones to inoculate wells of a UV-sterilized ELISA multi-well plate containing 200 μl of L-broth added with 100 mg/ml Amp, 20 μg/ml Kan and 1% glucose. After overnight growth at 37 °C, transfer colonies from this master plate in one step onto membrane B (which is layered on top of membrane A and L-agar plate) using a sterile 96-pin gridding device ("Q" rep, Genetix). This operation can be repeated several times on different membranes. Square plates can be used to efficiently use space on L-agar plates (*Figure 1(b)*) (Nunc Omnitray, Life Technologies).

**2** Dilute crude serum 1 : 100 in blocking solution containing $1 \times 10^{10}$ pfu/ml f1R1 phage and 5 µl/ml of infected bacterial cells protein extract (*Protocol 2*). Incubate for 30 min at room temperature. Competing antigen at the desired concentration can be added at this step.

**3** Following overnight growth at 37 °C, mark the relative orientation of the two membranes. Remove membrane A. The colonies on it remain largely viable and can be re-grown for propagation or plasmid isolation. Transfer membrane B to a Petri dish and incubate with the pre-incubated serum mixture for 2 h at room temperature.

**4** Extensively wash membrane B with PBST, and incubate for 1 h at room temperature with antihuman (Fc-specific) alkaline-phosphatase conjugated antibodies.

**5** Extensively wash the membrane as above, rinse in developing buffer and incubate for 2–15 min at room temperature in developing solution. Stop the reaction by rinsing filter with 20 mM EDTA solution.

[a] Inducing the expression of recombinant protein with IPTG generates phage displaying a high number of recombinant capsid proteins.

[b] Use of DH5αF′ bacterial cells is crucial to achieving the best results, as these cells ensure a good growth of Amp$^R$, Kan$^R$ colonies.

## 5.2 DNA-based screening

Screening can also be efficiently performed by a DNA-based protocol. Each phage-displayed peptide in a library is physically associated with its encoding gene. This link can be used to identify peptides common to or different from given sets of phage. Binding of serum antibodies to a phage-displayed peptide can be conveniently detected by hybridizing a probe derived from the pool of peptide-coding sequences enriched by the serum to the peptide-coding sequence of the individual clone. This protocol is advantageous when parallel selection has been performed and limited amounts of serum samples are available.

# Protocol 8

## DNA-based screening

### Equipment and reagents

- LB (see *Protocol 6*)
- Tween-20 (Sigma)
- XL1-Blue TB cells (see *Protocol 1*)
- L-agar + Amp/Glu (see *Protocol 6*)
- UV-sterilized microplate (Immuno plate Maxisorp, Nunc)
- PCR multi-well plate Optiplate (Perkin Elmer)
- Primer F24: 5′-TTGCAGGGAGTCAAAGG-CCGCTTT-3′ (64 pmol)
- Primer E24: 5′-GCTACCCTCGTTCCGAT-GCTGTCT-3′ (64 pmol)
- dNTPs (see *Protocol 1*)
- DMSO (Dimethyl sulfoxide) (Sigma)
- *Taq* polymerase buffer (50 mM KCl, 10 mM Tris-HCl pH 9.0, 0.1% Triton X-100, 1.5 mM MgCl$_2$)

- *Taq* polymerase (Boehringer Mannheim)
- 96-pin gridding device ('Q' rep, Genetix).
- BBP solution (0.1% bromophenol blue solution)
- 10 × SSC (Saline-sodium citrate)
- Nylon membrane (Hybond N+, Amersham Biosciences)
- Dot-Blot apparatus (BioRad)
- Denaturing solution (0.5 M NaOH, 1.5 M NaCl), freshly prepared
- Neutralizing solution (0.5 M Tris-HCl pH 7.5, 1.5 M NaCl, 1 mM EDTA)
- 0.4 M NaOH
- 5 × SSC: (See *Protocol 6*)
- Qiagen plasmid DNA purification kit

- M13 primer: 5'-GTTTTCCCAGTCACGAC-3' (80 pmol)
- pC89 primer: 5'-AGAGATTACGCCAAGCC-3' (80 pmol)
- 20% PAGE (polyacrylamide gel electrophoresis)
- $\alpha[^{32}P]dATP$ and $\alpha[^{32}P]dCTP$
- Klenow enzyme
- G-50 Sephadex (Amersham Biosciences)
- Sodium dodecyl sulphate (SDS) (Sigma)
- Denhardt's solution (according to *Protocol 6*)
- Sonicated salmon sperm DNA, as described by Sambrook *et al.* (10)
- PhosphorImager ® (Molecular Dynamics)

## A. Master plate preparation

1  Make suitable dilutions of a supernatant of the pool of selected clones grown in repressed conditions (1% glucose, no IPTG) in LB + 0.05% Tween-20, in order to obtain a final concentration around $10^3$ TU/ml. Aliquot in an Eppendorf tube 100 μl of the phage dilution. Incubate phage 30 min at 70 °C with agitation.[a]

2  Add 200 μl of XL1-Blue TB cells to the phage dilution. Mix and allow infection for 1 min at 37 °C with gentle agitation.[b] Plate the infection mix on L-agar + Amp/Glu plates. Incubate overnight at 37 °C.

3  Individually transfer 96 clones from the Petri dish to the single wells of a UV-sterilized microplate filled with 200 μl of L-agar + Amp/Glu. Include in the microplate negative (e.g. no cells, cells transformed with wt phagemid, etc.) and positive controls. Incubate the microplate overnight at 37 °C.

## B. Target clone preparation

1  Amplify the peptide-coding sequence of each target clone as follows. Aliquot in each well of a PCR multi-well plate 80 μl of a reaction mixture containing: 0.8 μM F24 and E24 primers, 250 μM dNTPs, 7.5% DMSO, 1× *Taq* polymerase buffer, 1.25 units *Taq* polymerase.

2  Seal the PCR plate and run the PCR process on a GeneAmp9600 for 30 cycles: 10 s at 94 °C, 10 at 60 °C, 10 at 72 °C.

3  Spin the sealed PCR plate briefly to remove drops on top of the wells. Carefully remove the sealing film, avoiding cross-contamination between wells.

4  Add 10 μl dye of a BBP solution in each well. Seal the PCR plate again and mix on a plate shaker for 20 min at room temperature.

## C. Filter preparation

1  Wet a nylon filter (Hybond N + , 0.45 µm) and two 3 MM paper sheets with 10 × SSC. Place the wet 3 MM sheets in a dot blot apparatus, layer the nylon membrane on the top and then set the upper part of the apparatus.

2  Spot approximately 1 µl of the PCR amplification product diluted in 5 × SSC on the filter using the dot blot apparatus with vacuum suction.

3  Disassemble the dot blot apparatus and transfer the nylon membrane (DNA side up) on a 3 MM paper wetted with denaturing solution and incubate 7 min at room temperature.

4  Transfer the filter on a 3 MM paper wetted with neutralizing solution and incubate 3 min at room temperature. Blot dry with 3 MM filter paper. Air dry.

5  Fix the DNA samples by placing the nylon membrane DNA-side up on 3 MM paper wetted with the 0.4 M NaOH solution and incubate twice for 20 min at room temperature.

6  Rinse the membrane by immersion in 5 × SSC for 5 min.

## D. Probe

1  Add XL1-Blue TB cells to a representative aliquot of the pool of phage to be used as probe (usually, pools of phage derived from two subsequent rounds of panning have complexity lower than $10^5$). Mix and allow infection for 1 min at 37 °C with gentle agitation. Plate the infection mix onto L-agar + Amp/Glu plates. Incubate overnight at 37 °C.

2  Harvest Amp$^R$ bacterial clones, extract and purify the plasmid dsDNA according to the Qiagen protocol.

3  Assemble a PCR reaction containing: 50 ng template dsDNA of the phage pool, 0.8 µM M13 and pC89 primers, 250 µM dNTPs, 7.5% DMSO, 1 × *Taq* polymerase buffer, 1.25 units *Taq* polymerase in a final volume of 100 µl.

4  Program amplification for 30 cycles using the following thermal conditions: 10 s at 94 °C, 10 at 45 °C, 10 at 72 °C.

5  Digest the amplification products with *Eco*RI and *Bam*HI restriction enzymes, separate the digested fragment by 20% PAGE and then purify from the gel the fragment corresponding to the peptide-coding sequences.

6  Radiolabel 3 pmol of this purified product with α[$^{32}$P]dATP and α[$^{32}$P]dCTP by fill-in reactions using Klenow-fragment and then purified by G-50 Sephadex gel filtration, as described by Sambrook *et al.* (6).

## E. Hybridization

1  Pre-hybridize the membrane in 5 × SSC, 0.1% SDS, 5 × Denhardt's, 100 µg/ml sonicated salmon sperm DNA for 4 h (or overnight) at 65 °C in a water bath with gentle agitation.

**Protocol 8** continued

2  Remove the pre-hybridization mix and incubate overnight at 65 °C in 5 × SSC, 0.1% SDS, 5 × Denhardt's containing about $1 \times 10^6$ cpm/ml [$^{32}$P] labeled heat-denatured DNA probe.

3  Wash membrane briefly with 2 × SSC at room temperature, then wash the filter twice for 20 min at 70 °C in a water bath with gentle agitation in 0.1 × SSC/0.1% SDS.

4  Quantify the hybridization signal by a PhosphorImager® (Molecular Dynamics).[c]

[a] The above treatment is meant to reduce aggregation of phage particles, which in turn could generate mixed clones upon infection.

[b] Limited time of infection is meant to prevent production of new phage particles and co-infection of cells.

[c] Choose optimal exposure time in order to obtain values in the $10^4$–$10^5$ range.

## 6  Characterization of selected clones

### 6.1  ELISA

The binding properties of the clones singled out by the screening step are further characterized. A small-scale preparation of phage supernatant (*Protocol 9*) is sufficient to carry out several experiments. First, ELISA quantitatively measures reactivity of the clone with a large panel of positive and negative sera. This assay can be performed adopting the classical format, where phage are immobilized on the bottom of the well and human antibodies binding to phage-displayed peptides are detected by an antihuman conjugate (*Protocol 10*). In addition, binding of serum antibodies to the phage-displayed peptide can be detected by phage-ELISA: in this case serum antibodies are immobilized at the bottom of the well and binding of phage-displayed peptide is detected by anti-phage conjugate (*Protocol 11*). Sensitivity of phage-ELISA depends on the percentage of DS-Ab antibodies in the serum.

## Protocol 9

## Small-scale preparation of phage supernatant

### Reagents

- Phage buffer (see *Protocol 6*)
- XL1-Blue TB cells (see *Protocol 2*)
- L-agar + Amp/Glu plates (see *Protocol 6*)
- LB + Amp (see *Protocol 7*)
- M13K07 helper phage (Amersham Biosciences)
- IPTG (see *Protocol 6*)

### Method

1  Pick phage from single plaques using Pasteur pipettes and transfer to microtubes containing 200 µl of phage buffer.

2  Heat at 70 °C for 20 min and then centrifuge for 30 min in order to remove bacteria from the phage suspension.

3 Use dilutions of the phage-containing supernatant to infect XL1-Blue TB cells. Phage suspensions can then be stored at 4 °C.

4 Incubate 15 min without agitation, and then 30 min with vigorous shaking at 37 °C.

5 Plate onto L-agar + Amp/Glu plates and incubate overnight at 37 °C.

6 Inoculate single colonies in 2 ml of LB + Amp and grow at 37 °C with vigorous shaking (>350 rpm) until the cell density reaches $OD_{600} = 0.25$.

7 Incubate 15 min at 37 °C with very gentle agitation.

8 Super-infect with helper phage M13K07 (moi = 30) and add IPTG (0.1 mM final concentration), mix gently for 10 min and then incubate at 37 °C for 5 h with strong agitation.

9 Transfer cultures into micro-tubes and centrifuge for 30 min at maximum speed to remove the bacteria and store the phage supernatant at 4 °C. Titres are usually above $1 \times 10^{10}$ TU/ml.

# Protocol 10

## ELISA using phage supernatants and human sera

### Equipment and reagents

- Multi-well plates (Immuno plate Maxisorp, Nunc)
- Coating buffer (50 mM $NaHCO_3$, pH 9.6, 0.05% Thimerosal).
- Monoclonal antibody (m Ab) 57D1 (9). This antibody recognizes the N-terminal portion of the minor capsid protein (pIII) of M13 phage. It was generated by immunization of rats with a suspension of M13 phage particles. Supernatant from 57D1 hybridoma is precipitated using 50% $(NH_4)_2SO_4$, dissolved in PBS and dialyzed. Finally the antibody is immuno-purified on protein G-Sepharose.
- PBST (see Protocol 7)
- Blocking solution (PBS, 5% nonfat dry milk, 0.05% Tween-20, 0.02% Thimerosal)

- Phage supernatants, (see Protocol 9).
- Bacterial extract (see Protocol 2).
- PEG-concentrated f1R1-11.1 phage (9).
- Supernatant from unrelated rat hybridoma cells.
- Goat antihuman IgG Fc-specific AP-conjugated antibodies (Sigma) diluted 1 : 5000 in blocking solution.
- Substrate buffer (10% diethanolamine, 0.5 mM $MgCl_2$, 0.05% $NaN_3$, adjusted at pH 9.8 with HCl)
- Developing solution (1 mg/ml of p-nitrophenyl phosphate in substrate buffer)
- Automated ELISA reader

### Method

1 Aliquot in each well of multi-well plate 200 µl of anti-pIII mAb 57D1[a] at a concentration of 1 µg/ml in coating buffer 2. Incubate overnight at 4 °C.

2 Discard the coating solution and wash several times with PBST.

3  Add 250 µl/well of blocking solution and incubate plates for 60 min at 37 °C. Coated plates can then be used immediately or stored at − 20 °C for several weeks.

4  Add to each well a mixture of 100 µl of blocking solution and 100 µl of cleared phage supernatant prepared as described above. Allow the phage particles to bind for 60 min at 37 °C.

5  Pre-incubate human serum in 200 µl of blocking solution containing 2.5 µl of bacterial extract, about $5 \times 10^{10}$ pfu of PEG-concentrated f1R1-11.1 phage and 10 µl of supernatant from unrelated rat hybridoma cells for 30 min at room temperature. Use a 1 : 100 serum dilution when testing single clones and a 1 : 400 serum dilution when testing phage pools. This step eliminates the interference of serum antibodies directed against wild-type phage, bacterial contaminants and rat anti-pIII mAb.

6  Discard the phage supernatant from coated plate and extensively rinse wells with PBST.

7  Add 200 µl of the pre-incubated serum mix in each well and incubate for 2 h at room temperature.

8  Discard the serum mixture and extensively rinse wells with PBST. Add 200 µl of alkaline-phosphatase conjugated secondary antibody to each well and incubate 60 min at room temperature.

9  Discard the secondary antibody solution and extensively rinse wells with PBST. Briefly rinse wells with substrate buffer.

10  Add 200 µl of developing solution to each well. Incubate 15–60 min at 37 °C in the dark.

11  Record the results as the difference between $A_{405}$ and $A_{655}$ using an automated ELISA reader.

[a] The use of the anti-pIII mAb in ELISA with crude phage supernatants significantly reduces the background deriving from interaction of serum antibodies with bacterial proteins or other contaminants present in the culture. Alternatively, PEG-purified phage clones (see *Protocol 13*) can be directly coated on ELISA plates using $10^{10}$-$10^{11}$ phage particles in 100 µl of coating buffer per well, leaving the plates overnight at 4 °C. The optimal mAb concentration to be used for coating is experimentally determined for each batch of purified antibody through coating ELISA plates with serial dilutions of mAb 57D1.

## 6.2 Characterization of binding specificity by affinity purification

As a result of the high complexity of the phage library, it is possible that different peptides bind HCV-specific antibodies with the same binding specificity, that is different peptidic structures mimic the same determinant of the natural antigen. An effective and very sensitive way to obtain this information is to immuno-purify antibodies from a positive serum using a phage and then test the binding of these antibodies to the other phage: a positive result demonstrates that both phage bind antibodies with overlapping specificities.

## Protocol 11

## Phage-ELISA using human sera

### Equipment and reagents

- Coating buffer (see *Protocol 10*)
- PBST (see *Protocol 7*)
- PBBST (PBS, 1% Triton X-100, 3% BSA (Sigma))
- Rabbit antihuman Fc-specific IgG (Pierce): dilution 5.0 µg/ml in coating buffer[a]

- Anti-phage antibody HRP-conjugated (Amersham Biosciences): 1/10.000 dilution in blocking solution
- Developing solution (3,3′,5,5′-tetramethyl benzidine (TMB) liquid substrate system, Sigma)

### Method

1 Aliquot in each well 200 µl of rabbit antihuman Fc-specific IgG. Incubate overnight at 4 °C.

2 Wash the plate with PBST.

3 Add 250 µl of PBBST. Incubate for 2 h at 37 °C

4 Discard the PBBST and aliquot in each well 200 µl of the phage supernatant. Add in each well 0.25 µl of serum.[b] Incubate 60 min at 37 °C with shaking.

5 Wash plate with PBST.

6 Add 200 µl of goat anti-phage antibody HRP-conjugated. Incubate 30 min at 37 °C with shaking.

7 Wash the plate with PBST.

8 Add 200 µl of developing solution to each well. Incubate 15 min at room temperature in the dark. Stop the reaction by adding 25 µl of 2 M $H_2SO_4$. Shake 10 s in the ELISA reader before analysis.

9 Record results as the difference between $A_{450}$ and $A_{655}$ using an automated ELISA reader.

[a] Alternatively, protein A (Amersham Biosciences) diluted 0.5 µg/ml in coating buffer can be used.

[b] Serum can be pre-diluted using blocking solution.

When the natural antigen is available, a similar protocol can be used to identify the region of the antigen, which is mimicked by the peptide.

### 6.3 Cross-inhibition assay

An alternative method to identify peptides that bind serum antibodies with the same binding specificity is by competition: the binding of serum antibodies to a phage-displayed peptide is competed by an excess of a different phage-peptide. Binding inhibition is measured through an ELISA format. Also in this case, the same protocol lends itself to mapping the region of the antigen mimicked by the selected ligand.

## Protocol 12

## Large-scale preparation of phagemid clones

### Equipment and reagents

- LB + Amp (see *Protocol 7*)
- LB + Amp/Glu (LB (see *Protocol 6*), 50 μg/ml Amp, 1% glucose)
- IPTG (see *Protocol 6*)
- M13K07 helper phage (Stratagene)
- PEG 8000 (Sigma)
- NaCl
- TBS (see *Protocol 3*)
- SW40 polyallomer tubes (Beckman)

### Method

1  Inoculate a single phagemid-containing colony into 10 ml of LB + Amp/Glu and grow overnight at 37 °C (glucose represses the expression of the pVIII gene under the control of the lac promoter).

2  Dilute the overnight culture in 500 ml of LB + Amp to obtain an $OD_{600} = 0.05$ and grow up at 37 °C in a 2 L flask with vigorous shaking (at least 300 rpm) till $OD_{600} = 0.25$.

3  Incubate for 15 min at 37 °C with very gentle agitation. Add IPTG (0.1 mM final concentration) to induce the pVIII expression from the lac promoter and super-infect with helper phage M13K07 (moi = 30). Mix and leave for 15 min at 37 °C with no agitation to allow infection. Incubate the culture at 37 °C for 5 h with strong agitation.

4  Centrifuge the bacteria/phage suspension for 30 min, 5000 rpm at 4 °C and recover the supernatant.

5  Precipitate phage particles by adding PEG 8000 and NaCl solution (4% and 0.5 M final concentrations, respectively) and incubating on ice for 4 h at 4 °C.

6  Recover the phage pellet by centrifugation for 40 min, 5000 rpm at 4 °C and resuspend in 1/10 of the original volume with TBS.

7  Incubate the phage suspension for 30 min at 70 °C in a water bath to denature contaminating bacterial proteins, which are removed by centrifugation for 30 min at 10,000 rpm at 4 °C.

8  Precipitate the phage again by adding PEG 8000/NaCl to the resulting supernatant and incubating on ice for 2 h at 4 °C.

9  Centrifuge for 30 min, 10,000 rpm, 4 °C and resuspend phage pellet in 10 ml of TBS. Centrifuge again for 15 min at 8000 rpm to remove insoluble material.

10  Titrate the phage for TU on LB + Amp plates.

## Protocol 13

# Affinity purification of serum antibodies reacting with a phage-displayed peptide

### Equipment and reagents

- 60 mm diameter polystyrene Petri dish (Falcon 1007)
- Coating buffer (see *Protocol 10*).
- mAb 57D1 (9). This antibody recognizes the N-terminal portion of the minor capsid protein (pIII) of M13 phage. It was generated by immunization of rats with a suspension of M13 phage particles. Supernatant from 57D1 hybridoma is precipitated using 50% $(NH_4)_2SO_4$, dissolved in PBS and dialyzed. Finally the antibody is immuno-purified on protein G-Sepharose.
- PBST (see *Protocol 7*)
- Blocking solution (see *Protocol 10*)
- PEG-purified phage, prepared as described in *Protocol 12*
- Bacterial extract (see *Protocol 2*)
- PEG-purified F1-R1 phage (see *Protocol 12*)
- Glycine-HCl/BSA (100 mM glycine, pH 2.7 with HCl, 10 µg/ml BSA)
- 2 M Tris-HCl pH 9.4 (fresh solution)

### Method

1 Aliquot in a polystyrene Petri dish 4 ml of PEG-purified phage diluted in coating buffer ($\approx 1 \times 10^{11}$ TU/ml). Leave overnight at 4 °C on rocking platform.

2 Discard and wash extensively with PBST at room temperature.

3 Add 6 ml of blocking solution and incubate 1 h at 37 °C on a rocking platform.

4 Mix 40 µl serum, 100 µl bacterial extract, $4 \times 10^{12}$ pfu PEG-purified f1-R1 carrier phage in a total volume of 4 ml of blocking solution and incubate 60 min at room temperature.

5 Discard the blocking solution. Briefly rinse with PBST. Add the preincubated mix to the plate. Incubate overnight at 4 °C on rocking platform.

6 Discard the solution and wash the plate extensively with PBST at room temperature.

7 Discard the washing buffer. Elute antibodies by adding 4 ml of glycine-HCl/BSA. Incubate for 10 min at room temperature on rocking platform. Neutralize in a polypropylene tube containing 100 µl of 2 M Tris-HCl pH 9.4.

8 Phage-peptides or antigens can be probed with purified antibodies using an ELISA format similar to that described in *Protocol 11*.

## Protocol 14

## Cross-inhibition assay

### Equipment and reagents

- Multi-well plates (Immuno plate Maxisorp, Nunc)
- TBS (See *Protocol 3*)
- PEG-purified phage (see *Protocol 12*).
- PBST (see *Protocol 7*)
- Blocking solution (5% nonfat dry milk, 0.05% Tween-20, 0.02% NaN$_3$ in PBS)
- XL1-Blue bacterial extract (see *Protocol 2*).

- Alkaline-phosphatase conjugated, goat antihuman-IgG (Fc-specific) Ab (Sigma).
- Substrate buffer (10% diethanolamine, 0.5 mM MgCl$_2$, 0.05% NaN$_3$, adjusted at pH 9.8 with HCl.
- Developing solution (1 mg/ml of *p*-nitrophenyl phosphate in substrate buffer.
- Automated ELISA reader.

### Method

1  Aliquot in each well of multi-well plates 100 µl of TBS containing $2 \times 10^{10}$ PEG-purified phage particles.[a] Incubate for 12 h at 4 °C.

2  Wash several times with PBST and block with 250 µl/well of blocking solution for 2 h at room temperature.

3  Dilute the sera in blocking solution containing $1 \times 10^{12}$ M13 phage particles/ml, 50 µl/ml of XL1-Blue bacterial extract and $2 \times 10^{11}$ particles/ml of PEG-purified inhibitor phagemid clone.[b] Pre-incubate 3 h at room temperature.

4  Discard the blocking solution from the plate, add the pre-incubated serum mixture to the phage coated wells and incubate for 3 h at room temperature.

5  Wash extensively with PBST.

6  Add 100 µl/well of alkaline-phosphatase conjugated secondary antibody, diluted 1 : 5000 in blocking solution, and incubate at room temperature for 2 h.

7  Wash plates several times with PBST, and once more with substrate buffer.

8  Add 100 µl/well of a 1 mg/ml solution of *p*-nitrophenyl phosphate in substrate buffer to reveal alkaline phosphatase reaction.

[a] Phage competition can also be performed using the cleared phage supernatant. In this case the ELISA format is essentially that described in *Protocol 9*. In this case, phage f1R1–11.1 (9) should be used as helper in preparation of the competing phage, to avoid signal reduction due to its binding to the coated antibody. f1R1-11.1 is a f1R1 mutant bearing a substitution of the glutamic acid at position 5 of pIII with a glycine; such mutation impairs binding of the mAb 57-D1 to pIII.

[b] Alternatively, a competing antigen can also be used.

## 6.4  DNA sequencing

Here, we report a simple protocol that allows the template of a large number of clones to be prepared rapidly and efficiently. This procedure does not involve the preparation/purification of plasmid DNA and requires minimal hands-on work, thus considerably saving material and time. The protocol is composed of three essential steps:

- PCR amplification of the insert region from a bacterial colony.

- Rapid purification of the amplified material by gel filtration (purification of the PCR-amplification product by gel filtration efficiently eliminates the problems derived from carry over of PCR primers and dNTPs in the subsequent sequencing steps). Purification is performed in a format that allows the parallel processing of 96 clones (*Protocol 15*).

- Sequencing by standard methods (11,12).

## Protocol 15

## PCR amplified template

### Equipment and reagents

- MicroAmp™ reaction tube (Perkin Elmer)
- Elution buffer (20 mM Tris-Cl pH 8.5, 2 mM EDTA, 1% Triton X-100)
- GeneAmp9600 thermal cycler (Perkin-Elmer)
- Forward primer: 5'-AGAGATTACGC-CAAGCC-3' (7.5 pmol)
- Reverse primer: 5'-TCCCAGTCACGACGT-3' (7.5 pmol)
- dNTPs (see *Protocol 1*)
- NP-40 (Nonidet P-40)
- *Taq* buffer (50 mM KCl, 10 mM Tris-Cl

pH 8.3, 1.5 mM $MgCl_2$, 0.01% gelatin)
- Ampli-*Taq* DNA polymerase (Perkin-Elmer)
- L-agar + Amp plates (L-agar (see *Protocol 5*), 50 µg/ml Amp )
- Sephacril® S-300 (Amersham Biosciences)
- Silent Monitor™ membrane-bottomed microplate (Model SM120LP, 1.2 µm Loprodyne; Pall Bio Support)
- Vacuum filtration unit (mod. EVENT 4160, Eppendorf)

### Method

1  Pick a single Amp$^R$ phagemid colony (about 1 mm in diameter) with a sterile plastic pipette tip and transfer it into a 0.2 ml MicroAmp™ reaction tube containing 20 µl of elution buffer. Heat the sample 10 min at 95 °C in a GeneAmp9600 thermal cycler. (Clarification by centrifugation should be avoided, since heating makes cell debris to stick to the bottom of the tube.)

2  Keep the colony extract as a master copy of the clone. Alternatively, the colony can be transferred directly to the PCR reaction and a heating step added to the PCR program. In this case, a master copy of the clone is stabbed into a well of a UV-sterilized micro plate filled with L-agar + Amp.

3  Assemble a PCR reaction by adding 3 µl of colony extract to a 0.2 ml MicroAmp™ tube in a final volume of 50 µl containing 150 nM forward and reverse primers, 50 µM dNTPs, 0.5% NP-40, *Taq* buffer and 1 µl Ampli-*Taq* DNA polymerase. Run PCR process using the following program: 30 cycles of 94 °C for 10 s, 60 °C for 10 s, 72 °C for 10 s.

4   Load Sephacril® S-300 aqueous slurry in the wells of a Silent Monitor™ membrane-bottomed microplate to give a settled bed volume of about 400 µl of resin per well. Applying a vacuum to the bottom of the plate using a vacuum filtration unit.

5   Pipette the PCR-amplified material from each clone into the drained Sephacril®-containing wells. Collect flow-through in a 96-well microplate by vacuum aspiration. (The length of the PCR-amplified templates dictates the choice of the gel filtration medium: we found that Sephacril® S-300 gives the best results for our 330 base pail PCR-amplified fragment. Using the above procedure, both handling and assembly of the samples is easier and faster than using commercially available spin columns packaged in microcentrifuge tubes, with saving on material costs.)

6   Use the purified PCR-product for sequencing according to standard protocols.

## 7  Epitope maturation

The complexity of phage libraries has often been limited to $10^8$ by the low efficiency of DNA transformation in bacteria (3; however, see Chapter 2). Thus a library comprises only a very small number of all the possible sequences. This is why ligands selected from random peptide libraries are often of low affinity. An effective way to overcome this limitation and identify ligands with improved properties is through a process of "epitope maturation." Peptide sequences selected from the initial library are partially mutagenized to generate a population of variants. Screening this secondary library with serum samples identifies ligands that are better than the original lead sequence.

Here we report a site-directed mutagenesis protocol, which generates diversity from a single clone or population of clones. This process is designed to introduce mutations in the peptide-coding sequences and to extend the size of the same target sequences by adding random residues at both ends.

In the first step of the protocol, peptide coding sequences of a single or population of clones are amplified using specific primers, under conditions known to introduce mutation in the amplified sequences (error-prone PCR; 13, 14). The amplified products are then digested with BamHI restriction endo-nuclease and biotinylated vector sequences at the 3'-end of the peptide-coding region removed by streptavidin capturing. Single-stranded 5' extensions gener-ated by enzymatic restriction are removed by mung-bean nuclease treatment and the resulting products are ligated to the 3'-NNK adapter, with the following structure: an $(NNK)_5$ random sequence followed by a tag sequence which includes a BamHI restriction site. An $NH_2$ group at the 3'-end of the anti-sense strand directs the orientation of the adapter in ligation.

By using bio.for and bio.3'-NNK as primers, ligation products are specifically PCR-amplified under mutagenic conditions to introduce further diversity in the peptide-coding sequence. The amplified products are digested with EcoRI and

*Bam*HI restriction endonucleases and the biotinylated terminal fragments are removed by streptavidin capturing. Finally, the resulting products are cloned into the *Eco*RI/*Bam*HI sites of the pC89 vector *(Figure 2)* (7).

A similar strategy can be adopted to extend the foreign peptide sequence by adding random residues at its 5' end.

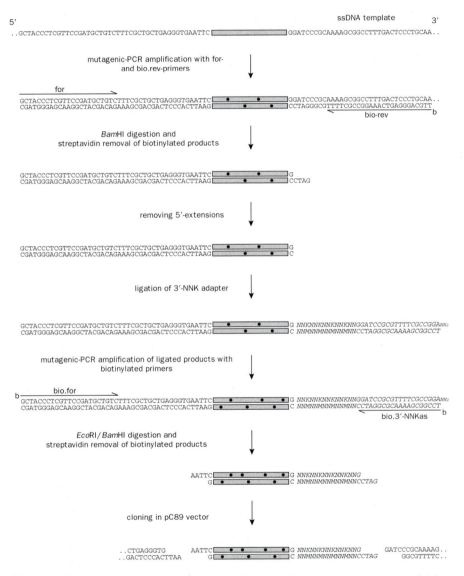

**Figure 2** Scheme of the protocol for epitope affinity maturation. Grey box indicates foreign peptide sequences, while dots denote mutation of the original sequence. Arrows denote PCR primers, NH$_2$ indicates 3'-end amino-linker. *b* identifies biotin label.

217

## Protocol 16

# Epitope maturation

## Equipment and reagents

- ssDNA purified as described by Qiagen
- Oligonucleotides

| | |
|---|---|
| for | 5′-GCTACCCTCGTTCCGATG CTGTCT-3′ (20 pmol) |
| bio.for | 5′-bGCTACCCTCGTTCCGATGCTGTCT-3′ (20 pmol) |
| rev | 5′-TTGCAGGGAGTCAAAGGCCGCTTT-3′ (20 pmol) |
| bio.rev | 5′-bTTGCAGGGAGTCAAAGGCCGCTTT-3′ (20 pmol) |
| 3′-NNKs | 5′-NNKNNKNNKNNKNNGGATCCGCGTTTTCGCCGGA-NH$_2$-3′ (100 pmol) |
| 3′-NNKas | 5′-TCCGGCGAAAACGCGGATCC-3′ (100 pmol) |
| bio.3′-NNKas | 5′-bTCCGGCGAAAACGCGGATCC-3′ (100 pmol) |
| 5′-NNKs | 5′-AGCGACGACTCCCAGAATTC-3′ (100 pmol) |
| bio.5′-NNKs | 5′-bAGCGACGACTCCCAGAATTC-3′ (100 pmol) |
| 5′-NNKas | 5′-NNMNNMNNMNNMNNGAATTCTGGGAGTCGTCGCT-NH$_2$-3′ (100 pmol) |

- dNTPs (see *Protocol 1*)
- dITP
- *Taq* polymerase buffer (50 mM KCl, 10 mM Tris-HCl pH 9.0, 0.1% Triton X-100, 6.5 mM MgCl$_2$)
- *Taq* polymerase (Promega)
- MnCl$_2$
- PCIA (phenol/CHCl$_3$/isoamyl-alcohol 25 : 24 : 1)
- CIA (CHCl$_3$/isoamyl-alcohol 24 : 1)
- 3 M sodium acetate (pH 5.2)
- Ethanol
- *Eco*RI (5′-targeted extension) or *Bam*HI (for 3′-targeted extension) restriction endonuclease (New England Biolabs)
- Streptavidin-coated magnetic micro-beads (Dynal)

- Mung bean nuclease buffer (New England Biolabs)
- Mung bean nuclease (New England Biolabs)
- Annealing buffer (10 mM Tris-HCl, 50 mM NaCl, 10 mM MgCl$_2$)
- T4 DNA ligase buffer (New England Biolabs)
- T4 DNA ligase (New England Biolabs)
- DH5αF′ *E. coli* strain strain (8)
- DH5αF′ competent cells, as described by Sambrook *et al.* (10)
- SOC medium, as described by Sambrook *et al.* (10)
- L-agar + Amp/Glu (see *Protocol 6*)
- LB (see *Protocol 6*)
- LB + Amp (see *Protocol 6*)

## Method

1  Assemble a first mutagenic PCR-amplification of peptide-coding sequences in a final volume of 50 μl containing 0.5 ng of the ssDNA template, 0.4 μM bio.for and rev primers (for 5′-targeted extension) or for and bio.rev primers (for 3′-targeted extension), 200 μM dNTPs, 200 μM dITP, *Taq* polymerase buffer, 0.5 mM MnCl$_2$, 7.5% DMSO and 2.5 units of *Taq* polymerase.

**2** Perform amplification in 30 cycles on a GenAmp 9600 using the following conditions: 10 s at 94 °C, 10 s at 65 °C, and 30 s at 72 °C, with a final segment of 7 min at 72 °C.

**3** Collect together the PCR products from 10 reactions, extract with PCIA, then CIA. Precipitate the DNA by adding 0.3 M sodium acetate, 10 mM MgCl₂ and 2.5 volumes of cold ethanol. Leave 5 min on ice, and then spin 30 min in bench top centrifuge in cold room. Remove the supernatant, wash the DNA pellet with cold 70% ethanol, dry at room temperature and re-dissolve in 40 µl H₂O.

**4** Digest the PCR product with *Eco*RI (5′-targeted extension) or *Bam*HI (for 3′-targeted extension) restriction endonuclease, according to manufacturer's instruction. Extract the reaction mix with PCIA, then with CIA. Precipitate the DNA by adding 0.3 M sodium acetate, 10 mM MgCl₂ and 2.5 volumes of cold ethanol. Leave 5 min on ice, and then spin 30 min in bench top centrifuge in cold room. Remove the supernatant, wash DNA pellet with cold 70% ethanol, dry at room temperature and redissolve in H₂O.

**5** Remove the biotinylated fragments by incubation with streptavidin-coated magnetic micro-beads according to manufacturer's instructions and concentrate the supernatant by ethanol precipitation.

**6** Resuspend the DNA fragments in mung bean nuclease buffer at a concentration of 0.1 µg/ml and incubate with 1 U/µg of DNA for 30 min at 30 °C. Stop the reaction by extracting the reaction mixture with PCIA/CIA, and ethanol-precipitate.

**7** Resuspend 100 pmoles of oligonucleotide 5′-NNKs and 5′-NNKas (for 5′-adapter) or 3′-NNKs and 3′-NNKas (for 3′-adapter) in 30 µl of annealing buffer and transfer to a 94 °C water bath that is allowed to slowly return to room temperature. Dilute the annealed mix to a final volume of 60 µl containing 300 µM dNTPs and 5 units Klenow polymerase, and incubate for 40 min at room temperature. Extract the elongated product by PCIA/CIA and ethanol-precipitate.

**8** Incubate 50 pmoles of random 5′- or 3′-adapter and 10 pmoles of *Eco*RI- or *Bam*HI-digested PCR fragments overnight at 16 °C in a 20 µl volume of T4 DNA ligase buffer added with 400 U of T4 DNA ligase.

**9** Use the ligation mixture as the template in a second mutagenic PCR amplification. Carry out the amplification reaction in a final volume of 50 µl containing 2 µl of the ligation mixture, 0.8 µM bio.5′-NNKs and bio.rev primers (for 5′-targeted randomization) or 0.8 µM bio.for and bio.3′-NNKas primers (for 3′-targeted randomization), adopting the same reaction conditions used in the first mutagenic PCR. Extract the reaction mixture with PCIA/CIA, and ethanol-precipitate as described above.

**10** Collect together the PCR-amplification products from five reactions, PCIA/CIA-extract, ethanol-precipitate, resuspended in the appropriate reaction buffer and digest with *Eco*RI and *Bam*HI restriction enzymes. The reaction mixture is then PCIA/CIA-extracted, ethanol-precipitated and redissolved in H₂O.

**Protocol 16** continued

11  Remove the biotinylated fragments by incubation with streptavidin-coated magnetic micro-beads according to manufacturer's instructions. Concentrate the supernatant by ethanol precipitation and resuspend in $H_2O$. Ligate about 2 µg (0.8 pmoles) of *EcoRI/Bam*HI-digested pC89 vector (7) to 2 pmoles of purified insert at 16 °C overnight in 20 µl of ligase reaction buffer containing 400 U T4 DNA ligase. The ligation mixture is PCIA/CIA-extracted, ethanol-precipitated and redissolved in 10 µl of $H_2O$.

12  The ligation product is used to transform competent bacterial cells by electroporation using a Gene Pulser apparatus according to the manufacturer's instructions. 3 µl aliquots of the ligated product are mixed with 100 µl DH5αF' competent cells. The transformed cells are immediately resuspended in a total of 9 ml of SOC medium, grown at 37 °C for 60 min and plated on four 230 × 230 mm square dishes containing LB-agar, 100 µg ampicillin/ml and 1% glucose. After overnight incubation at 37 °C, bacterial colonies are scraped from the plates and resuspended in 40 ml LB medium.

13  An aliquot of this cell suspension is used as starting inoculum in LB + Amp medium and incubated at 37 °C with vigorous shaking to $A_{600} = 0.25$. Cells are then infected with helper phage M13-K07, at a moi = 20–50, and 1 mM IPTG added. After incubation for 4 h at 37 °C with strong agitation, the bacteria are removed by centrifugation (8000 g for 10 min at 4 °C), the cleared phage supernatant collected and precipitated by adding 4% PEG (polyethylene glycol) 8000/0.5 M NaCl final concentrations. The phage pellet is resuspended in TBS.

## Acknowledgment

We thank Janet Clench for a linguistic revision of the manuscript.

## References

1.  Alter, H. J. (1995). *Blood*, **85**, 1681.
2.  Smith, G. P. and Petrenko, V. (1997). *Chem. Rev.*, **97**, 391.
3.  Cortese, R., Monaci, P., Luzzago, A., Santini, C., Bartoli, F., Cortese, I. *et al.* (1996). *Curr. Opin. Biotech.*, **7**, 616.
4.  Neuner, P., Cortese, R., and Monaci, P. (1998). *Nucl. Acids Res.*, **26**, 1223.
5.  Puntoriero, G., Meola, A., Lahm, A., Zucchelli, S., Bruni Ercole, B., Tafi, R. *et al.* (1998). *EMBO J.*, **17**, 3521.
6.  Felici, F., Castagnoli, L., Musacchio, A., Jappelli, R., and Cesareni, G. (1991). *J. Mol. Biol.*, **222**, 301.
7.  Hanahan, D. (1983). *J. Mol. Biol.*, **166**, 557.
8.  Model, P. and Russel, M. (1988). In *The bacteriophages*, Vol. 2 (ed. R. Calendar), p. 375, Plenum Press, New York.
9.  Dente, L., Cesareni, G., Micheli, G., Felici, F., Folgori, A., Luzzago, A. *et al.* (1994). *Gene*, **148**, 7.

10. Sambrook, J., Fritsch, T., and Maniatis, T. (eds.) (1989). *Molecular cloning: a laboratory manual.* Cold Spring Harbor Laboratory Press, New York.
11. Rapley, R. (ed.) (1996). *PCR sequencing protocols.* Humana Press, Totowa, NJ.
12. Howe, C. J. and Ward, E. S. (eds.) (1989). *Nucleic acids sequencing.* IRL Press, Oxford.
13. Leung, D. W., Chen, E., and Goeddel, D. V. (1989). *Technique*, **1**, 11.
14. Cadwell, R. C. and Joyce, G. F. (1992). *Methods Appl.*, **2**, 28.

Chapter 12

# Interaction cloning using cDNA libraries displayed on phage

## Laurent Jespers
MRC Laboratory of Molecular Biology, Hills Road, Cambridge,
CB2 2QH, UK.

## Marc Fransen
Katholieke Universiteit Leuven, Faculteit Geneeskunde, Campus
Gasthuisberg, Departement Moleculaire Celbiologie, Afdeling
Farmacologie, Herestraat 49, B-3000 Leuven, Belgium.

## 1 Introduction

Expression cloning of complementary DNA (cDNA) is an extremely valuable tool for the isolation and characterization of genes. Whereas nucleic acid probe-based screening relies on preliminary sequence information related to the gene of interest, screening of an expressed cDNA library only requires a natural ligand or a specific (monoclonal or polyclonal) antibody as probe (1, 2). *In vitro* screening of cDNA libraries expressed from λ phage or plasmids involves affixing the translated products to a membrane prior to challenging with a labeled probe. This transfer can denature the cDNA-encoded proteins and may therefore compromise detection. In contrast, the yeast-based two-hybrid system operates entirely *in vivo* by reconstituting a transcriptional activator to detect protein–protein interactions via the expression of a reporter gene (3). Overall, these systems are not easily amenable to the screening of large libraries and this has triggered the development of several systems based on selection rather than screening. In view of sensitivity and selectivity, selection for a specific interaction through iterative enrichment steps is less demanding than screening. However, cDNA cloning via selection of expressed clones requires a physical linkage between each cDNA (for propagation) and its gene product (for detection). One of the first selection systems designed for expression cloning of cDNA was developed for isolation of mammalian cell transformants expressing specific cell receptors (4). Recently, the advent of phage display technology (for reviews, see 5–6) as a powerful tool for selection of specific binding peptides or proteins from large libraries ($\geq 10^8$ clones) has provided the opportunity to develop this system for expression cloning of cDNA.

## 2  Vectors for display of cDNA libraries on phage

### 2.1  Filamentous bacteriophage

Since full-length cDNAs contain translational stop codons at their 3'-end and lack a suitable ribosome-binding site for procaryotic expression (if they are derived from eucaryotic cells), the most practical way to ensure expression of cDNAs on phage is to fuse them through their 5'-ends to a carrier gene. Traditionally, phage display involves fusion to the N-terminal end of the coat protein III (pIII) or VIII (pVIII) (5, 6) (see Chapter 1), and it was widely accepted that direct fusion to the C-terminal end of either coat protein is not compatible with phage assembly. Three specialized vectors have therefore been developed to circumvent this limitation. One approach involves the fusion of pIII with the leucine zipper of c-Jun while the translated cDNA products are fused to the C-terminal end of the leucine zipper of c-Fos (pJuFo vector) (7). Upon phage assembly, the zipper domains associate at the tip of the phage, thereby linking the encapsidated cDNA with its encoded product. A second strategy, Direct Interaction Rescue, exploits the functional modularity of the minor coat protein pIII: cDNAs are expressed as fusion to the C-terminal end of the infectivity domain of pIII, while the protein-probe is attached to the C-terminal part of pIII, which is embedded in the phage core (8). Only phages that are able to establish a stable interaction between the probe and a specific cDNA-encoded product can restore their infectivity and be selected throughout propagation in *Escherichia coli*. The third strategy—which is described in detail in this chapter—circumvents pIII altogether and is based on the finding that the C-terminal end of the minor coat protein VI (pVI) is surface-exposed (9) and therefore suitable for phage display of expressed gene *VI* (*gVI*)–cDNA fusions through a glycine-rich hexapeptide (10). Protein VI is thought to stabilize the phage structure by providing a link between pIII and the phage core (11, 12) (see Chapter 1). All three systems have proven successful for interaction cloning of cDNAs from libraries by selection on protein ligands and/or polyclonal antisera. Recently, a group has reported that, contrary to common expectations, polypeptides can be functionally displayed on the filamentous phage surface by fusion to the C-terminal end of pIII (13) or pVIII (14) through optimized linker sequences (ranging from 8 to 10 amino acids). In theory, such surprising functional fusions should also be amenable for expression cloning of cDNA but proof-of-principle still remains to be demonstrated.

### 2.2  Lytic bacteriophage

In filamentous phage systems, the coat proteins are directed to the periplasm prior to incorporation into phage particles. While this pathway may be appropriate for proteins that require disulfide bond formation (e.g. secreted proteins or extracellular domains of receptors), it may not be compatible with cDNA-encoded products that normally fold in a reducing environment. Therefore, phage particles that assemble intracellularly, such as those of phages λ, T4 and T7, have been explored for display of cDNAs encoding intracellular proteins.

In phage lambda, fusions have been reported at the C-terminal end of the capsid protein D (15) or the tail protein V (16). The former display protein was used to display and select cDNA fragments prepared from Hepatitis C virus genome (17), human brain, and mouse embryo (18). In phage T4, the C-terminal ends of the coat proteins WAC (19) and SOC (20) have been targeted for covalent display of proteins, while the T7 display system is based on the solvent accessibility of the C-terminal end of the capsid protein 10A for display. Recently, the utility of the latter system for selection of cDNAs has been demonstrated in several reports (21–23). A cloning kit exploiting this promising approach is commercially available (Novagen: see Chapter 1, section 8).

# 3 Cloning of cDNA libraries in gVI-based display vectors

Over the past few years, the conventional preparation of cDNA libraries has been greatly simplified by the availability of reliable cDNA synthesis and cloning kits. Because the excellent manuals provided with these kits make the construction of cDNA libraries straightforward, these procedures will not be discussed here. However, an alternative procedure in which cDNA inserts from a pre-made cDNA library are first amplified by polymerase chain reaction (PCR) (see *Protocol 1*) and subsequently transferred into the gVI phagemid expression vectors (see *Protocol 2*) will be described in detail.

## 3.1 Description of the pG6 vectors

The pG6 vectors described in this chapter (*Figure 1*) are phagemid vectors designed for monovalent display of fusion proteins: wild-type pVI from helper phage competes with the pVI–cDNA fusion expressed by the phagemid vector for incorporation into nascent phage particles. With the exception of an additional *Sal*I site in the polylinker region, the phagemids pG6A, pG6B, and pG6C are virtually identical to the pDONG6 set of vectors (10) which allow cloning of cDNA libraries at the 3′-end of *gVI*. Transcription from the *lac* promoter generates a bicistronic mRNA: the first cistron encodes a fusion between *lacZ* and the 3′-end of *gIII*, while the second cistron comprises the *gVI*–cDNA fusion. Translation of the second cistron is controlled by the ribosome-binding site of *gVI*, that is located within the 3′-end of *gIII*. Although the isopropyl β-10-thiogalactolide (IPTG)-inducible *lac* promoter controls the expression of *gVI*-cDNA, we have exploited the leakiness of the *lac* promoter (in the absence of glucose) to govern the expression of *gVI* fusion for display on phage.

## 3.2 Preparation of cDNA inserts from pre-made λGT11 libraries by PCR

The pG6 vectors are designed for directional cloning of cDNAs primed with a tagged $(dT)_n$ oligonucleotide. By using three vectors with alternate reading frames, each cDNA has the possibility of being translated correctly until the

(a)

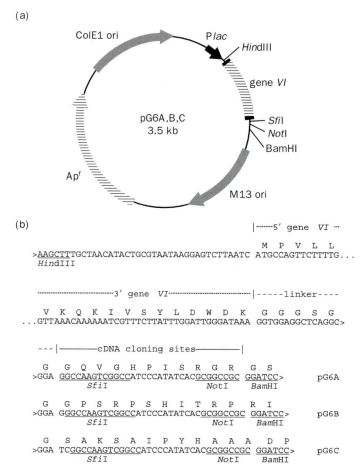

(b)

```
                                         |········5' gene  VI ···
                                           M   P   V   L   L
     >AAGCTTTGCTAACATACTGCGTAATAAGGAGTCTTAATC ATGCCAGTTCTTTTG...
      HindIII


      ··········─────────3' gene  VI·······················|-----linker----

          V   K   Q   K   I   V   S   Y   L   D   W   D   K   G   G   G   S   G
     ...GTTAAACAAAAAATCGTTTCTTATTTGGATTGGGATAAA GGTGGAGGCTCAGGC>


      ---|─────────cDNA cloning sites─────────|

          G   G   Q   V   G   H   P   I   S   R   G   R   G   S
     >GGA GGCCAAGTCGGCCATCCCATATCACGCGGCCGC GGATCC>         pG6A
             SfiI                       NotI   BamHI

          G   G   P   S   R   P   S   H   I   T   R   P   R   I
     >GGA GGGCCAAGTCGGCCATCCCATATCACGCGGCCGC GGATCC>        pG6B
             SfiI                       NotI   BamHI

          G   S   A   K   S   A   I   P   Y   H   A   A   A   D   P
     >GGA TCGGCCAAGTCGGCCATCCCATATCACGCGGCCGC GGATCC>       pG6C
             SfiI                       NotI   BamHI
```

**Figure 1** Circular map (a) and sequences (b) of the pG6 family of vectors for phage display of cDNA-encoded products. The vectors are based on pUC119 into which gene *VI* of bacteriophage M13 has been inserted. The set of pG6 vectors allows cloning of cDNAs as *Sfi*I–*Not*I fragments in each reading frame downstream of gene *VI*. Figure reproduced with the permission of Nature America, Inc.

ribosome encounters the natural stop codon. The construction of randomly primed cDNA libraries might be more appropriate for interaction cloning in view of the modular assembly of many proteins. Indeed, functional domains rather than full-length proteins, especially of cytoplasmic origin, may be more amenable for nontoxic, efficient expression as pVI fusions. However, cloning of randomly primed cDNA library in these vectors would ideally require the introduction of translational stops in each reading frame after the *Not*I site of each vector. Moreover, only 50% of the transformants would contain a cDNA insert in the correct orientation for expression.

*Protocol 1* describes a PCR-based procedure to prepare adequate amounts of cDNA inserts from pre-made cDNA libraries for the generation of *gVI*–cDNA phagemid libraries. Although *Protocol 1A* describes the amplification of cDNA inserts from a pre-made, directionally cloned *Sfi*I–*Not*I λGT11 cDNA library, this protocol can easily be adapted to amplify the cDNA inserts from other cDNA libraries by using appropriate vector primers containing the restriction sites for the enzymes *Sfi*I and *Not*I. Ten identical PCR reactions are usually set up to establish a *gVI*-cDNA library of considerable size. Although it is difficult to offer specific guidelines, as a rule of thumb, mammalian cDNA libraries consisting of $5 \times 10^5$ to $1 \times 10^6$ independent clones statistically contain at least one copy of every mRNA. An essential aspect is that the template cDNA library should faithfully represent the sequence, size, and complexity of the mRNA population from which it was generated. To increase the fidelity of the DNA synthesis it is recommended to use a DNA polymerase with $3'$–$5'$ exonuclease proofreading activity.

After purification, the amplified DNA is sequentially digested with *Sfi*I and *Not*I (see *Protocol 1B*). To remove small cDNA fragments, the digested cDNA inserts are fractionated through a 0.8% (w/v) agarose gel by electrophoresis (see *Protocol 1C*). These short fragments would otherwise be preferentially inserted and decrease the number of long cDNA clones in the library. The target cDNAs (ranging in size from 500 to 3000 bp) are recovered and concentrated by using commercially available systems to purify DNA from agarose slices.

## Protocol 1

## Preparation of cDNA inserts from premade λGT11 libraries

### Equipment and reagents
### General

- Sterile $H_2O$
- Sterile microcentrifuge tubes
- Sterile pipettes and pipette tips

- Microcentrifuge
- Benchtop centrifuge

*For part A*

- Sterile 0.2 ml thin-walled polypropylene microcentrifuge tubes
- *Pfu* DNA polymerase (Stratagene)
- $10 \times Pfu$ buffer as provided by the supplier of the *Pfu* DNA polymerase
- dNTP mixture (10 mM each)
- λGT11 forward primer (New England Biolabs, #1218; 5′-GGTGGCGACGACCCTG-GAGCCCG-3′)

- λGT11 reverse primer (New England Biolabs, #1222; 5′-TTGACACCAGACCAAC-TGGTAATG-3′)
- Pre-made *Sfi*I–*Not*I λGT11 cDNA library (titer $\geq 10^9$ pfu (plaque forming units)/ml)
- Mineral oil
- Automated thermal cycler

**Protocol 1** continued

### For part B

- QIAquick PCR purification kit (Qiagen)
- Restriction endonucleases and the corresponding $10 \times$ restriction endonuclease buffers

### For part C

- Agarose
- TAE buffer ($50 \times$ stock): 2 M Tris-acetate, 50 mM EDTA (pH 8.0)
- 0.5 mg/ml ethidium bromide ($1000 \times$ stock)
- $6 \times$ gel-loading buffer (0.25% (w/v) bromophenol blue, 0.25% (w/v) xylene cyanol FF, and 30% (v/v) glycerol in water)
- Agarose gel electrophoresis tank
- UV transilluminator
- Sterile 15 ml polypropylene tubes
- Sterile razor blades
- Geneclean (Qbiogene) or Wizard PCR prep DNA purification system (Promega)
- DNA Dipstick kit (Invitrogen)

## A. Amplification of cDNA inserts by PCR

**1** Add the following components to a 0.2-ml microcentrifuge tube on ice (perform 10 reactions)

| | |
|---|---|
| $H_2O$ | 36.5 μl |
| $10 \times Pfu$ buffer | 5.0 μl |
| 10 mM dNTP | 1.5 μl |
| λGT11 forward primer (10 μM) | 2.5 μl |
| λGT11 reverse primer (10 μM) | 2.5 μl |
| Pre-made *Sfi*I–*Not*I λGT11 cDNA library | 1.0 μl |
| *Pfu* DNA polymerase (2.5 U/μl) | 1.0 μl |

**2** Mix and overlay the reaction mixtures with 50 μl of mineral oil.

**3** Denature the template at 95 °C for 2 min and perform 25 PCR cycles as follows: (a) denature at 95 °C for 1 min, (b) anneal at 50 °C for 1 min, and (c) extend at 72 °C for 3 min (lengthen the time for the last extension step to 7 min).

## B. Purification and restriction enzyme digestion of the amplified cDNA inserts

**1** Purify the PCR products obtained in *part A* with the QIAquick PCR purification kit. Use one spin column per two PCR reactions and elute the purified PCR products in 30 μl of $H_2O$.

**2** Pool the eluted PCR products and set up the restriction digest as follows:[a]

| | |
|---|---|
| $H_2O$ 25 μl | |
| Purified PCR product | 150 μl |
| $10 \times$ restriction endonuclease buffer | 20 μl |
| Restriction endonuclease (10 U/μl) | 5 μl |

**3** Digest at the appropriate temperature for 2 h.

## C. Size fractionation and recovery of the amplified and digested cDNA inserts

**1** Gel-purify the restriction enzyme digest on a 0.8% (w/v) agarose/TAE gel (apply a voltage of 1–5 V/cm).

**2** Carefully excise the target region (from 500 to 3000 bp) by using a sterile razor blade and transfer the gel slice to a 15-ml polypropylene tube. Note that an increased running time will result in a larger volume of the target region.

**3** Purify the fragments from the agarose gel using, for example, the Geneclean or Wizard PCR prep DNA purification system. Recover the amplified cDNA inserts in 50 µl of $H_2O$.

**4** Estimate the DNA concentration by using the DNA Dipstick kit.

[a] Since the optimal buffer compositions as well as the optimal incubation temperatures are different for *Sfi*I and *Not*I, it is advised to perform the reactions sequentially with an intermediate purification of the cDNA fragments (see *Protocol 1 B*, step 1–2).

### 3.3 Ligation and transformation

*Protocol 2* details the large-scale ligation of the amplified cDNA inserts into the pG6 set of vectors. To determine the optimal conditions, it is useful to perform test ligations as described in the footnote. To increase the transformation efficiency, clean up of the ligation mixture is recommended (see *Protocol 2*, steps 4–7). Currently, electroporation is the most efficient method for transforming *E. coli* with plasmid DNA. A detailed procedure to transform Top10F′ cells by electroporation is described in *Protocol 3*, which also includes a recommended procedure for harvesting and storing *gVI*–cDNA libraries.

## Protocol 2

# Large-scale ligation and purification of ligated DNA

### Equipment and reagents

As *Protocol 1 (general)* with the following reagents:

- T4 DNA ligase and $10 \times$ T4 DNA ligase buffer (New England Biolabs)
- DNA thermal cycler as incubator at 16 °C and 70 °C
- Phenol equilibrated to pH $\geq 8.0$ with Tris buffer (USB)
- 24/1 (v/v) chloroform/isoamyl alcohol
- 70% ethanol
- Isopropanol
- 5 M $NaClO_4$ (pH unadjusted)

- Ultrafree-MC 30,000 NMWL cellulose microfilter (Millipore)

- Phagemid vectors (pG6A, pG6B, pG6C) and cDNA inserts (see *Protocol 1*) digested with *Sfi*I and *Not*I

## Method

1  Set up the following ligation reaction:[a]
   - vector DNA (pG6A, pG6B, pG6C) (10 µg)  x µl
   - cDNA insert (3–5 µg)  y µl
   - 10 × ligase buffer  40 µl
   - T4 DNA ligase (6 Weiss U/ul)  5 µl
   - H$_2$O to an end volume of  400 µl

2  Incubate overnight at 16 °C.

3  Incubate for 30 min at 70 °C to denature the T4 DNA ligase.

4  Add one volume of phenol and vortex for 45 s. Centrifuge for 10 min in a micro-centrifuge at 14,000g. Transfer the upper aqueous layer to a new microcentrifuge tube.

5  Add one volume of 24/1 (v/v) chloroform/isoamyl alcohol and vortex for 45 s. Centrifuge in a microcentrifuge at 14,000g for 5 min. Transfer the upper aqueous layer to a new microcentrifuge tube.

6  Repeat step 5.

7  (a) Precipitate the DNA by adding 0.1 volume of 5 M NaClO$_4$ and 1 volume of isopropanol. Vortex briefly and spin the sample in a microcentrifuge at 14,000g for 30 min at 4 °C. Carefully discard the supernatant. Wash the pellet twice with 250 µl of 70% ethanol (follow each washing step by a 5-min centrifugation at 14,000g at room temperature). Briefly dry the pellet and resuspend in 40–50 µl of H$_2$O. Store the DNA at − 20 °C.

   (b) Alternatively, use microfiltration and concentration for purification: add the upper aqueous layer onto a microfilter. Spin at 3000g for 15–20 min in a microcentrifuge. Remove and discard the eluate. Add 350 µl H$_2$O and spin again at 3000 g for 15–20 min. Repeat this step three times. Remove the upper part of the microfilter, cut the protruding ring out, invert and place in a fresh microcentrifuge tube. Vortex and spin for 5 s at 14,000g to recover the DNA (in 40–50 µl) at the bottom of the microcentrifuge tube. Store the DNA at − 20 °C.

[a] Optimal conditions for ligation are normally obtained with a total DNA concentration ≥ 30 µg/ml and a vector : cDNA molar ratio ranging from 1 : 2 to 1 : 3. Since it is difficult to quantify the cDNA concentration accurately, it is useful to perform test ligations with a constant amount of vector (e.g. 200 ng) and twofold dilutions of cDNA insert in a final volume of 10 µl (with 1 µl of T4 DNA ligase). After overnight ligation at 16 °C, dilute the ligation reaction fivefold with water and use 2 µl for electroporation of competent Top10F′ cells (see *Protocol 3*). Count the number of transformants on LBAT plates and evaluate the percentage of clones carrying an insert by PCR screening using the primers pG6F (5′-CTCCCGTCTAATGCGCT-TCCCTG-3′) and pG6R (5′-GCTGCAAGGCGATTAAGTTGGGT-3′).

# Protocol 3

## Electroporation and storage of the library in Top10F′ cells

### Equipment and reagents

As *Protocol 1 (general)* with the following reagents:

- Top10F′ electrocompetent cells (Invitrogen)
- Bio-Rad Gene Pulser Plus
- 0.2 cm electroporation cuvettes, prechilled
- pUC19 (New England Biolabs) at 15 ng per μl
- LB: Luria Bertani broth
- LBAT: LB with 50 μg/ml ampicillin and 20 μg/ml tetracycline

- LBAT plates: LBAT supplemented with 15 g agar per litre (small (90 mm diameter) plates and square (243 × 243 mm) plates (Nunc))
- A sterile spreader
- Sterile 15-ml polypropylene culture tubes (Falcon)
- Polypropylene freezing tubes (Corning)
- Sterile glycerol
- Purified ligated DNA (see *Protocol 2*)

### Method

1   Add 2–3 μl of the purified ligation reaction (see *Protocol 2*, max. 0.3 μg DNA) to 80 μl of Top10F′ electrocompetent cells and mix (gently) by flicking. For the control of the electroporation efficiency, use 30 ng (2 μl) of pUC19.

2   Transfer the cell + DNA suspension to a prechilled 0.2 cm cuvette and place on ice for 3 min.

3   Set up the Gene Pulser Plus to give a 2.5-kV pulse, using the 25 μF capacitor and the pulse controller set to 200 Ωs.

4   Place the cuvette, after drying with a paper towel, in the electroporation chamber and apply one pulse (the registered time constant should be 4.5–5.0 ms).

5   Immediately add 1 ml of LB to the cuvette, resuspend the cells, and transfer the cell suspension to a 15-ml polypropylene culture tube.

6   Allow the cells to express the antibiotic resistance marker by shaking for 1 h at 37 °C.

7   Combine all electroporation samples and determine the library size by plating the appropriate dilutions of a 100-μl aliquot on small LBAT plates. The control electroporation with pUC19 should yield approximately $3 \times 10^7$ transformants.

8   Spin the combined electroporation samples at 2500g for 15 min at 20 °C. Resuspend the pellet in 0.5 × the starting volume of LBAT. Spread 1 ml of this cell suspension per square LBAT plate.

9   Incubate the plates overnight at 37 °C.

10  Harvest the library by pouring 10 ml of LBAT on each square plate and scraping the cells with a sterile spreader. Transfer the suspended cells to a 50 ml polypropylene tube.

11 Repeat step 10 with 5 ml of LBAT to remove the remaining cells from the plates.

12 Measure the optical density (OD) of the cell suspension at 600 nm. If needed, dilute the concentrated cell stock to 40 $OD_{600\,nm}$ units/ml with LBAT.

13 To prepare glycerol stocks, add 5 ml of sterile glycerol to 5 ml of the concentrated cell stock, mix well and divide into 1-ml aliquots in polypropylene freezing tubes. Snap-freeze and store at $-70\,°C$. The final cell concentration will be 20 $OD_{600}$ units/ml, which corresponds to approximately $1.6 \times 10^{10}$ cells/ml.

## 3.4 Evaluation and storage of the *gVI*-cDNA library

To ensure that a negative result upon screening is not the result of a technical error during the construction of the library, it is necessary to verify the quality of the library. Therefore, it is recommended to determine the number of independent library clones (*Protocol 3*, step 7), the percentage of clones with insert, the average insert size, and the insert size range. A more rigorous quality control is to randomly pick 10 clones, sequence the 5'- and 3'-ends of each cDNA insert and analyze the obtained sequences by BLASTN (24) (see Section 5.3.1.) database searches.

# 4 Affinity selection of *gVI*–cDNA fusion phage

Each round of affinity selection comprises the rescue and purification of phage-mid particles from the primary or enriched *gVI*–cDNA library (see *Protocol 4*), and the panning procedure itself (see *Protocols 5-6*). Two different approaches for eluting bound phage are described. *Protocol 5* provides a procedure to elute phage-bait complexes from streptavidin-coated beads by reducing the disulfide bond between the biotin moiety and the bait. In *Protocol 6*, a harsher but less specific phage elution procedure using a low pH buffer is described. After each round of panning, the selected library is plated out to estimate the number of selected phage.

## 4.1 Rescue and purification of the phagemid libraries

At the beginning of each panning cycle, (enriched) Top10F' cells have to be superinfected with helper phage to generate and to purify *gVI*–cDNA phagemid particles to perform a subsequent panning. *Protocol 4* details these procedures. In panning procedures where no special buffers are required and the affinity of the phage for the bait is high (e.g. for immunoaffinity selections), it is not always necessary to purify the phage particles (see *Protocol 4B*) obtained by superinfection.

## Protocol 4

# Rescue and purification of *gVI–cDNA* phagemid libraries

### Equipment and reagents

As *Protocol 1 (general)* with the following reagents:

#### For part A

- LBAT (see *Protocol 3*)
- Sterile 100 ml and 1 L shake flasks
- R408 helper phage (Stratagene)
- Glycerol stock of the primary or enriched *gVI–cDNA* library in Top10F′ cells (see *Protocol 3*)

- Shaking incubator
- Spectrophotometer

#### For part B

- Sterile 50 ml polypropylene tubes (Falcon)
- Sterile glycerol
- Phosphate buffered saline (PBS): 50 mM sodium phosphate, 140 mM NaCl, pH 7.4. Sterilize by autoclaving

- Polyethylene glycol (PEG)/NaCl: 20% (w/v) polyethylene glycol 6000, 2.5 M NaCl, and sterile filtered
- Spectrophotometer
- 0.45 μm filter units (Corning; cat. number 431223)

### A. Rescue of phagemid libraries

1 Add 50 μl of the glycerol stock ($8 \times 10^8$ cells) to 20 ml of LBAT in a 100-ml shake flask and grow at 37 °C with shaking (260 rpm) for 2–3 h (or when $OD_{600\,nm}$ reaches 0.5–0.7).

2 Add the R408 helper phage at a multiplicity of infection of 10 : 1.

3 Incubate the cells for 30 min without shaking at 37 °C.

4 Transfer the infected cells to 100 ml of prewarmed LBAT in a 1L flask.

5 Grow overnight with shaking (260 rpm) at 30 °C.

### B. Purification of phagemid libraries

1 Transfer the cultures to 50 ml polypropylene tubes and centrifuge at 3500g for 30 min at 4 °C.

2 Filtrate the supernatants through a 0.45 μm filter (use fresh 50-ml polypropylene tubes) and add 1/5 volume of PEG/NaCl.

3 Mix well and incubate for 1 h on ice.

4 Centrifuge at 3500g for 20 min at 4 °C. Discard the supernatant and centrifuge again at 4000g for 3 min at 4 °C. Remove as much of the supernatant as possible.

5 Resuspend each pellet in 0.5 ml of PBS supplemented with 5% (v/v) sterile glycerol, combine and transfer to a fresh microcentrifuge tube.

6 Spin at 14,000g in a microcentrifuge to remove cell debris and transfer the supernatant to a fresh microcentrifuge tube.

7 Measure phage titer by determining absorbance at 260 nm ($A_{260}$). One absorbance unit corresponds to $4.4 \times 10^{13}$ Transducing Units (TU)/ml.

8 For storage, divide the phage stock into 250-µl aliquots in fresh microcentrifuge tubes. Snap-freeze and store at $-20\,°C$.

## 4.2 Preparation of biotinylated bait

One way to standardize panning with different ligands is to modify the baits with a common tag such as biotin and to use streptavidin-coated enzyme-linked immunosorbent assay (ELISA) plate wells or beads for panning. Two useful (amine-reactive) biotinylation reagents are biotin disulfide N-hydroxysuccinimide ester and biotinamidocaproate N-hydroxysuccinimide (Sigma). In contrast to the latter reagent, the former reagent introduces a cleavable disulfide bond between the biotin moiety and the bait, and is the preferred reagent for specific elution (with 50 mM dithiothreitol) of bound phage during biopanning (see *Protocol 5*). The level of biotin incorporated can be evaluated with the HABA method (24).

## 4.3 Panning procedure with biotinylated bait

*Protocol 5* describes a standard procedure to affinity-purify *gVI*–cDNA phagemid particles using biotinylated bait. In this protocol, nonspecific streptavidin-binding phage particles are first removed by preclearing the phage particle stock with streptavidin-coated magnetic beads (see *Protocol 5*, step 2). After the third round of panning, individual clones can be analyzed by ELISA (see *Protocol 8*) or additional selection rounds can be performed. However, note that successful display requires that the pVI fusion proteins translocate to the periplasm, fold correctly and incorporate into the phage particles. As a result, binders expressing pVI-fusion proteins interfering with these processes may be diluted out in the pool of "noninterfering" binders as selection progresses.

## 4.4 Immunoaffinity panning procedure

In *Protocol 6* a procedure is described for the isolation of *gVI*–cDNA phagemid particles by using antisera. Only minute amounts (in the µl range) of the antisera are required.

## Protocol 5

# Selection of *gVI*–cDNA phage libraries using biotinylated antigen

## Equipment and reagents

As *Protocol 1 (general)* with the following additional reagents:

- Streptavidin-coated magnetic beads and Magnetic Particle concentrator (Dynal)
- PBS (see *Protocol 4B*)
- PBS2M: PBS supplemented with 2% (w/v) skimmed milk powder (Difco)
- PBS4M: PBS with 4% (w/v) skimmed milk powder (Difco)
- PBST: PBS supplemented with 0.1% (v/v) Tween-20
- Dithiothreitol (DTT)
- Sterile 0.1 M glycine-HCl, pH 2.2
- Sterile 1 M Tris-HCl, pH 8.0

- End-over-end rotator
- Plating Top10F' cells
- Sterile 15-ml polypropylene culture tubes (Falcon)
- Incubator at 37 °C
- LBAT (see *Protocol 3*)
- LBAT plates (see *Protocol 3*)
- Sterile 250-ml shake flask
- *gVI*-cDNA phagemid library (see *Protocol 4*)
- Biotinylated bait

## Method

1. Resuspend 2 mg streptavidin beads (200 µl) in 2 ml of PBS2M blocking buffer and incubate for 1 h at room temperature on a rotator. Rinse the beads 3 times with PBST.

2. Mix $2.5 \times 10^{12}$ TU of the cDNA-phage library (about 250 µl of the phage stock prepared according to *Protocol 4*) with 1 mg blocked beads resuspended in 250 µl of PBS4M. Incubate for 1 h at room temperature on a rotator. This step should remove most, if not all, streptavidin-binding phage particles.

3. Pellet the beads using the concentrator and transfer the supernatant to a fresh microcentrifuge tube. Add biotinylated bait (final concentration 250 nM) and incubate for 3 h at 4 °C on a rotator.

4. Add 1 mg blocked beads (prepared in step 1) to the phage + bait mix and incubate for 30 min at 4 °C on a rotator.

5. Pellet the beads using the concentrator, and rinse the beads 10 times with PBST, using a short spin in a microcentrifuge followed by the concentrator to keep the beads at the bottom of the microcentrifuge tube.

6. (a) Elute the phage particles by resuspending the beads in 0.5 ml of PBS with 50 mM DTT.[a] Incubate for 10 min at room temperature, then pellet the beads with the concentrator and transfer the supernatant to a fresh microcentrifuge tube. Keep the solution on ice.
   (b) Alternatively: resuspend the beads in 400 µl of 0.1 M glycine-HCl, pH 2.2. Incubate for 10 min at room temperature, then pellet the beads with the concentrator and transfer the supernatant to a fresh microcentrifuge tube. Neutralize the solution by adding 100 µl of 1 M Tris-HCl, pH 8.0. Keep the solution on ice.

7  Add half of the eluted phage[b] (250 µl) to 10 ml of plating Top10F′ cells.[c] Incubate for 30 min at 37 °C.

8  Take an aliquot of the infected culture (100 µl), perform serial dilutions and plate them out on LBAT agar plates. Grow overnight at 37 °C and count the colonies.

9  Spin the remaining of the infected culture at 2500g for 10 min at 20 °C. Remove the supernatant and resuspend the cell pellet in 2 ml LBAT. Spread 1 ml of the cell suspension per LBAT square plate. Incubate the plates overnight at 37 °C.

10  Harvest and store the selected clones as described in *Protocol 3*, steps 10–13. If a further round of selection is required, inoculate 20 ml of LBAT in a 100-ml flask with 1 $OD_{600}$ cells ($8 \times 10^8$ cells) and proceed with the rescue and purification of phagemid particles (see *Protocol 4*).

[a] This protocol is suitable for elution of bound phage when there is an interceding disulfide bound between the biotin moiety and the bait molecule.

[b] The remaining 250 µl of phage can be stored at 4 °C to repeat the infection and amplification steps if needed.

[c] For the preparation of plating Top10F′ cells, inoculate 50 ml of LB supplemented with 20 µg/ml tetracycline in a 250-ml shake flask with a single Top10F′ colony (grown on a LB agar plate with 20 µg/ml tetracycline), and grow for 6 h at 37 °C. Chill the bacterial culture to 4 °C before use. Plating Top10F′ cells can be stored for periods up to 5 days at 4 °C.

## Protocol 6

# Selection of *gVI*–cDNA phage libraries by immunoaffinity

### Equipment and reagents

As *Protocol 1 (general)* with the following additional reagents:

- PBST (see *Protocol 5*)
- Bovine serum albumin (BSA) (Sigma)
- Antiserum and *gVI*–cDNA phage library
- Sterile 15-ml polypropylene tubes (Falcon)
- End-over-end rotator
- Protein A-Sepharose beads (Amersham Biosciences)
- Sterile 0.1 M glycine-HCl, pH 2.2 with 0.1% (w/v) BSA

- Sterile 2 M Tris base (pH unadjusted)
- Plating Top10F′ (see *Protocol 5, footnote c*)
- Incubator at 37 °C
- LB (see *Protocol 3*)
- LBAT plates (see *Protocol 3*)
- Sterile 16- or 18-mm-diameter culture tubes
- Shaking incubator at 37 °C

### Method

1  Add $10^{11}$–$10^{12}$ TU of the *gVI*–cDNA phage library to a 15-ml polypropylene tube containing 2.5 µl of antiserum diluted in 10 ml of PBST with 1% (w/v) BSA.

2  Incubate this mixture with gentle rotation for 1 h at room temperature.

3  Add 50 μl of a 1/1 (v/v) slurry of Protein A-Sepharose beads (preblocked with 5% (w/v) BSA in PBST) to this mixture and incubate for an additional 1 h at room temperature on a rotator.

4  Collect the protein A-Sepharose beads by centrifugation (500g for 5 min) and wash the beads 10 times with 10 ml of PBST.

5  Resuspend the beads in 1 ml of PBST, transfer this mixture to a microcentrifuge tube and spin the beads down (10,000g for 10 s). Discard the supernatant.

6  Resuspend the protein A-Sepharose beads in 0.5 ml of 0.1 M glycine-HCl, pH 2.2 with 0.1% (w/v) BSA.

7  Incubate the samples for 10 min at room temperature on a rotator.

8  Briefly centrifuge the samples (10,000g for 10 s) and transfer the supernatant to a fresh microcentrifuge tube.

9  Neutralize the solution by adding 30 μl of 2 M Tris. Keep the solution on ice.

10  For phage propagation, add half of the eluted phage (265 μl) to 10 ml of plating Top10F' cells (see *Protocol 5, footnote c*). Incubate for 30 min at 37 °C, and proceed as in *Protocol 5*, steps 8–10.

## 5  Analysis of selected clones

An essential step in the whole panning procedure is to select only those clones that specifically interact with the bait. Measures to eliminate false-positive clones depend on the nature of the bait used. Binders can be distinguished from non-binders in a phage-ELISA assay (see *Protocol 8*). To discriminate specific binders from nonspecific binders, suitable control baits (e.g. pre-immune serum) should be tested alongside with the panning baits. To reduce the possibility of analyzing duplicated clones, it is advisable to determine the insert size of interesting clones by PCR using the primers pG6F and pG6R (see *Protocol 2*, footnote a).

## Protocol 7

## Small-scale rescue and purification of *gVI*-cDNA phagemid particles

### Equipment and reagents

As *Protocol 1 (general)* with the following additional reagents:

- Sterile 96 well culture plate with lid (Costar)
- LBAT (see *Protocol 3*)
- R408 helper phage (Stratagene)
- Shaking incubator
- 96-prong replicaplater
- PEG/NaCl (see *Protocol 4B*)
- PBS (see *Protocol 4B*)

Protocol 7 continued

## Method

1  Inoculate colonies into 100 µl of LBAT in a 96-well culture plate and grow overnight at 37 °C with shaking (180 rpm).

2  Using a 96-prong replicaplater, inoculate a second 96-well culture plate wherein wells are filled with 200 µl of LBAT. Grow at 37 °C until mid-log phase (approximately 1.5 h).[a]

3  To each well, add 10 µl of a $2 \times 10^{11}$ TU/ml R408 helper phage stock.

4  Place the culture plates for 30 min at 37 °C without shaking.

5  Grow overnight at 37 °C[b] with shaking.

6  Collect the cells by centrifugation at 800g for 20 min at 4 °C.

7  Transfer the supernatant to a new 96-well culture plate and add 40 µl of PEG/NaCl.

8  Leave the plate on ice for 1 h.

9  Collect the phagemid particles by centrifugation at 800 g for 20 min at 4 °C. Carefully discard the supernatant and resuspend the pellet in 50 µl of PBS.

[a] The first 96-well culture plate can be used for long-term storage of the clones by adding sterile glycerol to a final concentration of 15% (v/v) and storing the plate at − 80 °C.

[b] Lowering the overnight incubation temperature to 30 °C may improve the expression level of functional cDNA-encoded proteins on the phage coat without decreasing the yield of the phagemid particles.

### 5.1  Phage ELISA

To perform a phage ELISA (see *Protocol 8*), the phagemid particles have first to be rescued from the isolated bacterial Top10F′ clones. In *Protocol 7* a small-scale rescue and purification procedure of phagemid particles is described. Depending on the use of specific buffers or the affinity of the phage for the bait, the precipitation step (steps 7–9) can be omitted. Phage-ELISA screening is generally not recommended before the third round of panning. The percentage of positive clones is in most cases too low.

# Protocol 8

# Phage ELISA

## Equipment and reagents

As *Protocol 1 (general)* with the following additional reagents:

- Streptavidin-coated microtiter plates (Pierce)
- PBST and PBS2M (see *Protocol 5*)
- Biotinylated antigen selected (BAS) phages (see *Protocol 5*) and/or immuno affinity selected (IAS) phages (see *Protocol 6*)

- Biotinylated ligand (for BAS phages only)
- Biotinylated capture antibody (Roche Molecular Biochemicals), bait antiserum and pre-immune serum (for IAS phages only)
- Horse radish peroxidase (HRP)/anti-M13 conjugate (Amersham Biosciences)
- 3,3', 5,5'-tetramethyl benzidine (TMB) substrate kit (Pierce)
- 2 M $H_2SO_4$

## Method

1  Rinse each well of the strepavidin-coated microtiter plate 3 times with 200 μl of PBST.

2  Dilute the biotinylated ligand (BAS phage-ELISA) or the biotinylated capture antibody (IAS phage-ELISA) in PBS2M to a final concentration of 10–50 nM. Add 100 μl of this solution to each microtiter plate well.

3  Incubate the microtiter plate for 1 h at room temperature.

4  Rinse each well 5 times with 200 μl of PBST.

5  Dilute the phage solutions in PBS2M to a final concentration of $1 \times 10^{11}$ TU/ml. Supplement the PBS2M used for the dilution of the IAS fusion phages first with bait antiserum or with pre-immune serum.[a] Transfer 100 μl of this suspension to each well.

6  Incubate the microtiter plate for 2 h at room temperature.

7  Rinse each well 5 times with 200 μl of PBST.

8  Add 100 μl of the HRP/anti-M13 conjugate (1 : 5000 diluted in PBS2M).

9  Incubate the microtiter plate for 1 h at room temperature.

10  Rinse each well 5 times with 200 μl of PBST.

11  Add 100 μl of the TMB substrate solution to each well.

12  Incubate the microtiter plate for 15–30 min (or until the desired color develops)[b] at room temperature.

13  Stop the reaction with 100 μl of 2 M $H_2SO_4$.

14  Read the absorbance at 450 nm.

[a] To eliminate nonspecific phages, it is strongly recommended to use the bait antiserum side by side with the corresponding pre-immune serum. Most antisera can, depending on their titer, be diluted 500- to 10,000-fold.

[b] The color changes from clear to brilliant blue. A high HRP concentration will yield a greener solution. The reaction should be stopped before any wells display a green color.

## 5.2 Sequence analysis and full-length cloning of the selected cDNAs

The ultimate goal of each biopanning experiment is to infer the function of the protein encoded by the affinity-selected cDNA insert and to obtain the corresponding full-length cDNA. To reduce the number of duplicated clones that will be analyzed, it is highly recommended to determine first the cDNA insert size of the isolated gVI-cDNA phagemid clones. The gVI-cDNA phagemid vectors from distinct positive Top10F' clones can be isolated by using standard plasmid isolation procedures and the corresponding cDNA inserts can be sequenced by using standard methods. Information on the selected cDNAs can then be obtained by scanning public databases.

### 5.2.1 *In silico* biology

Most researchers use the basic local alignment search tool (BLAST) family of programs developed by the National Center for Biotechnology Information (NCBI) (25) (http://www.ncbi.nlm.nih.gov/BLAST/). Because the reading frame of the cDNA insert can be determined, it is recommended to search the predicted protein sequence against one of the protein databases by using BLASTP. Sequences already listed in one of the DNA or protein databases can provide information about the function or phenotype of the corresponding proteins. Because in most cases the obtained cDNA insert does not represent a full-length cDNA, the information available in databases such as the expressed sequence tags (ESTs) database (dbEST) and its minimally redundant version (Unigene) can frequently be used to compose the full-length cDNAs. This *in silico* cloning approach is made easier by the Web tool ESTBlast (http://www.hgmp.mrc. ac.uk/ESTBlast/).

### 5.2.2 Cloning full-length cDNAs

Many of the ESTs that exist in the dbEST database are derived from clones from the Integrated Molecular Analysis of Genome Expression (IMAGE) Consortium or The Institute for Genomic Research (TIGR). Most of these clones are available from a number of distributors (e.g. the American Type Culture Collection, the UK HGMP Resource Center, the Resource Center/Primary Database). Therefore, the easiest way to obtain the *in silico* cloned full-length cDNA is often to order the corresponding clone from one of the distributors. In cases in which the full-length cDNA sequence cannot be assembled using the *in silico* approach, the quickest way to obtain more sequence information is to perform 5'- and/ or 3'-Rapid Amplification of cDNA Ends (RACE). RACE is a PCR-based technique to amplify DNA sequences from a messenger RNA template between a defined internal site and unknown sequences of either the 3'- or the 5'-end of the mRNA. A number of companies offer convenient RACE- kits to generate full-length cDNAs which, in a final step, can be subcloned by using a PCR cloning kit.

## 6 Applications

To date, five groups (10, 26–29) have reported the construction and successful use of *gVI*-cDNA fusions for expression cloning. Jespers *et al.* (10) isolated cDNAs encoding protease inhibitors from a *Ancylostoma caninum* hookworm cDNA library by biopanning on trypsin and factor Xa, and a collagen-binding protein was identified by Viaene *et al.* (26) using a cDNA library derived from the related *Necator Americanus* hookworm. Fransen *et al.* (27) reported the cloning of four different cDNAs coding for peroxisomal enzymes by affinity selection using immobilized antisera directed against peroxisomal subfractions. The same selection approach (using either a specific monoclonal antibody or a polyclonal antiserum) was also perfomed for functional cloning of cDNAs encoding β2-microglobulin and constant regions from human immunoglobulins from a *gVI*-colorectal cancer cDNA library (28), and cDNAs encoding various antigens (such as p53, centromere-F, int-2, and pentraxin I) targeted by serum antibodies from patients with breast cancer (29). Taken together, these results indicate that cDNAs encoding for protein fragments ranging from 28 to 252 amino acids can be selected from *gVI*–cDNA libraries comprising about $1 \times 10^6$ clones. In contrast to the pJuFo vector, *gVI*-based vectors appear to be less prone to deletions in later rounds of selection (28).

Since the export pathway of pVI to the periplasm is not precisely known (pVI lacks a signal sequence), it is unclear whether the *gVI*–cDNA fusion system may introduce a bias between secreted and cytoplasmic proteins for efficient display on phage.

A successful selection heavily relies on the affinity of the specific cDNA-encoded product for the bait. In the above mentioned examples, strong interactions ($K_D$ in nM range) involving inhibitor-protease or antibody–antigen complexes yielded successful selections. Valency of display is also a key factor. The pG6 phagemid vectors are designed for "monovalent display" on phage, and model experiments have shown that the display levels of various molecules such as a protease inhibitor, the human $CH_3$ domain and alkaline phosphatase are systematically lower (by a factor ranging from 5- to 160-fold) when fused to pVI than when fused either directly or via the c-Jun/c-Fos heterodimer to pIII (10, 28). As a result, the threshold of affinity-selectable pVI-cDNA fusions might be higher than that of other cDNA display systems based on pIII (or pVIII) fusions. Overall, phage-based systems for expression cloning of cDNA are inherently less sensitive (but more specific) than the two-hybrid system in which the compartmentalization of matching partners within a cell greatly reduces the affinity threshold for reconstitution of a functional transcriptional activator (for review, see (30)). The use of one system instead of another should be decided case by case depending on parameters such as sensitivity ($K_D$ in nM or μM range), selectivity (size of the library, unique, or multiple cDNA candidates), protein localization (cytoplasm, membrane, periplasm, or secreted), and binding conditions (buffer, pH, temperature, metal ions, detergents) for selection.

## Acknowledgments

The work done in Leuven is supported by a "Geconcerteerde Onderzoeksacties" grant from the Flemish government.

## References

1. Helfman, D. M., Fiddes, J. R., Thomas, G. P., and Hughes, S. (1983). *Proc. Natl. Acad. Sci. USA*, **80**, 31.
2. Young, R. A. and Davis, R. W. (1983). *Proc. Natl. Acad. Sci. USA*, **80**, 1194.
3. Fields, S. and Song, O. K. (1989). *Nature*, **340**, 245.
4. Seed, B. and Aruffo, A. (1987). *Proc. Natl. Acad. Sci. USA*, **84**, 3365.
5. Clackson, T. and Wells, J. (1994). *Trends Biotechnol.*, **12**, 173.
6. Smith, G. P. and Petrenko, V. A. (1997). *Chem. Rev.*, **97**, 391.
7. Crameri, R., Jaussi, R., Menz, G., and Blazer, K. (1994). *Eur. J. Biochem.*, **226**, 53.
8. Gramatikoff, K., Georgiev, O., and Schaffner, W. (1994). *Nucl. Acids Res.*, **22**, 5761.
9. Makowski, L. (1992). *J. Mol. Biol.*, **228**, 885.
10. Jespers, L. S., Messens, J. H., De Keyser, A., Eeckhout, D., Van Den Brande, I., Gansemans, Y. G. *et al.* (1995). *Biotechnology*, **13**, 378.
11. Endemann, H. and Model, P. (1995). *J. Mol. Biol.*, **250**, 496.
12. Rakonjac, J. and Model, P. (1998). *J. Mol. Biol.*, **282**, 25.
13. Fuh, G. and Sidhu, S. S. (2000). *FEBS Lett.*, **480**, 231.
14. Fuh, G., Pisabarro, M. T., Li, Y., Quan, C., Lasky, L. A., and Sidhu, S. S. (2000). *J. Biol. Chem.*, **275**, 21486.
15. Mikawa, G. Y., Maruyama, I. N., and Brenner, S. (1996). *J. Mol. Biol.*, **262**, 21.
16. Maruyama, I. N., Marayuma, H. I., and Brenner, S. (1994). *Proc. Natl. Acad. Sci. USA*, **91**, 8273.
17. Santini, C., Brennan, D., Mennuni, C., Hoess, R. H., Nicosia, A., Cortese, R. *et al.* (1998). *J. Mol. Biol.*, **282**, 125.
18. Santi, E., Capone, S., Mennuni, C., Lahm, A., Tramontano, A., and Nicosia, A. (2000). *J. Mol. Biol.*, **296**, 497.
19. Efimov, V. P., Nepluev, I. V., and Mesyanzhinov, V. V. (1995). *Virus Genes*, **10**, 173.
20. Ren, Z. J., Lewis, G. K., Wingfield, P. T., Locke, E. G., Steven, A. C., and Black, L. W. (1996). *Protein Sci.*, **5**, 1.
21. Yamamoto, M., Kominato, Y., and Yamamoto, F. (1999). *Biochem. Biophys. Res. Commun.*, **255**, 194.
22. Zozulya, S., Lioubin, M., Hill, R. J., Abram, C., and Gishizky, M. J. (1999). *Nat. Biotechnol.*, **17**, 1193.
23. Sche, P. P., McKenzie, K. M., White, J. D., and Austin, D. J. (1999). *Chem. Biol.*, **6**, 707.
24. Green, N. M. (1965). *Proc. Natl. Acad. Sci. USA*, **80**, 4045.
25. Altschul, S. F., Gish, W., Miller, W., Myers, E. W., and Lipman, D. J. (1990). *J. Mol. Biol.*, **215**, 403.
26. Viaene, A., Crab, A., Meiring, M., Pritchard, D., and Deckmyn, H. (2001). *J. Parasitol.*, **87**, 619.
27. Fransen, M., Van Veldhoven, P. P., and Subramani, S. (1999). *Biochem. J.*, **340**, 561.
28. Hufton, S. E., Moekerk, P. T., Meulemans, E., de Bruine, A., Arends, J.-W., and Hoogenboom, H. R. (1999). *J. Immunol. Methods*, **231**, 39.
29. Sioud, M. and Hansen, M. (2001). *Eur. J. Immunol.*, **31**, 716.
30. Allen, J. B., Walberg, M. W., Edwards, M. C., and Elledge, S. J. (1995). *Trends Biotechnol.*, **20**, 511.

# Chapter 13

# Phage antibody libraries

## Andrew R. M. Bradbury

Biosciences Division, Los Alamos National Laboratory, Los Alamos,
New Mexico, USA and International School for Advanced
Studies (SISSA), Via Beirut 2-4, Trieste 34012, Italy.

## James D. Marks

Departments of Anesthesia and Pharmaceutical Chemistry,
University of California, San Francisco, Rm 3C-38, San Francisco
General Hospital, 1001 Potrero, San Francisco, CA 94110, USA.

## 1 Introduction

The use of phage display to generate antibodies recognizing specific antigens is
one of the more successful uses of the technology. This application requires a
number of crucial steps common to all molecular diversity technologies: the
creation of diversity, coupling of phenotype (antibody protein) to genotype (antibody gene), selection, amplification, and analysis. Antibodies with affinities com-
parable to those obtained using traditional hybridoma technology can be
selected from either large naïve antibody libraries or from immune phage antibody libraries. The affinity of initial isolates can be further increased, to levels
unobtainable in the immune system, by using the selected antibodies as the basis
for construction and selection of libraries where the antibody gene is diversified
(see Chapter 14). In this chapter, we will discuss the issues in making phagemid
based single chain Fv (scFv) libraries on phage coat protein pIII and selecting and
characterizing antigen specific antibodies. Many of the protocols, however, can
be modified for alternative formats.

## 2 Phage antibody library construction

In general two kinds of libraries can be constructed: immune or naive. Immune
libraries have been made from immunoglobulin (Ig) variable region (V) genes
derived from either immunized humans (1–3) or mice (4–6) and contain V-genes
heavily biased toward antibodies recognizing the immunogen. As a result, the
affinity of the antibodies isolated is far higher than that which can be isolated
from a naïve library of the same size. Such libraries have also been used to select
antibodies against antigens which were not used in the immunization (2). Naïve

libraries are intended to be unbiased, and so antibodies can be selected against any antigen. They have been derived from either natural unimmunized human rearranged V-genes (7–11), or synthetic human V-genes (12–15). In general, the affinity of the antibodies selected is proportional to the size of the library, with $K_d$s ranging from $10^{-6}$ to $10^{-7}$ M for smaller libraries (8, 12) to $10^{-9}$ M for larger ones (7, 9, 11, 13, 15). This finding is consistent with theoretical considerations (16).

A comparison of antibody libraries made from either natural or synthetic V-genes reveals that libraries made from natural V-genes tend to yield antibodies of higher affinities than those obtained from similarly sized libraries made from synthetic V-genes. This may be due to the use of complementarity determining region (CDR) residues in synthetic libraries which are not tolerated by the antibody structure, or from the creation of structures which are not able to bind at high affinity. The use of polymerase chain reaction (PCR) to create synthetic V-regions in this way can also lead to frameshifts and stop codons in the region of variability. This can be overcome by the use of trinucleotides instead of single bases to create the oligonucleotides (17), although this technology is expensive and not yet generally available. The problem of poorly folding synthetic scFv can be overcome by preselecting both $V_H$ and $V_L$ genes for correct folding with a ligand such as protein A or protein L which recognizes conformational V-region epitopes (I. M. Tomlinson, unpublished data), although the library is then limited to $V_H$ and $V_L$ genes known to bind these proteins. One advantage of the use of synthetic V-regions is that V-genes can be chosen or engineered which are known to express well in bacteria, although single amino acid changes within CDRs can cause great differences in bacterial expression levels. One library has been made using synthetic trinucleotide techniques which allow synthetic sequences to be inserted into any of the CDRs using flanking restriction sites (15): these may be particularly useful for subsequent affinity maturation of the antibody (see Chapter 14).

Although in theory there are many different choices to be made when constructing a library, most published work has tended to follow a tried and tested formula. In particular, filamentous phage based phagemid vectors have been used almost exclusively, with pIII being the display protein used. Most libraries have been made using the scFv (18) format, although a few have been made in the Fab format (10, 13). Fabs consist of two chains, the $V_H$–$C_H$1 and the $V_L$–$C_L$, which need to assemble. scFv fragments on the other hand are single proteins with the two V-domains joined covalently by a flexible linker (19). These formats are illustrated at the protein level in *Figure 1* and at the genetic level in *Figure 2*. In general, Fab fragments are more difficult to assemble on phage, more likely to be degraded, have lower yields as soluble antibody fragments, and are more likely to cause problems of DNA instability in the phage(mid) due to the larger size of the encoding DNA. All these problems arise from the fact that Fabs contain two protein chains which need to assemble, each one of which is the same size as the single scFv chain. However, Fabs do not appear to suffer from the problems of dimerization or aggregation which afflict scFv fragments, and one published naïve Fab library (10) appears to be as effective as many of the published scFv

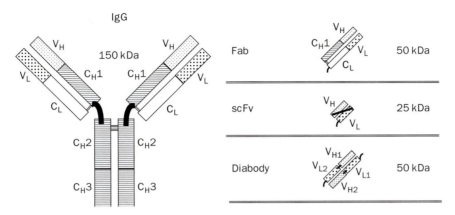

**Figure 1** Antibody structure and recombinant forms used for phage display. The single chain Fv (scFv), Fab, and diabody contain the antigen binding $V_H$ and $V_L$ domains from immunoglobulin G (IgG) and can be displayed on phage. In the case of the Fab and the diabody, one of the chains is secreted as a pIII fusion and the other is secreted as native non-fusion protein. Thick lines indicate peptide connections, and thin double lines indicate disulfide bonds. In the case of the diabody, the $V_H$ and $V_L$ encoding a single specificity are found on different chains, but come together at the protein level, as shown.

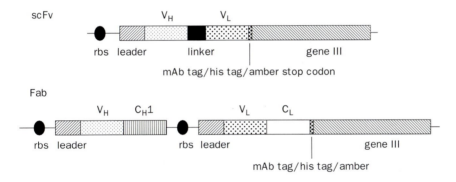

**Figure 2** Genetic structure of displayed recombinant antibodies. The genetic structures of scFv and Fab fragments are shown. The Fab fragment is expressed from a bicistronic construct, with one chain secreted into the periplasm and the other chain expressed as a pIII fusion. rbs, ribosome binding site; leader, secretion signal; linker, scFv linker; mAb tag, epitope tag recognized by monoclonal antibody for detection; his tag, hexahistidine tag for purification.

libraries. One of the major problems with scFv is their tendency to form dimers, in which the $V_H$ of one scFv interacts with the $V_L$ of another, this problem being aggravated by shorter linkers. The formation of such dimers has been exploited to create dimeric antibodies (termed diabodies) in which each member of the dimer expresses a different antibody specificity (20), a concept which has been extended to the creation of a phage antibody library in which each phage carries two antibody specificities (21) (see *Figure 1*).

245

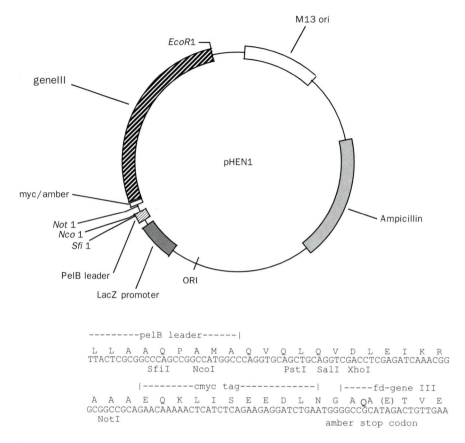

```
---------pelB leader------|
  L  L  A  A  Q  P  A  M  A  Q  V  Q  L  Q  V  D  L  E  I  K  R
TTACTCGCGGCCCAGCCGGCCATGGCCCAGGTGCAGCTGCAGGTCGACCTCGAGATCAAACGG
        SfiI     NcoI                    PstI  SalI XhoI

    |---------cmyc tag-------------|    |-----fd-gene III
  A  A  A  E  Q  K  L  I  S  E  E  D  L  N  G  A  A (E) T  V  E
GCGGCCGCAGAACAAAAACTCATCTCAGAAGAGGATCTGAATGGGGCCGCATAGACTGTTGAA
  NotI                                          amber stop codon
```

**Figure 3** The pHEN1 phage display vector. In this early phage display vector (8), the pelB leader directs secretion of scFv-pIII fusions to the periplasm. scFv are cloned into the vector as *Sfi*1-*Not*1 fragments. The c-myc epitope tag is used for detection of expressed scFv and the amber codon allows simple switching between displayed scFv and secreted native scFv (see text). The vector polylinker sequence is shown below the vector map.

In this section we will discuss the issues in making phagemid based scFv libraries on pIII. Virtually all antibody libraries, unlike peptide libraries, are constructed in phagemid vectors. One of the first vectors used, pHEN1, is shown in *Figure 3*. Subsequent vector improvements, including the presence of hexahistidine tags and alternative epitope tags, do not fundamentally change the nature of the vector. Phagemid vectors are used for two reasons: first, significantly larger libraries can be constructed due to the higher transformation efficiency of phagemid compared with phage vectors; and second, use of phagemid display results in primarily monovalent display of the antibody fragment (see Chapters 1 and 4), allowing better selection of the antibody based on binding constant. The use of phagemid vectors necessitates rescue of phage by the use of helper phage, which supplies the other phage proteins in trans.

## 2.1 The source of V-region diversity

Natural V-regions can be considered to be those found in lymphocytes (see later for a discussion of sources), which may or may not have undergone antigen stimulation. As a result, some of the V-genes amplified will be rearranged germline genes, whereas others will contain mutations induced by the encounter of the B cell with antigen. Such genes are amplified using primers which recognize the 5′ end of the V-genes and the 3′ end of the J-genes or the 5′ end of the $C_L$ or $C_H1$ domain (*Figure 4*) (22). In general, the majority of V-regions which are amplified from natural sources are functional and do not contain stop codons or frameshifts. This is because the transcription of such nonfunctional V-regions is suppressed in B cells. It should be noted that the same is not true of hybridomas, in which nonfunctional V-regions are often encountered (23).

Synthetic V-regions have usually been made by PCR and involve the addition of random CDR3's and framework region 4 to the 3′ ends of cloned immunoglobulin genes (*Figure 5*) using long oligonucleotide primers (13, 15, 24). In *Figure 5*, CDR3 is indicated as being created by the use of NNS in the amplifying primer. This creates codons for all amino acids with a minimal number of stop codons (see Chapter 2). Such synthetic V-genes are based on natural V-gene segments but have synthetic elements within the CDR3. The CDR3 can be made with a wide variety of lengths, depending upon the oligonucleotides used for amplification.

Many different V-region primer sets have been published for amplification of full-length natural V-genes from humans (8, 25–30) and from mice (31, 32). Those listed in *Table 1* represent a set of human primers used to construct two naïve human phage libraries (8, 11). Those listed in *Table 2* (taken from (30)) have been developed based on the sequences in VBASE (http://www.mrc-cpe.cam.ac.uk/vbase) and used by one of the authors to create a large functional phage antibody library (9). In general, the tendency has been to create primer sets which recognize as many different V-genes as possible. This may not be the best strategy, however, since there is a difference between V-gene families found in randomly picked clones and those in selected antibodies (9, 11). This could be due to differences among V-gene families with respect to display levels

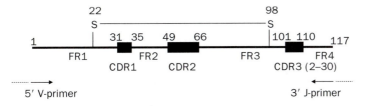

**Figure 4** Amplifying and cloning natural V-genes. Naturally rearranged V-genes are amplified by PCR using primers which anneal to the 5′ and 3′ end of the rearranged V-gene (indicated in bold). Such primers may also have restriction sites appended to their 5′ ends (indicated with a dotted line). The structure of the rearranged V-genes is shown, with disulfide bonds, CDR and framework regions (FR). Numbers indicate the amino acid number at each position. CDR3 (2–30) indicates that CDR3 in natural antibodies can vary in length from 2 to 30 amino acids.

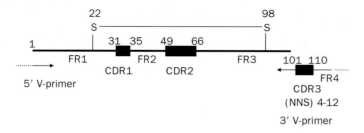

**Figure 5** Amplifying and cloning synthetic V-genes. Synthetic V-genes are constructed from unrearranged V-gene segments, with diversity supplied by degeneracy within an oligonucleotide that encodes the CDR3. In this figure, the CDR3 length is indicated as ranging from 4 to 12 amino acids, provided by an oligonucleotide degeneracy NNS (where N is any nucleotide and S is C or G). Other lengths and forms of oligonucleotide degeneracy are also possible. Other abbreviations are as for *Figure 4*.

**Table 1** Primers for creation of human scFv libraries

*Primers for 1st strand cDNA synthesis*

Human heavy chain constant region primers

| | |
|---|---|
| HuIgG1-4CH1FOR | 5′–GTC CAC CTT GGT GTT GCT GGG CTT–3′ |
| HuIgMFOR | 5′–TGG AAG AGG CAC GTT CTT TTC TTT–3′ |

Human κ constant region primers
HuC$_\kappa$FOR

| | |
|---|---|
| | 5′–AGA CTC TCC CCT GTT GAA GCT CTT–3′ |

Human λ constant region primers
HuC$_\lambda$FOR

| | |
|---|---|
| | 5′–TGA AGA TTC TGT AGG GGC CAC TGT CTT–3′ |

*Primers for primary amplifications of $V_H$, $V_\kappa$, and $V_\lambda$ genes*

Human $V_H$ Back Primers

| | |
|---|---|
| HuVH1aBACK | 5′–CAG GTG CAG CTG GTG CAG TCT GG–3′ |
| HuVH2aBACK | 5′–CAG GTC AAC TTA AGG GAG TCT GG–3′ |
| HuVH3aBACK | 5′–GAG GTG CAG CTG GTG GAG TCT GG–3′ |
| HuVH4aBACK | 5′–CAG GTG CAG CTG CAG GAG TCG GG–3′ |
| HuVH5aBACK | 5′–GAG GTG CAG CTG TTG CAG TCT GC–3′ |
| HuVH6aBACK | 5′–CAG GTA CAG CTG CAG CAG TCA GG–3′ |

Human $V_\kappa$ Back Primers

| | |
|---|---|
| HuVκ1aBACK | 5′–GAC ATC CAG ATG ACC CAG TCT CC–3′ |
| HuVκ2aBACK | 5′–GAT GTT GTG ATG ACT CAG TCT CC–3′ |
| HuVκ3aBACK | 5′–GAA ATT GTG TTG ACG CAG TCT CC–3′ |
| HuVκ4aBACK | 5′–GAC ATC GTG ATG ACC CAG TCT CC–3′ |
| HuVκ5aBACK | 5′–GAA ACG ACA CTC ACG CAG TCT CC–3′ |
| HuVκ6aBACK | 5′–GAA ATT GTG CTG ACT CAG TCT CC–3′ |

Human $V_\lambda$ Back primers

| | |
|---|---|
| Huλ1BACK | 5′–CAG TCT GTG TTG ACG CAG CCG CC–3′ |

**Table 1** (*Continued*)

| | |
|---|---|
| Huλ2BACK | 5′–CAG TCT GCC CTG ACT CAG CCT GC–3′ |
| Huλ3aBACK | 5′–TCC TAT GTG CTG ACT CAG CCA CC–3′ |
| Huλ3bBACK | 5′–TCT TCT GAG CTG ACT CAG GAC CC–3′ |
| Huλ4BACK | 5′–CAC GTT ATA CTG ACT CAA CCG CC–3′ |
| Huλ5BACK | 5′–CAG GCT GTG CTC ACT CAG CCG TC–3′ |
| Huλ6BACK | 5′–AAT TTT ATG CTG ACT CAG CCC CA–3′ |
| Human J_H Forward Primers | |
| HuJH1-2FOR | 5′–TGA GGA GAC GGT GAC CAG GGT GCC–3′ |
| HuJH3FOR | 5′–TGA AGA GAC GGT GAC CAT TGT CCC–3′ |
| HuJH4-5FOR | 5′–TGA GGA GAC GGT GAC CAG GGT TCC–3′ |
| HuJH6FOR | 5′–TGA GGA GAC GGT GAC CGT GGT CCC–3′ |
| Human J_κ Forward Primers | |
| HuJκ1FOR | 5′–ACG TTT GAT TTC CAC CTT GGT CCC–3′ |
| HuJκ2FOR | 5′–ACG TTT GAT CTC CAG CTT GGT CCC–3′ |
| HuJκ3FOR | 5′–ACG TTT GAT ATC CAC TTT GGT CCC–3′ |
| HuJκ4FOR | 5′–ACG TTT GAT CTC CAC CTT GGT CCC–3′ |
| HuJκ5FOR | 5′–ACG TTT AAT CTC CAG TCG TGT CCC–3′ |
| Human J_λ Forward Primers | |
| HuJλ1FOR | 5′–ACC TAG GAC GGT GAC CTT GGT CCC–3′ |
| HuJλ2-3FOR | 5′–ACC TAG GAC GGT CAG CTT GGT CCC–3′ |
| HuJλ4-5FOR | 5′–ACC TAA AAC GGT GAG CTG GGT CCC–3′ |
| *Primers to create scFv linker DNA* | |
| Reverse J_H primers | |
| RHuJH1-2 | 5′–GCA CCC TGG TCA CCG TCT CCT CAG GTG G–3′ |
| RHuJH3 | 5′–GGA CAA TGG TCA CCG TCT CTT CAG GTG G–3′ |
| RHuJH4-5 | 5′–GAA CCC TGG TCA CCG TCT CCT CAG GTG G–3′ |
| RHuJH6 | 5′–GGA CCA CGG TCA CCG TCT CCT CAG GTG C–3′ |
| Reverse V_κ primer for scFv linker | |
| RHuVκ1aBACKFv | 5′–GGA GAC TGG GTC ATC TGG ATG TCC GAT CCG CC–3′ |
| RHuVκ2aBACKFv | 5′–GGA GAC TGA GTC ATC ACA ACA TCC GAT CCG CC–3′ |
| RHuVκ3aBACKFv | 5′–GGA GAC TGC GTC AAC ACA ATT TCC GAT CCG CC–3′ |
| RHuVκ4aBACKFv | 5′–GGA GAC TGG GTC ATC ACG ATG TCC GAT CCG CC–3′ |
| RHuVκ5aBACKFv | 5′–GGA GAC TGC GTG AGT GTC GTT TCC GAT CCG CC–3′ |
| RHuVκ6aBACKFv | 5′–GGA GAC TGA GTC AGC ACA ATT TCC GAT CCG CC–3′ |
| Reverse V_λ primer for scFv linker | |
| RHuVλBACK1Fv | 5′–GGC GGC TGC GTC AAC ACA GAC TGC GAT CCG CCA CCG CCA GAG–3′ |
| RHuVλBACK2Fv | 5′–GCA GGC TGA GTC AGA GCA GAC TGC GAT CCG CCA CCG CCA GAG–3′ |
| RHuVλBACK3aFv | 5′–GGT GGC TGA GTC AGC ACA TAG GAC GAT CCG CCA CCG CCA GAG–3′ |

**Table 1** (*Continued*)

| | |
|---|---|
| RHuVλBACK3bFv | 5′–GGG TCC TGA GTC AGC TCA GAA GAC GAT CCG CCA CCG CCA GAG–3′ |
| RHuVλBACK4Fv | 5′–GGC GGT TGA GTC AGT ATA ACG TGC GAT CCG CCA CCG CCA GAG–3′ |
| RHuVλBACK5Fv | 5′–GAC GGC TGA GTC AGC ACA GAC TGC GAT CCG CCA CCG CCA GAG–3′ |
| RHuVλBACK6Fv | 5′–TGG GGC TGA GTC AGC ATA AAA TTC GAT CCG CCA CCG CCA GAG–3′ |

*Primers with appended restriction sites for reamplification of scFv gene repertoires*

| | |
|---|---|
| HuVH1aBACKSfi | 5′–GTC CTC GCA ACT GCG GCC CAG CCG GCC ATG GCC CAG GTG CAG CTG GTG CAG TCT GG–3′ |
| HuVH2aBACKSfi | 5′–GTC CTC GCA ACT GCG GCC CAG CCG GCC ATG GCC CAG GTC AAC TTA AGG GAG TCT GG–3′ |
| HuVH3aBACKSfi | 5′–GTC CTC GCA ACT GCG GCC CAG CCG GCC ATG GCC GAG GTG CAG CTG GTG GAG TCT GG–3′ |
| HuVH4aBACKSfi | 5′–GTC CTC GCA ACT GCG GCC CAG CCG GCC ATG GCC CAG GTG CAG CTG CAG GAG TCG GG–3′ |
| HuVH5aBACKSfi | 5′–GTC CTC GCA ACT GCG GCC CAG CCG GCC ATG GCC CAG GTG CAG CTG TTG CAG TCT GC–3′ |
| HuVH6aBACKSfi | 5′–GTC CTC GCA ACT GCG GCC CAG CCG GCC ATG GCC CAG GTA CAG CTG CAG CAG TCA GG–3′ |
| HuJκ1BACKNot | 5′–GAG TCA TTC TCG ACT TGC GGC CGC ACG TTT GAT TTC CAC CTT GGT CCC–3′ |
| HuJκ2BACKNot | 5′–GAG TCA TTC TCG ACT TGC GGC CGC ACG TTT GAT CTC CAG CTT GGT CCC–3′ |
| HuJκ3BACKNot | 5′–GAG TCA TTC TCG ACT TGC GGC CGC ACG TTT GAT ATC CAC TTT GGT CCC–3′ |
| HuJκ4BACKNot | 5′–GAG TCA TTC TCG ACT TGC GGC CGC ACG TTT GAT CTC CAC CTT GGT CCC–3′ |
| HuJκ5BACKNot | 5′–GAG TCA TTC TCG ACT TGC GGC CGC ACG TTT AAT CTC CAG TCG TGT CCC–3′ |
| Hu Jλ1FORNot | 5′–GAG TCA TTC TCG ACT TGC GGC CGC ACC TAG GAC GGT GAC CTT GGT CCC–3′ |
| Hu Jλ2-3FORNot | 5′–GAG TCA TTC TCG ACT TGC GGC CGC ACC TAG GAC GGT CAG CTT GGT CCC–3′ |
| Hu Jλ4-5FORNot | 5′–GAG TCA TTC TCG ACT TGC GGC CGC ACY TAA AAC GGT GAG CTG GGT CCC–3′ |

*Primers for DNA sequencing and PCR fingerprinting*

| | |
|---|---|
| LMB3 | 5′–CAG GAA ACA GCT ATG AC–3′ |
| fdseq | 5′–GAA TTT TCT GTA TGA GG–3′ |
| Linkseq | 5′–CGA TCC GCC ACC GCC AGA–3′ |
| pHENseq | 5′–CTA TGC GGC CCC ATT CA–3′ |

The primers correspond to those used to create libraries described in (8) and (11). Although they may not amplify all V-genes, they have produced excellent libraries. They are adapted for cloning into pHEN1 (see *Figure 3*).

**Table 2** Alternative primer set for creation of human scFv libraries

| | Primer name | V-gene family optimally recognized |
|---|---|---|
| *(A) BACK primers (5′)* | | |
| **V_H** | | |
| CAG GTG CAG CTG CAG GAG TCS G | VH4back | 4 |
| CAG GTA CAG CTG CAG CAG TCA | VH5back | 6 |
| CAG GTG CAG CTA CAG CAG TGG G | VH6back | 4 (DP63) |
| GAG GTG CAG CTG KTG GAG WCY | VH10back | 3 |
| CAG GTC CAG CTK GTR CAG TCT GG | VH12back | 1 |
| CAG RTC ACC TTG AAG GAG TCT G | VH14back | 2 |
| CAG GTG CAG CTG GTG SAR TCT GG | VH22back | 1, 2, 5, 7 |
| **V_λ** | | |
| CAG TCT GTS BTG ACG CAG CCG CC | VL1back | 1 |
| TCC TAT GWG CTG ACW CAG CCA C | VL3back | 3 |
| TCC TAT GAG CTG AYR CAG CYA CC | VL38back | 3 |
| CAG CCT GTG CTG ACT CAR YC | VL4back | 1, 4, 5, 9 |
| CAG DCT GTG GTG ACY CAG GAG CC | VL7/8back | 7, 8 |
| CAG CCW GKG CTG ACT CAG CCM CC | VL9back | 1, 5, 9, 10 |
| TCC TCT GAG CTG AST CAG GAS CC | VL11back | 3 (DPL16) |
| CAG TCT GYY CTG AYT CAG CCT | VL13back | 2 |
| AAT TTT ATG CTG ACT CAG CCC C | VL15back | 6 |
| **V_κ** | | |
| GAC ATC CRG DTG ACC CAG TCT CC | VK1back | 1 |
| GAA ATT GTR WTG ACR CAG TCT CC | VK2backts | 3, 6 |
| GAT ATT GTG MTG ACB CAG WCT CC | VK9back | 2, 3, 4, 6 |
| GAA ACG ACA CTC ACG CAG TCT C | VK12back | 5 |
| *(B) FOR primers (3′)* | | |
| **V_H** | | |
| TGA GGA GAC RGT GAC CAG GGT G | VH1/2for | JH1, JH2 |
| TGA GGA GAC GGT GAC CAG GGT T | VH4/5for | JH4, JH5 |
| TGA AGA GAC GGT GAC CAT TGT | VH3for | JH3 |
| TGA GGA GAC GGT GAC CGT GGT CC | VH6for | JH6 |
| GGT TGG GGC GGA TGC ACT CC | IGMfor | CH1 Cμ |
| SGA TGG GCC CTT GGT GGA RGC | IGGfor | CH1 Cλ |
| **V_λ** | | |
| TAG GAC GGT SAS CTT GGT CC | VL1/2for | Jλ1, Jλ2, Jλ3 |
| GAG GAC GGT CAG CTG GGT GC | VL7for | Jλ7 |
| **V_κ** | | |
| TTT GAT TTC CAC CTT GGT CC | VK1for | Jκ1 |
| TTT GAT CTC CAS CTT GGT CC | VK2/4for | Jκ2, Jκ4 |
| TTT GAT ATC CAC TTT GGT CC | VK3for | Jκ3 |
| TTT AAT CTC CAG TCG TGT CC | VK5for | Jκ5 |

These primers have been described by Sblattero *et al.* (9). The portion of the primer annealing to the V-gene is given in nucleic acid triplets (corresponding to amino acids). For cloning, restriction sites should be appended to the 5′ end of the primer according the vector used.

on phage or toxicity to *Escherichia coli*. Use of primers biased toward V-gene families and members found in selected clones might be beneficial. Such restriction of V-genes present in a library to a small subset of those which express well in bacteria does not appear to compromise selection ability (15).

In this chapter, methods to create naïve scFv phage antibody libraries from natural sources will be described. The same protocols can also be used to construct libraries from immunized humans or patients with diseases where an immune response is mounted. Alternatively, different primer sets can be used to generate libraries from immunized rodents or hybridomas (31, 32).

The steps required to construct scFv phage antibody libraries are:

- Isolation of RNA from a V-gene source (*Protocol 1*)
- First strand complementary DNA (cDNA) synthesis (*Protocol 2*)
- V-region PCR and construction of a scFv gene repertoire (*Protocols 3–6*)
- Restriction digestion of the phage display vector and scFv gene repertoire (*Protocol 7*)
- Ligation of vector and scFv gene repertoire and transformation of *E. coli* (*Protocols 8–10*)

## 2.2 The source of natural V-genes

When making a phage antibody library from natural V-genes a number of different sources of B cells (and hence V-genes) can be considered. In general, the choice is between V-genes which have not undergone somatic mutation (naïve and generally immunoglobulin M (IgM)) and those which have (immune and generally immunoglobulin G (IgG)), there being a gradient of naivity going from IgM to IgG. Early experiments with naïve libraries of diversities less than $10^8$ members suggested that IgM-derived V-genes were better than IgG-derived V-genes (8). However, a far larger library (7) used unfractionated V-genes from bone marrow, peripheral blood, and tonsil, and was used successfully to isolate antibodies against many antigens. Analysis of the V-genes derived from this library shows that most of the selected antigen-binding V-genes contained many mutations, suggesting that they were derived either from IgG or from the mutated IgM compartment. Other large naïve libraries (9–11, 13) have invariably used IgM-derived V-genes.

Lymphocytes from peripheral blood, spleen, tonsils, and bone marrow have been used to make libraries, with no systematic study of the advantages or disadvantages, although it would be expected that tissues removed for chronic infections (tonsil) would have V-genes heavily biased toward infectious organisms. The procedure described in *Protocol 1* generates high quality RNA; alternatively, we have had good results using the RNAgents kit from Promega for RNA generation. The procedure is essentially identical whatever the source of lymphocytes.

## Protocol 1

# RNA isolation from peripheral blood lymphocytes (PBLs)[33]

### Equipment and reagents[a]

- 30 ml Corex tubes, acid washed and silated
- Prechilled centrifuge and swinging bucket rotor
- Ficoll (Amersham Biosciences)
- Ice cold phosphate buffered saline (PBS), pH 7.4
- Lysis buffer: 5 M guanidine monothiocyanate, 10 mM ethylene diamine tetracetic acid (EDTA), 50 mM Tris-HCl, pH 7.5, 1 mM dithiothreitol (DTT), filter with Millipore 0.45 micron
- 4 M and 3 M lithium chloride, autoclaved

- RNA solubilization buffer: 0.1% SDS, 1 mM EDTA, 10 mM Tris-HCl, pH 7.5, autoclaved
- 3 M sodium acetate, pH 4.8, diethylpyrocarbonate (DEPC) treated (add 0.2 ml DEPC/100 ml solution) and autoclaved
- DEPC treated water (0.2 ml DEPC/100 ml water), autoclaved
- Phenol equilibrated in Tris (Sigma)
- Chloroform : Isoamyl alcohol (24 : 1) (Sigma)
- 100% ethanol at $-20\,°C$

### Method

1. Collect fresh human blood and separate white blood cells over Ficoll immediately.

2. Wash the PBLs 3 times with ice cold PBS.

3. Collect the PBLs in a 50-ml plastic centrifuge tube and add 7 ml of lysis buffer (will lyze up to $5 \times 10^8$ PBLs) and vortex vigorously to lyze the cells.

4. Add seven volumes (49 ml) of 4 M lithium chloride and incubate at $4\,°C$ for 15–20 h (overnight).

5. Transfer the suspension to 30 ml Corex tubes and centrifuge at 6500 rpm for 2 h at $4\,°C$ in a swinging bucket rotor.

6. Discard the supernatant and wipe the lips of the tubes with Kimwipes. Pool the pellets by resuspending them in 3 M lithium chloride (approximately 15 ml). Centrifuge the resuspended pellets for 1 h at 6500 rpm.

7. Discard the supernatant and dissolve the pellets in 2 ml of RNA solubilization buffer. Freeze the suspension thoroughly at $-20\,°C$.

8. Thaw the suspension by vortexing for 20 s every 10 min for 45 min.

9. Extract once with an equal volume of phenol and once with an equal volume of chloroform.

10. Precipitate the RNA by adding 1/10 volume 3 M sodium acetate, pH 4.8 and two volumes of $-20\,°C$ ethanol. Mix the solution thoroughly and incubate it overnight at $-20\,°C$.

**11** Centrifuge the RNA at 12,000 rpm for 30 min in a swinging bucket rotor. Suspend the pellet in 0.2 ml of DEPC treated water. Transfer the dissolved DNA to a 1.5-ml microcentrifuge tube and reprecipitate it by adding 1/10 volume 3 M sodium acetate, pH 4.8 and two volumes of −20°C ethanol. Store the RNA as an ethanol precipitate until ready to use.

[a] Use disposable plasticware when possible. All glassware, including Corex tubes, should be baked overnight at 180°C. Use separate reagents (phenol, chloroform, ethanol) for RNA work.

It is not necessary to prepare poly-A RNA to prepare cDNA: total RNA works very well. Immune libraries should be created using IgG-based primers, while naïve libraries are best prepared using IgM primers (see *Table 1*). Alternatively, the RNA may be primed with random hexamers with subsequent isolation of IgM or IgG specific V-genes by PCR using specific 3′ primers.

## Protocol 2
## First strand cDNA synthesis[a]

### Equipment and reagents

- 10 × RT buffer: 1.4 M KCl, 500 mM Tris-HCl, pH 8.1 at 42°C, 80 mM MgCl2
- RNAsin (Promega)
- AMV Reverse Transcriptase (Promega)
- 70% ethanol
- 100 mM DTT
- 20 × dNTPs (each at 5 mM) (New England Biolabs)
- IgG, IgM, $C_\kappa$, and $C_\lambda$ primers or random hexamer primers (10 pmol/µl) (see *Table 1*)
- Silated 1.5 ml centrifuge tubes
- RNA (from Protocol 1)

### Method

**1** For first strand cDNA synthesis, prepare the following reaction mixture in a silated 1.5-ml microcentrifuge tube:[b]

| | |
|---|---|
| 10 × RT buffer | 5 µl |
| 20 × dNTPs | 5 µl |
| 100 mM DTT | 5 µl |
| HuIgMFOR primer | 2 µl |
| HuC$_\kappa$FOR primer | 2 µl |
| HuC$_\lambda$FOR primer | 2 µl |
| RNAsin | 80 U (2 µl) |

All primers are 10 pmol/µl

2   Take an aliquot (1–4 μg) of RNA in ethanol, place in a sterile 1.5-ml microcentrifuge tube and centrifuge it for 5 min in a microcentrifuge. Wash the pellet once with 70% ethanol, dry it, and resuspend in 24.5 μl of DEPC treated water (see *Protocol 1*).

3   Heat the solution to 65 °C for 3 min to denature the RNA, quench it on ice for 2 min, and add it to the first strand reaction mixture. Add 2.5 μl AMV reverse transcriptase and incubate the solution at 42 °C for 1 h.

4   Boil the cDNA reaction mixture for 3 min, centrifuge it for 5 min in a microcentrifuge, transfer the cDNA-containing supernatant to new silated tube and use it immediately for PCR amplification of V-genes (*Protocol 3*).

[a] This protocol has been used successfully in our laboratories. There are also many kits available which are suitable for cDNA synthesis.

[b] The heavy chain primer used in this protocol are for naïve library generation. For immune libraries, use an IgG specific primer.

After cDNA synthesis, V regions are amplified from the cDNA using specific primers which recognize relatively conserved elements at the 5′ and 3′ end of rearranged V genes.

## Protocol 3

## Amplification of $V_H$ and $V_L$ genes from first strand cDNA

### Equipment and reagents

- Vent DNA polymerase and 10 × buffer (New England Biolabs)
- 20 × dNTPs (each at 5 mM) (New England Biolabs)
- V-region specific forward (3′) ($J_H$FOR, $J_\kappa$FOR, and $J_\lambda$FOR) and back (5′) ($V_H$BACK, $V_\kappa$FOR, and $V_\lambda$FOR) PCR primers (*Table 1*)[a]

- PCR thermocycler
- Mineral oil (Sigma)
- Geneclean kit (Qbiogene, Inc.)
- First-strand cDNA (from *Protocol 2*)

### Method

1   Make up 50 μl PCR reaction mixes[a, b] in 0.5-ml microcentrifuge tubes containing:

| | |
|---|---|
| Millipore water | 31.5 μl |
| 10 × Vent polymerase buffer | 5.0 μl |
| 20 × dNTPs | 2.5 μl |
| Forward primer(s) | 2.0 μl |
| Back primer(s) | 2.0 μl |
| cDNA reaction mix from *Protocol 2* | 5.0 μl |

2   Heat the reactions to 94 °C for 5 min in a thermal cycling block. If this does not have a heated lid, add one drop of light mineral oil to cover the reaction.

**3** Add 2.0 µl (2 units) Vent DNA polymerase.

**4** Cycle 30 times to amplify the V-genes at 94 °C for 1 min, 60 °C for 1 min, and 72 °C for 1 min.

**5** Purify the PCR fragments by electrophoresis on a 1.5% agarose gel, extract from the gel using Geneclean according to the manufacturer's instructions. Resuspend each product in 20 µl of water. Determine the concentration of the DNA by analysis on a 1% agarose gel compared to markers of known size and concentration.

[a] $V_H$, $V_\kappa$, and $V_\lambda$ genes are amplified in separate PCR reactions using the appropriate Back and Forward primers. Back primers are an equimolar mixture of either the 6 $V_H$Back, 6 $V_\kappa$Back, or 7 $V_\lambda$Back primers (final total concentration of primer mixture equals 10 pmol/µl). Forward primers are an equimolar mixture of either the 4 $J_H$, 5 $J_\kappa$, or 4 $J_\lambda$ primers (final total concentration of primer mixture equals 10 pmol/µl). Alternatively, amplifications can be done with each back primer individually with the mixture of FOR primers. This has the tendency to bias the library slightly toward those V-genes which are less highly expressed in the PBL population.

[b] If first strand synthesis in *Protocol 2* was performed using random hexamers, amplification of IgM or IgG V-genes can be performed using *Protocol 3* and the IgM or IgG specific primers listed in *Table 1*. *Protocol 3* should then be repeated using the V-gene primers as described, but using the gel purified PCR products from this amplification as substrate DNA instead of cDNA.

## 2.3 Assembly and cloning of V-gene repertoires

To create phage antibody libraries, $V_H$ genes need to be juxtaposed to $V_L$ genes in the same plasmid, for both the scFv or Fab formats (see *Figure 2*). Although the discussion below refers specifically to the creation of scFv, it equally applies to the creation of Fabs. In each case the region between the V-genes is involved in connecting the two genes. For scFv fragments, this region encodes a polypeptide linker, which covalently joins the two V-genes, whereas in the case of Fabs, this region will be a segment of DNA containing a ribosome-binding site and a leader sequence.

There are three basic ways to join $V_H$ and $V_L$ genes: PCR assembly (for which protocols are provided in this chapter), sequential cloning, and recombination. Early libraries (8) were created using PCR splicing by overlap extension (assembly), but many subsequent ones have been generated by making separate $V_H$ and $V_L$ libraries and then assembling them either by cloning one into the other using standard methods, or by carrying out PCR assembly from cloned material.

PCR assembly methods can involve the splicing of either two or three fragments (*Figure 6*). In two-fragment assembly, V-genes are amplified with primers which overlap with one another, so that splicing them together by PCR creates the linker region. This is only possible if the linker region is not repetitive, and so is not suitable for the $(Gly_4Ser)_3$ linker typically used in scFv libraries (*Figure 7*), nor for the assembly of Fab fragments since the linker region is too large to be derived from the overlap of two primers. When building a library of scFv fragments connected by the Gly–Ser linker, either the region of overlap must extend

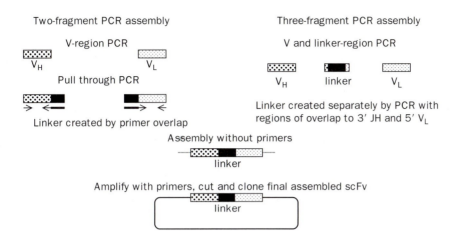

**Figure 6** Assembling scFv genes from V-gene repertoires by PCR. Assembly of scFv can be performed using either two or three fragments as shown. In the case of 3-fragment assembly, the linker is created in such a way that it contains a region of homology to all 3' $J_H$ and 5' $V_L$ genes. This anchors the assembly and ensures that the linker remains the correct length and is important when the linker is repetitive, such as with the (Gly$_4$Ser)$_3$ scFv linker. If a non-repetitive linker is used, a two-fragment assembly can be carried out, as shown, with the region of overlap between the two V-genes containing an overlapping region of the linker.

to the opposite chain, or the linker must be first created separately as a pool of fragments, each of which has homology to a different $V_H$ and $V_L$ gene. This linker can then be spliced to $V_H$ and $V_L$ genes in a three-fragment assembly (*Figure 6*).

In both two- and three-fragment assemblies, several amplification cycles without the addition of external primers are first performed, that involve an initial annealing of the regions of overlap followed by an extension. This process joins $V_H$ genes randomly to $V_L$ genes via the linker, and the assembled scFv can then be amplified by subsequent PCR in which external primers are added. The reaction is easier if separate $V_H$ and $V_L$ libraries are first made and reamplified to create assembly fragments.

Construction of libraries by sequential cloning involves the ligation of $V_H$ and $V_L$ gene repertoires consecutively into an appropriate display vector using standard methods. The origin of the V-genes used can be either primary PCR products (from cDNA), or precloned V-genes isolated by PCR or restriction digestion. In general, the use of precloned V-genes is easier, since PCR fragments tend to be difficult to clone.

Regardless of the approach used to construct libraries, it is important to use restriction enzymes which cut infrequently within V-regions. *Asc*I, *Bss*HII, *Bst*BI, *Mlu*I, *Not*I, *Pml*I, *Sal*I, *Sfi*I, *Sna*BI, and *Srf*I are all suitable, cutting no more than one germline V-gene.

The protocols below describe the procedure for PCR amplification of $V_H$ and $V_L$ genes and their assembly into scFvs by the three-fragment PCR method.

Variations on this, with intermediate cloning of $V_H$ and $V_L$ libraries prior to assembly by either PCR or cloning are relatively straightforward and involve standard molecular biological techniques (34) together with the protocols outlined in this chapter. The first step in a three-fragment assembly is preparation of the scFv linker, which is prepared by PCR using as a template any cloned scFv gene with a $V_\kappa$ gene and a $(Gly_4Ser)_3$ linker. The amplified linker is generated with primers that anneal to the 5′ and 3′ ends of the cloned linker and have overhangs encoding 24 nucleotides of perfect complementarity with the various $J_H$ (5′ end) or $V_L$ (3′ end) genes. Since there are four different $J_H$ primers and six different $V_\kappa$ primers, a total of 24 separate PCR reactions is required to generate the $V_H$–$V_\kappa$ linker. Similarly, since there are four different $J_H$ primers and seven different $V_\lambda$ primers, a total of 28 separate PCR reactions is required to generate the $V_H$–$V_\lambda$ linker.

```
      G   G   G   G   S   G   G   G   G   S   G   G   G   G   S
GGTGGAGGCGGTTCAGGCGGAGGTGGTTCTGGCGGTGGCGGATCG
```

**Figure 7** The $(Gly_4Ser)_3$ scFv linker. Many phage antibody libraries use the 15 amino acid $(Gly_4Ser)_3$ linker to connect $V_H$ and $V_L$ in the order $V_H$–$V_L$. It has been shown that linkers of this length can have a tendency to form dimeric scFv, in which the $V_H$ of one scFv pairs with the $V_L$ of another scFv. By increasing the length of the linker to 20 amino acids, this tendency can be reduced, although not eliminated.

## Protocol 4

## Preparation of scFv linker DNA

### Equipment and reagents

- PCR reagents and equipment as described in *Protocol 3*
- 1 µg of any vector containing an scFv with the $(Gly_4Ser)_3$ linker
- $RHuJ_H$, $RHuV_\kappa$, and $RHuV_\lambda$ primers at 10 pmol/µl (see *Table 1*)

### Method

1  Make up a master mix for amplification in 0.5-ml microcentrifuge tubes. The master mix is shown for the $V_H$–$V_\kappa$ linker. For the $V_H$–$V_\lambda$ linker master mix, $n = 29$.

|  | Individual reaction | Master mix (µl) ($n = 25$) |
|---|---|---|
| Water | 37 | 925 |
| $10 \times$ Vent buffer | 5.0 | 125 |
| $20 \times$ dNTPs | 2.5 | 62.5 |
| Vent DNA polymerase | 0.5 | 12.5 |

Protocol 4 continued

**2** Place into 24 0.5-ml tubes 2 μl each of RHuJ$_H$ and RHuV$_κ$ primers for each of the 24 possible combinations of RHuJ$_H$ and RHuV$_κ$ primers. For the V$_H$–V$_λ$ linker use the RhuV$_λ$ primers instead of RHuV$_κ$ primers and make up 28 tubes. If this does not have a heated lid, add one drop of light mineral oil to cover the reaction (Sigma).

**3** Put 50 μl of the master mix into a negative control tube.

**4** Add 1.0 μl of the scFv gene template (~1 ng) per sample to the master mix.

**5** Add 46 μl of master mix to each of the 24 tubes generated in step 2.

**6** Preheat the PCR block to 94 °C.

**7** Cycle 25 times to amplify the linker sequence at 94 °C for 1 min, 55 °C for 1 min, and 72 °C for 1 min.

**8** Purify the PCR fragments by electrophoresis on a 2.0% TBE agarose gel, excise the PCR product, extract it with Geneclean and resuspend the DNA in 50 μl of water.

## Protocol 5

# PCR assembly of V$_H$ and V$_L$ genes with scFv linker to create scFv gene repertoires

### Equipment and reagents

- PCR reagents and equipment as described in *Protocol 3*
- Geneclean kit (Qbiogene, Inc.)
- V-region specific forward (J$_κ$FOR and J$_λ$FOR) and back (V$_H$BACK) PCR primers (*Table 1*)
- scFv linker DNA (see *Protocol 4*)

### Method

**1** Set up the following PCR reaction mixtures in 0.5-ml microcentrifuge tubes. Make up two "reaction tubes," one containing the V$_κ$ repertoire DNA and one containing the V$_λ$ repertoire DNA.

| | Reaction tube | Control 1 | Control 2 |
|---|---|---|---|
| scFv linker DNA (100 ng/μl) | 1.0 μl | 1.0 μl | |
| V$_H$ repertoire DNA (100 ng/μl) | 3.5 μl | | 1.0 μl |
| V$_L$ repertoire DNA (V$_κ$ or V$_λ$) (100 ng/μl) | 3.0 μl | | 1.0 μl |
| Water | | 6.5 μl | 5.5 μl |

**2** To each tube add:

| | |
|---|---|
| Water | 33 μl |
| 10 × Vent buffer | 5.0 μl |
| 20 × dNTPs | 2.5 μl |

3  Heat the reaction tubes to 94 °C for 5 min in a thermal cycling block. If this does not have a heated lid, add one drop of light mineral oil to cover the reaction (Sigma).

4  Add 2.0 µl (2 units) of Vent DNA polymerase.

5  Cycle 7 times at 94 °C for 1 min, 65 °C for 1 min, and 72 °C for 2 min to randomly join the fragments.

6  After seven cycles, hold the temperature at 94 °C while adding 1 µl of each set of flanking primers (an equimolar mixture of the 6 $V_H$BACK primers and an equimolar mixture of the 5 $J_\kappa$FOR or 4$J_\lambda$FOR primers, giving a final total concentration of each primer mix of 10 pmol/µl).

7  Cycle 25 times to amplify the fragments at 94 °C for 1 min, 60 °C for 1 min, and 72 °C for 1 min.

8  Analyze 3 µl of the PCR reactions to determine success of the splicing. The assembled scFv should be 0.8–0.9 kb and there should not be a product of this size in the negative control reactions.

9  Purify the assembled gene repertoires by electrophoresis on a 1.0% agarose gel, followed by extraction from the gel using Geneclean. Resuspend each product in 20 µl of water. Determine the DNA concentration by analysis on a 1% agarose gel compared with markers of known size and concentration.

For library creation, the spliced scFv gene repertoires must next be reamplified to append restriction sites to permit cloning into the phagemid vector.

# Protocol 6

# Reamplification of scFv gene repertoires to append restriction sites for cloning

## Equipment and reagents

- PCR reagents and equipment as described in *Protocol 3*, except substitute *Taq* polymerase buffer for Vent polymerase buffer. Any brand of *Taq* polymerase in its appropriate buffer can be used.

- V-region specific forward (3′) ($J_\kappa$FORNot and $J_\lambda$FORNot) and back (5′) ($V_H$BACKSfi) PCR primers (*Table 1*)
- Wizard PCR purification kit (Promega)
- Assembled scFv repertoires (from *Protocol 5*)

## Method

1  Make up two 50 µl PCR reaction mixes (one for the $V_H$–$V_\kappa$ scFv repertoire and one for the $V_H$–$V_\lambda$ scFv repertoire) in 0.5 ml microcentrifuge tubes containing:

| | |
|---|---|
| Water | 36.5 µl |
| 10 × *Taq* buffer | 5.0 µl |

| | |
|---|---|
| 20 × dNTPs | 2.5 µl |
| Forward primer[a] | 2.0 µl |
| Back primer[b] | 2.0 µl |
| scFv gene repertoire (~10 ng) | 1.0 µl |

2  Heat the reaction mixes to 94 °C for 5 min in a thermal cycling block. If this does not have a heated lid, add one drop of light mineral oil to cover the reaction.

3  Add 1.0 µl (5 units) *Taq* DNA polymerase.

4  Cycle 25 times to amplify the scFv genes at 94 °C for 1 min, 55 °C for 1 min, and 72 °C for 1 min.

5  Purify the PCR product using the Wizard PCR purification kit.

[a] Forward primer is an equimolar mix of the 5 $J_\kappa$FORNot or 4 $J_\lambda$FORNot, at a final concentration of 10 pmol/µl.

[b] Back primer is an equimolar mix of the 6 $V_H$BACKSfi primers, at a final concentration of 10 pmol/µl.

In order to clone the scFv gene repertoires into pHEN1, they must first be digested with the restriction enzymes *Nco*I and *Not*I. We have found that the repertoires can be digested by these enzymes simultaneously with good efficiency. However, digestion of PCR products can be inefficient. In this protocol this is mitigated by increasing the length of the primer beyond the restriction site, and carrying out prolonged digestions with clean DNA.

## Protocol 7

## Restriction digestion of the scFv gene repertoires and pHEN1 phage display vector

### Equipment and reagents

- *Nco*I and *Not*I restriction enzymes and manufacturer's buffer 3 (New England Biolabs) (NEB 3 buffer)[a]

- 100 × acetylated bovine albumin serum (BSA) (New England Biolabs, supplied with *Not*I)

- 37 °C incubator or heat block with heated lid

- Geneclean kit (Qbiogene, Inc.)

- pHEN1 DNA (8), prepared by cesium chloride purification

- Reamplified scFv repertoires (from *Protocol 6*)

## Method

1   Make up two 100 µl reaction mixes to digest the scFv repertoires, one for the $V_H$-$V_\kappa$ scFv repertoire and one for the $V_H$-$V_\lambda$ scFv repertoire:

| | |
|---|---|
| scFv DNA (1–4 µg) | 50 µl |
| Water | 33 µl |
| 10 × NEB 3 buffer | 10 µl |
| Acetylated BSA (100×) | 1.0 µl |
| *Nco*I (10 U/µl) | 3.0 µl |
| *Not*I (10 U/µl) | 3.0 µl |

2   Incubate the reaction at 37 °C overnight. Use of a 37 °C incubator or heat block with heated lid prevents evaporation. Otherwise add a layer of mineral oil (Sigma) as used for PCR reactions.

3   Gel purify the gene repertoire on a 1% agarose gel and extract the DNA using the Geneclean kit. Resuspend the product in 30 µl of water. Determine the DNA concentration by analysis on a 1% agarose gel with markers of known size and concentration.

4   Prepare the vector by digesting 4 µg of cesium chloride purified pHEN1 in a reaction exactly as described above for the scFv gene repertoire, but substituting plasmid DNA for scFv DNA.[b]

5   Purify the digested vector DNA on a 0.8% agarose gel. Extract the cut vector from the gel using the Geneclean kit the same way as for the scFv insert. Resuspend the product in 30 µl of water. Determine the DNA concentration by analysis on a 1% agarose gel compared with markers of known size and concentration.

[a] For other vectors, use different restriction enzymes as appropriate

[b] Efficient digestion is important since a small amount of undigested vector leads to a very large background of nonrecombinant clones. Use of vector DNA prepared by techniques other than cesium chloride will give lower transformation efficiencies!

After digestion, the repertoires are gel purified and ligated into digested pHEN1.

# Protocol 8

# Ligation of scFv and vector DNA

## Equipment and reagents

- T4 DNA ligase and 10 × buffer (NEB)
- Phenol/chloroform (Sigma)
- 100% and 70% ethanol at −20 °C
- Digested vector and scFv repertoires from *Protocol 7*
- TE (10 mM Tris pH 8, 1 mM EDTA)

## Method

1 Make up two 100 µl ligation mixtures as described below, one for the $V_H$-$V_\kappa$ scFv repertoire and one for the $V_H$-$V_\lambda$ scFv repertoire, as well as the two control reactions.[a] The controls determine how much of the vector is uncut (control 2) and how much is cut once and can be religated in the absence of insert (control 1). The proportions are maintained, and the number of colonies obtained (for the same ligation volume) should be less than 10% of the library:

|  |  | Control 1 | Control 2 |
|---|---|---|---|
| 10 × ligation buffer | 10 µl | 0.5 | 0.5 |
| Millipore water | 32 µl | 2.3 | 2.5 |
| Digested pHEN 1 (100 ng/µl) | 40 µl | 2 | 2 |
| scFv gene repertoire (100 ng/µl) | 14 µl |  |  |
| T4 DNA ligase (400 u/µl) | 4 µl | 0.2 |  |

2 Incubate the mixtures overnight at 16 °C.

3 Increase the volume of the reaction mixes to 200 µl with TE. Extract the DNA with an equal volume of phenol/chloroform. Ethanol precipitate the DNA and wash the pellet twice in 70% ethanol.

[a] The amounts of DNA indicated are chosen to give the theoretical optimal molar ratio of insert to vector of 2 : 1. Given that the ratio of sizes of scFv (800 bp) to vector (4500 bp) is approximately 6 : 1, this translates into a ratio of insert to vector of 1 : 3 in weight terms.

Ligation mixtures are introduced into *E. coli* TG1 using electroporation, by far the most efficient way to introduce DNA into bacteria. Although electrocompetent cells are available commercially, higher efficiencies (up to $1.0 \times 10^{10}$ colonies/µg supercoiled pUC) are obtained using "home-made" cells prepared fresh as described in *Protocol 9*. Although this TG1 protocol is specifically included because it has been successfully used for antibody libraries, an alternative and optimized bacterial strain and set of protocols for high-efficiency electroporation of bacterial cells is described in Chapter 2.

# Protocol 9

# Preparation of electrocompetent *E. coli* TG1

## Equipment and reagents[a]

- *E. coli* strain TG1 (Stratagene)
- 37 °C incubator (New Brunswick Scientific)
- Sorvall RC5 centrifuge (or similar), GS3 rotor (both prechilled to 4 °C)
- Prechilled 500-ml polypropylene centrifuge tubes

- 2-L baffled flasks
- Minimal media agar plate [10.5 g $K_2HPO_4$, 4.5 g $KH_2PO_4$, 1 g $(NH_4)_2SO_4$, 0.5 g $Na_3C_6H_5O_7 \cdot 2H_2O$, 15 g agar with water to 1 L. Autoclave and allow to cool until the flask is hand hot, and then add 1 ml 1 M $MgSO_4$, 0.5 ml 1% vitamin B1 (thiamine), and 10 ml 20% dextrose].

- $2 \times$ TY media (Add 16 g bacto-tryptone, 10 g yeast, 5 g NaCl to 1l of distilled water and autoclave).
- 1 M Hepes, (N-[2-hydroxy ethyl]piperazine-N'-[2-ethanesulfonic acid]) pH 7.4
- 1 M Hepes, pH 7.4, 10% glycerol
- 10% glycerol

## Method

1   Inoculate a minimal media plate with *E. coli* TG1, and incubate overnight at 37 °C.

2   Inoculate 15 ml of $2 \times$ TY in a 100-ml flask with *E. coli* TG1 from the minimal plate. Shake the culture overnight at 37 °C.

3   Inoculate two 2-L baffled flasks containing 500 ml of $2 \times$ TY with 5 ml of the overnight TG1 culture. Grow in a 37 °C shaker for ~1.5 h, until the $A_{600}$ is no more than 0.5.

4   Transfer the bacteria to four 500-ml prechilled centrifuge bottles (~250 ml/bottle). Balance the bottles and keep them on ice for 20 min.

5   Centrifuge the bottles at 3000 g, 4 °C, for 15 min. Prechill all rotors before use.

6   Discard the supernatant and resuspend the cells in each bottle in the original volume (~250 ml) of ice cold 1 mM Hepes solution. All solutions should be made fresh.

7   Centrifuge the bottles again at 3000 g, 4 °C for 15 min.

8   Resuspend the cells in half the original volume (~125 ml) of ice cold 1-mM Hepes, now using two centrifuge bottles.

9   Centrifuge the bottles again at 3000 g, 4 °C for 15 min.

10  Resuspend all the cells in total volume of 20 ml of 10% glycerol/1-mM Hepes.

11  Centrifuge the bottles again at 3000 g, 4 °C for 15 min.

12  Prepare an ethanol/dry ice bath, if the cells are to be stored rather than used fresh.

13  Resuspend the cells in a total volume of 2 ml 10% glycerol.

14  Use the cells fresh, or freeze aliquots of cells using the dry ice bath.[b] Use the cells promptly after testing efficiency (within a week).[c]

[a] It is critical that all glass and plastic equipment, medium, and buffer are absolutely soap and detergent-free. (It is best to use a set of dedicated equipment that is washed only with water and autoclaved).

[b] Cells can be quick frozen in 25–200 µl aliquots and stored at − 80 °C, with an approximately 10-fold decrease in transformation efficiency.

[c] Test that cells are free of contaminant plasmids, phagemids, or phages, by growing on appropriate antibiotic plates and in top agar with lawn bacteria.

DNA used for electroporation must be salt free. The presence of salt in the bacteria/DNA mixture causes arcing (a short circuit) which must be avoided. In general, 2 µl of a standard ligation can be used after a heat inactivation step of 10 min at 70°C. For library construction, the DNA should be purified by phenol/chloroform extraction and ethanol precipitation (as described in *Protocol 8*) or using proprietary kits (e.g. Geneclean or Qiagen).

## Protocol 10

## Electroporating the ligation and archiving the library

### Equipment and reagents

- Electrocompetent *E. coli* strain TG1 (from *Protocol 9*, or purchased from Stratagene)
- 37°C incubator (New Brunswick)
- Electroporator and 0.2 cm path cuvettes (Gene Pulser™ BioRad)
- 16°C water bath
- SOC media (Dissolve 20 g bacto-tryptone, 5 g yeast extract, and 0.5 g NaCl in 950 ml water; add 10 ml of 250 mM KCl, adjust to pH 7.0 with 5 N NaOH and make up to 1 liter with distilled water. Autoclave the solution. When it is cool to the touch, add 5 ml of 2 M MgCl$_2$ and 20 ml of 1 M glucose)

- 100 and 150 mm TYE plates with 100 µg/ml ampicillin and 2% glucose (TYE/amp/glu) (Dissolve 16 g bacto-tryptone, 10 g yeast extract, 5 g NaCl, and 15 g bacto-agar in distilled water and adjust the volume to 1 L. Autoclave the solution. When it is cool to the touch, add ampicillin and glucose, mix, and pour into plates.)
- 2 × TY media with 100 µg/ml ampicillin and 2% glucose (2 × TY/amp/glu) (see *Protocol 9* for 2 × TY media; add ampicillin and glucose when the autoclaved solution is cool to the touch.)
- 50% Glycerol

### Method

1  Set the electroporator at 200 Ω (resistance), 25 µF (capacitance), and 2.5 kV.

2  Thaw the electrocompetent bacteria on ice or use freshly prepared electrocompetent cells. Place the electroporation cuvettes, cuvette holder, cells and DNA on ice.

3  Mix 80 ng purified, salt-free DNA with 50 µl bacteria. Incubate the mixture on ice for 1 min.

4  Place the mixture in a 0.2 cm path electroporation cuvette, taking care not to leave air bubbles between the electrodes. Tap the cuvette gently to transfer all liquid to the bottom of the cuvette. Move quickly to keep the cuvette cold.

5  Place the cuvette into the electroporator, making sure the sides of the cuvette are dry (to avoid arcing). Pulse, and immediately add 1.0 ml of SOC medium and mix. The time constant should be in the 4.5–5.5 ms range. If the time constant is less than 4.3 s, repeat the DNA precipitation.

6  Grow the bacteria from each electroporation in 1 ml SOC media at 37°C for 1 h, shaking at 250 rpm.

**7** Combine all the electroporation cultures and plate serial dilutions onto 100 mm TYE/amp/glu plates to determine the size of the library.

**8** Centrifuge the remaining bacterial culture at 2000 g for 10 min at 4 °C. Resuspend the pellet in 500 μl for every four electroporations and plate 125 μl aliquots onto 150 mm TYE/amp/glu plates or plate 500 μl on Nunclon Square Dishes (500 cm$^2$) containing TYE/amp/glu. Incubate the plates overnight at 30 °C.

**9** Scrape the bacteria from the plates by adding 10 ml 2 × TY/amp/glu media containing 15% glycerol (sterilized by filtration through 0.2 μm filter). Store the library stock at − 70 °C.[a,b]

[a] Typical transformation efficiencies are 1–10 × 10$^9$/μg pUC19, and 1–100 × 10$^7$/μg ligated vector with insert. The background of ligated vector without insert should be less than 10% of the ligated vector with insert—this will depend upon the efficiency of the ligation reaction and the quality of the vector used—see controls 1 and 2 in *Protocol 8*.

[b] Ligations and electroporations should be continued until there are at least 10$^9$ separate colonies for naïve libraries, and 10$^7$ separate colonies for immune libraries.

## 3 Preparing phage antibodies for selection on antigen

After transformation, the library is obtained as a bacterial lawn. This is used to prepare glycerol stocks which are kept frozen at −80 °C, as described in *Protocol 10*. This is the primary library and represents a very valuable resource. The best libraries are prepared from this primary stock. However, primary library can be amplified by plating it on large culture plates, with some loss of diversity. In any procedure in which phage are produced or bacteria as amplified, the inoculum should always be at least 5 times greater than the library size to preserve diversity.

In most phage display vectors, the expression of the antibody fragment is driven by the lacZ promoter. Although this is a leaky promoter, and probably could be improved, it works quite well provided a few guidelines are followed. Before phage are rescued, bacteria containing the library are grown to logarithmic phase in the presence of glucose. The glucose inhibits the activity of the lacZ promoter and hence reduces expression of both pIII and antibody. This has two important effects: the toxicity due to pIII and antibody is reduced, thus preventing library bias; and the inhibition of pIII expression permits bacterial pilus expression, allowing infection by helper phage. If glucose is omitted, the efficiency of infection drops dramatically, sharply reducing the effective size of the library. After infection by helper phage, the glucose must be removed, to permit pIII, and hence antibody expression. The pIII expression is then driven by the low basal level of the lac promoter, which is sufficient for effective display. Inducing this promoter with isopropyl β-D-thiogalactoside (IPTG) is counterproductive, and leads to very low display levels, due to the toxicity of the pIII-antibody fusion protein and degradation of the pIII-antibody fusion protein. For scFv and Fab libraries, 30 °C is typically chosen for the overnight growth phase as this seems to result in better antibody display levels. Prior to selection, phage are prepared from the supernatant of the culture by PEG-precipitation. This removes soluble antibody fragments that arise from incomplete suppression or proteolytic cleavage, which can compete with the phage for binding. PEG precipitation also removes any other contaminants which may interfere with selection. Subsequent purification by cesium chloride centrifugation is recommended if long-term storage of phage is contemplated.

*Protocol 11* describes the preparation of phage from library stocks. Prior to beginning this protocol, calculate the number of bacteria per milliliter in the library stock by carrying out titrations (plate serial dilutions on TYE/amp/glu plates. The number of bacteria for a particular absorbance at 600 nm ($A_{600}$) is lower for a library than for untransfected bacteria, perhaps because library bacteria containing plasmids are larger, or because dead bacteria are present. Once titrations have been related to the $A_{600}$, the absorbance can be used to calculate bacterial numbers. For phage preparation (library rescue) the size of the initial inoculum from the library stock bacteria should be 5 times more than the library size. After inoculation into the culture media, the bacterial concentration should not exceed an $A_{600}$ of 0.05. The use of 5 times more bacteria than the library size ensures representation of all library members. An initial $A_{600}$ of less than 0.05 for the culture ensures multiple doubling times prior to the addition of helper phage, increasing the efficiency of helper phage infection. The minimal culture volume for the rescue must be determined based on library size and a maximal initial $A_{600}$ of 0.05. For example, a library containing $1.0 \times 10^{10}$ members would require an initial inoculum of $5.0 \times 10^{10}$ bacteria. Since an $A_{600}$ of $1.0 = 1.0 \times 10^9$ bacteria, to keep the $A_{600}$ of the initial culture $< 0.05$, a volume of 1 L would be required. This culture volume should be equally distributed into 250 ml of $2 \times$ TY/amp/glu in 2-L flasks. The glucose inhibits the expression of pIII, which allows pilus expression.

## Protocol 11

## Preparing phage antibodies from library glycerol stocks

### Equipment and reagents

- 500 ml sterile polypropylene centrifuge bottles[a]
- 2-L culture flasks
- RC5 centrifuge and GS3 rotor (Sorvall)
- $2 \times$ TY/amp/glu media (see *Protocol 10*)
- TYE amp/glu plates (see *Protocol 10*)

- TYE plates containing 25 µg/ml kanamycin and 2% glucose (TYE/kan/glu) (see *Protocol 10*)
- $2 \times$ TY media containing 100 µg/ml ampicillin and 25 µg/ml kanamycin ($2 \times$ TY/amp/kan)

- Helper phage (VCSM13, R408 or M13K07—Stratagene, NEB, or Amersham Biosciences).[b]
- 20% Polyethylene glycol (PEG) 6000, 2.5 M NaCl (PEG solution)
- Phosphate buffered saline (PBS), pH 7.4
- Library glycerol stock (from *Protocol 10*).

## Method

1  Inoculate an appropriate number (see the discussion above) of 2-L culture flasks containing 250 ml $2 \times$ TY/amp/glu with the library glycerol stock. The $A_{600}$ of the culture media should be $< 0.05$. Plate 1 μl from each culture flask to determine the size of the initial inoculum.

2  Grow the culture at $37\,°C$ with shaking (300 rpm) to an $A_{600}$ of 0.5 (approximately 1.5–2.5 h).[c, d]

3  Add helper phage to a helper phage: bacteria ratio of 10–20 : 1. Incubate the culture at $37\,°C$ for 30 min, standing in a waterbath with occasional mixing, then incubate at $37\,°C$ for 30 min with shaking.

4  Remove 1 μl from each flask, dilute in 1 ml of $2 \times$ TY and plate 1 and 10 μl on TYE/amp/glu and TYE/kan/glu plates to determine the efficiency of helper phage infection. The number of colonies on each plate should be similar, with the proportion of kanamycin resistant colonies greater than 10% of the number of ampicillin resistant colonies.

5  Centrifuge the bacterial cultures in 500 ml bottles in a GS3 rotor at 3000 rpm. Remove the supernatant.

6  Resuspend the bacterial pellet(s) in $2 \times$ TY/amp/kan and distribute the suspended culture into 2-L flasks containing no more than 250 ml of media.[e] Grow with shaking (300 rpm) overnight at $25–30\,°C$.

7  Centrifuge the bacteria at 10,000 rpm in 500 ml bottles in a GS3 rotor for 30 min. Transfer the supernatant to a new 500 ml bottle.

8  Add 1/10–1/5 volume of PEG solution to the supernatant and leave on ice for 1 h. This precipitates the phage, which should be visible as a clouding of the supernatant.

9  Pellet the phage in 500 ml bottles by centrifuging in a GS3 rotor at 4000 rpm for 15 min at $4\,°C$.[f] Discard the supernatant. Centrifuge the "dry" pellet again for 30 s to bring down the last drops of supernatant, and remove the liquid. Resuspend the pellet in 1/10 volume of PBS.

10  Centrifuge the bottles at 10,000 rpm for 15 min to pellet bacterial debris and transfer the supernatant to a new tube.

11  Repeat steps 7–9 to further purify the phage, but resuspend in a final volume 1/50 of the original culture volume. Phage prepared in this way can be used immediately for selection, or stored at $-80\,°C$ in aliquots of $10^{13}$ phage per aliquot.[g]

[a] Keep a dedicated set of centrifuge bottles and culture flasks for library rescue to avoid contamination by previously isolated binding phage antibodies.

Protocol 11 continued

ᵇ M13K07 is usually used, although no direct comparison has been made between the different helper phages.

ᶜ Glucose is used in the culture media to repress the lac promoter and prevent scFv and p3 expression, which are toxic to *E. coli*.

ᵈ A temperature of 37 °C is essential for pilus expression, which is requisite for helper phage infection. It is important not to overgrow the culture (by ensuring that $A_{600} < 0.5$), or infection with helper phage will be inefficient due to loss of the pilus.

ᵉ Use a volume at least 2 times the initial culture volume to allow bacterial growth. Glucose is not present in the culture medium to allow leaky expression of the antibody-pIII fusion protein.

ᶠ It is important not to spin the phage too fast, as they will be difficult to resuspend, and titer will decrease.

ᵍ Phage can be further purified by filtration through 0.45 μm sterile filter (Minisart NML; Sartorius). While this is useful when long-term storage at 4 °C is envisaged, it can lead to a significant loss of titer.

The best selection results are obtained when fresh phage are used, since there is a slow proteolysis of the displayed antibody with time, even at −80 °C. Proteolysis is significantly reduced by purifying phage using cesium chloride gradient centrifugation (*Protocol 12*). After such purification displayed proteins are far more stable. The expected yield is $1–5 \times 10^{11}$ phage per ml original culture volume, although this may decrease by 50% after cesium chloride purification.

# Protocol 12

## Phage purification by cesium chloride centrifugation[35]

### Equipment and reagents

- L8-M Beckman ultracentrifuge or equivalent
- Swinging bucket SWTi41 rotor or equivalent
- Cesium chloride (Sigma)
- PBS
- Purified phage preparation (e.g. from *Protocol 11*)

### Method

1 Add 0.45 g of cesium chloride for each milliliter of final phage suspension derived from the output of *Protocol 11*, and mix to dissolve.

2 Centrifuge the solution in a SWTi41 swinging bucket rotor (or equivalent) for 17 h at 15 °C. Phage will be concentrated in fractions 1 and 2 (see *Figure 8*). The concentration is higher in fraction 1, but fraction 2 has a greater volume. Harvest these fractions with a pasteur pipette or plastic syringe. The first time this protocol is carried out, it is wise to collect all bands and test them separately.

3 Dialyse the harvested solutions against PBS overnight.

4  Titer the phage from the different fractions by infection of *E. coli* DH5αF' or TG1, and keep those with the highest titers. Usually phage will be concentrated in one band, with lower amounts in other bands (if visible). Add glycerol to 10%, aliquot the phage to $10^{13}$ phage per sample and store at $-80\,^{\circ}$C. Further quality control can be carried out by western blot, using either an anti-tag or anti-pIII antibody.

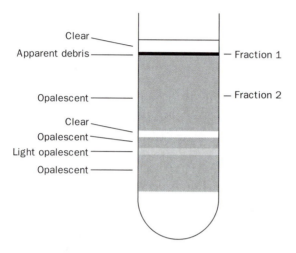

**Figure 8** Using cesium chloride to purify phagemids. The appearance of the centrifugation tube after centrifugation is shown. Phage are predominantly concentrated in the debris-ike band (fraction 1) and the opalescent band below it (fraction 2). Both bands should be collected.

# 4  Selecting antigen specific antibodies from phage libraries

Antigen specific antibodies are isolated from phage antibody libraries by subjecting the phage to recursive rounds of selection on antigen. Ideally only one round of selection should be required, but the fact that filamentous phage often bind nonspecifically to support surfaces limits the enrichment that can be achieved per round. In practice 2–5 rounds are necessary. Many different selection methods have been devised, including biopanning on antigen coated onto solid supports, columns or BIAcore sensorchips (13, 36), selection using biotinylated antigen (37), peptides (38), panning on fixed procaryotic (39) or mammalian (40) cells, tissue culture cells (41–43), fresh cells (44, 45), subtractive selection using sorting procedures such as FACS (46) and MACS (47), enrichment on tissue sections or pieces of tissue (48), selection for internalization (49) and, in principle, selection in living animals, as reported for peptide phage libraries (50) (see Chapter 10). More complex selection methods are "Pathfinder/Proximol" (51) and methods based on the restoration of phage infectivity via the antibody–antigen interaction (52, 53). In general, selection methods can be divided into two broad classes: those which attempt to isolate antibodies against known antigens, and those which use phage antibodies as a research tool to target unknown antigens (e.g. against cell surface antigens). This chapter will concentrate on the former type of selections.

Selection is most efficient when purified antigen is available. Although small quantities of antigen are required for selection, far greater amounts are needed for screening. In general, approximately 500 µg–1 mg of antigen is enough to complete a generous series of selections and screenings. This process can be accomplished with as little as 100 µg of antigen, if the antigen is used at low concentrations and can be stored and reused. Nonspecific phage-binding can be inhibited by using blocking agents such as fat free milk powder, gelatin, casein, BSA, or Tween 20.

Since there are so few copies of each individual phage antibody in a naïve library, the first round of selection should be as non-stringent as possible. In general, increasing stringency at this stage is counterproductive, as it reduces the number of phage eluted but does not increase the proportion of positive phage (I. M. Tomlinson, personal communication and (40)). After the first round of selection, the number of each individual phage antibody is far higher and subsequent washes can be far more stringent.

The most common selection method involves coating plastic immunotubes (Nunc) with the antigen. These tubes have a capacity of 650 ng antigen/cm$^2$, but have the potential problem that epitopes become partially denatured, leading to the selection of antibodies which do not recognize the native antigen. Although uncommon, this can be avoided by either first coating tubes with an antigen specific antibody or by using soluble biotinylated antigen in solution (37) with capture of the antigen and bound phage with magnetic streptavidin or avidin beads.

There are two ways to favor the selection of antibodies with higher affinity. One is to increase the stringency in the washing steps after the first round and the other is to reduce the antigen concentration during subsequent rounds of selection. The latter approach relies on the use of selection in solution using biotinylated antigen, since antigen coated to solid supports can lead to selection on the basis of avidity (37). Typically, for large unbiased antibody libraries, a starting antigen concentration of 300–500 nM is used, and this is reduced fivefold with each subsequent round of selection. Selecting from secondary libraries for affinity maturation is the subject of Chapter 14.

All other aspects being equal, a selection involving two antibodies with differing affinities will favor the antibody with the higher affinity. However, most aspects are not equal, and selection involves a number of factors influencing biological fitness other than affinity. In particular, display levels (a function of antibody folding efficiency, susceptibility to proteolysis and tendency to aggregate) and toxicity to the bacterial host are the two factors which most adversely affect selection efficiency. It is for this reason that if selections are continued for many rounds, diversity usually becomes reduced to a single clone, which sometimes may not even bind to the antigen and has a selective advantage due to deletion (54) (see Chapter 1).

Antibodies that display better will tend to be selected for two reasons: first, a greater proportion of phage will have antibody on their surface (normally only 1–10% of phagemid display an antibody fragment (55)); and second, because a greater proportion of phage will display two copies of antibody, contributing to an avidity effect. It is clear that antibodies that are toxic for their bacterial hosts will be discriminated against in any selection process.

Phage antibodies can be eluted from antigen in a number of different ways: acidic solutions such as 0.1 M HCl or 0.2 M glycine pH 2.5 (56, 57) basic solutions such as 0.1 M triethylamine (8), 0.1 M DTT when biotin is linked to antigen by a disulphide bridge (58), proteolytic cleavage of a site (e.g. using trypsin, Factor X or genenase) inserted between antibody and gene III (59), competition with excess antigen (60) or antibodies to the antigen (61), or by adding bacteria. In a comparison of different methods using a number of antibodies recognizing a single antigen with widely varying affinities, the most effective method was shown to be 0.1 M HCL (62). Bacterial elution, in particular, was not found to be effective in eluting high affinity phage.

The phage titer usually increases between rounds of selection and this indicates that selection is occurring. However, examples of successful selection with no change in output titer and failed selections with increases in titres have both been observed. In general, after the first round of selection, approximately 1% of phage bind antigen. This increases to 100% after 2–4 rounds of selection. However, the number of different antibodies typically decreases with each round of selection.

## Protocol 13

# Selection of phage antibodies on immobilized antigen

### Equipment and reagents

- 75 × 12 mm Nunc immunotubes (Maxisorb, catalog number 4-44202; available from VWR)
- PBS
- 2% skimmed milk powder in PBS (2% MPBS)
- 0.1% Tween 20 in PBS (PBS/Tween)
- 100 mM triethylamine
- 1 M Tris-HCl, pH 7.4

- Exponentially growing *E. coli* TG1 or DH5αF′
- 2 × TY media (see *Protocol 9*)
- 2 × TY/amp/glu media containing 10% glycerol (2 × TY/amp/glu/glycerol) (see *Protocol 10*)
- 100 and 150 mm TYE/amp/glu plates (see *Protocol 10*)
- Purified phage antibody library (from *Protocol 11 or 12*)

## Method

1   Coat an immunotube with 2 ml of antigen (10–1000 µg/ml) in PBS by incubating overnight at room temperature. Carbonate buffer can also be used.

2   Wash the tube 3 times with PBS and block nonspecific binding sites with 4 ml of 2% MPBS at 37°C for 1 h on a rotating turntable.

3   Wash the tube 3 × with PBS, add 1 ml of $10^{12}$–$10^{13}$ phage antibodies diluted in 1 ml of 4% MPBS[a] and incubate it for 30 min on a rotating turntable followed by a further 1.5 h standing.

4   Wash the tube once with PBS/Tween and twice with PBS. Each wash is performed by pouring buffer in and out of the tube.

5   Elute the bound phage by adding 1.0 ml of 100 mM triethylamine, capping the tube, and rotating it on a turntable for 8 min.[b]

6   Transfer the eluant to a new tube and neutralize it immediately by adding 0.5 ml of 1.0 M Tris-HCl, pH 7.4 and mixing.

7   Add half of the eluant (0.75 ml) to 10 ml of exponentially growing E. coli TG1 or DH5αF′ ($A_{600}$ 0.5) and incubate the culture without shaking at 37°C for 30 min.

8   Plate 1 µl of the culture on a 100 mm TYE/amp/glu plate to titer the eluted phage.[c]

9   Centrifuge the remainder of the culture at 4000 rpm for 15 min, resuspend in 0.5 ml 2 × TY and plate on two 150 mm TYE/amp/glu plates.

10  Prepare phagemid particles for the next round of selection by infection with helper phage as described in *Protocol 11*, but reducing the size of the initial culture volume to 10 ml (step 1) and the expansion culture volume to 50 ml (step 6). With the smaller volumes, a benchtop centrifuge can be used. Resuspend the phage preparation in a final volume of 1.0 ml PBS and titer the phage on E. coli TG1.

11  Repeat the selection, starting with step 1, for a total of 2–4 rounds. Increase the stringency of washes after the first round by washing 20 times in PBS/Tween and 20 times in PBS before phage elution.[d] Increased stringency can be provided by carrying out one or two 30-min washes with PBS and PBS/Tween.

[a] For a first round selection, $10^{13}$ phage should be used, for subsequent rounds $10^{12}$ are sufficient.

[b] Longer incubations with TEA may render the phage noninfectious. Alternative methods of elution include 1 ml of 100 mM HCl (10 min), or 0.5 ml trypsin (1 mg/ml in PBS for 10–60 min at 20–37°C) provided there is a trypsin cleavage site (Arginine or Lysine) between the antibody and pIII. Phage are resistant to trypsin for prolonged periods.

[c] The titer of the input and output phage can be used to calculate the recovery ratio (input/ouput ratio) which should increase each round of selection. In general, the expected first round output is $10^6$–$10^7$ phage particles.

[d] This reduces the output to 10,000–500,000 phage, but with a proportion of positives which can be as high as 100% after the second round.

An alternative to selecting antibodies on immobilized antigen is to select the antibodies in solution. This solves problems related to antigens that change conformation when directly coated onto solid surfaces. In addition, it is also easier to select on the basis of binding affinity, as phage displaying multiple antibodies do not have a selective advantage over those that do not. It is also easier to control the antigen concentration, which also facilitates selection for higher affinity antibodies. In order to select phage antibodies in solution, antigen is biotinylated using any of the kits sold by immunochemical reagent companies. After incubating the library with biotinylated antigen, the antigen and bound phage antibodies are separated from unbound phage antibodies using either streptavidin or avidin magnetic beads. After suitable washing, phage antibodies are eluted with either acid or alkaline solutions. One problem with this method is that antibodies recognizing streptavidin are also isolated. This can be avoided by depleting the library of streptavidin binders prior to selection, alternating between using avidin and streptavidin in selection rounds (these two biotin binding proteins are very different in structure and do not cross react), or by labeling the antigen with biotin disulfide N-hydroxy-succinimide ester (NHS-SS biotin) and eluting with DTT (58).

## Protocol 14

## Selection using soluble biotinylated antigen

### Equipment and reagents

- Streptavidin magnetic beads (Dynabeads M280, Dynal)
- Avidin magnetic beads (MPG Avidin, Catalog number MAVD0502, Dynal)
- Magnetic 1.5-ml tube holder (Dynal)
- PBS
- 2% skimmed milk powder in PBS (2% MPBS)
- Biotinylation kit (EZ-Link Sulfo-NHS-LC-Biotinylation kit, catalog number 21430, Pierce)

- PBS containing 0.1% Tween 20 (PBS/Tween)
- 100 mM HCl
- 1 M Tris, pH 7.4
- Exponentially growing *E. coli* TG1
- TYE amp/glu plates (see *Protocol 10*)
- 2 × TY media (see *Protocol 9*)
- 2 × TY/amp/glu media (see *Protocol 10*)
- Purified phage antibody library (from *Protocol 11 or 12*)

### Method

1  Biotinylate the antigen using the kit as recommended by the manufacturer. If DTT will be used for elution, use NHS-SS biotin.

2  Block a 1.5 ml microcentrifuge tube with 2% MPBS for 1 h at room temperature and then discard the blocking buffer.

3  Block 100 µl streptavidin-magnetic beads with 1 ml 2% MPBS for 1 h at room temperature in a 1.5 ml microcentrifuge tube. After blocking, collect the beads by pulling to one side with the magnetic tube holder. Discard the buffer.

4  Pre-deplete the library of streptavidin binders by incubating $10^{12-13}$ phage ($10^{13}$ for a first round selection, $10^{12}$ for subsequent rounds) in 2% MPBS with 100 µl streptavidin-magnetic beads for 1 h in a volume of 1 ml. Pull the beads to one side with the magnetic tube holder. Collect the phage antibodies and discard the beads.

5  Add the 1 ml of pre-depleted phage antibodies to the blocked tube and then add biotinylated antigen (100–500 nM). Incubate for 1 h, rotating on a turntable at room temperature.

6  Add 100 µl blocked streptavidin-magnetic beads to the tube and incubate on the rotator at room temperature for 15 min. Place the tube in the magnetic rack for 30 s. The beads will migrate toward the magnet.

7  Aspirate the tubes, leaving the beads on the side of the microcentrifuge tube. This is best done with a 200 µl pipette tip on a Pasteur pipette attached to a vacuum source. Wash the beads (1 ml per wash) with PBS/Tween 7 times, followed by MPBS 2 times, then once with PBS. Transfer the beads after every second wash to a fresh 1.5 ml tube to facilitate efficient washing.

8  Elute the phage by adding 1 ml of 100 mM triethylamine, capping the tube, and rotating end-over-end for 8 min. Draw the beads to one side of the tube with the magnet and transfer the solution to an eppendorf tube containing 500 µl 1 M Tris pH 7.4.[a]

9  Proceed as described in *Protocol 11*, step 7 onward to prepare phage for the next round of selection. Depending on the titer of the eluted phage, the antigen concentration can be reduced 10-fold for the next round of selection.

[a] If NHS-SS biotin was used to label the antigen, add instead 100 µl 1 mM DTT, cap the tube, and rotate end-over-end for 5 min. Draw the beads to one side with the magnet and transfer the solution to another eppendorf tube.

# 5  Identification and characterization of antigen-binding antibodies

When selecting on purified protein, the best way of assessing selected antibodies is by enzyme-linked immunosorbant assay (ELISA) in a high throughput micro-titer format. Although this may be unrelated to the final use of the antibody, it provides an initial screen which allows subsequent analysis to be focused on promising antibodies. The design of most phagemid display vectors permits two ELISA formats to be carried out: either phage ELISA in which bound phage are detected using labeled anti-phage secondary reagents, or soluble scFv ELISAs in which soluble scFv are detected by virtue of epitope tags attached at their N- or C-termini. In both cases the expression of the antibody is driven by the lacZ promotor. Display on phage is dependent on leaky expression in the absence of glucose while for the induction of soluble antibody fragments IPTG is used. Tags which have been used for detection include 9E10 (8, 63), SV5 (9, 64), and FLAG (65). In general, both ELISA formats give similar results, although we have identified antibodies which give signals in the phage format, but not the soluble format. This may be due to frameshifts occurring within the antibody which allow the production of sufficient scFv to be detected when displayed on phage, but not when expressed as native scFv, as has been previously described for peptide libraries (66). Most presently used phagemid vectors permit simple switching from phage bound to soluble scFv by including a suppressible amber codon (TAG) between the scFv gene and gIII (see Chapter 4). In a non-suppressor strain (e.g. HB2151) the amber codon will be read as a stop codon and only soluble scFv will be produced. In a suppressor strain (e.g. TG1, DH5αF′) the TAG codon is read as a glutamine as well as a stop codon and so both soluble scFv and scFv-pIII fusion are produced. While it may be desirable to carry out the switch from phage bound to soluble scFv by infecting a suppressor strain, this is in fact not necessary, and soluble scFv expression can also be easily accomplished in a suppressor strain since suppression is usually no more than 50%. Crude phage or scFv preparations are suitable for most preliminary analyses. However, subsequent analyses are facilitated by the ability to purify antibodies easily. This is best done by using phage vectors which include hexahistidine tags (9, 67) or subcloning the scFv genes into a vector which fuses the hexahistidine tag to the scFv C-terminus.

The percentage of positive clones at different rounds of selection will depend upon the quality of the library, the nature of the antigen, and the physical selection process, including antigen density, whether soluble or fixed antigen is used and the washing stringency. We have found that by following the immunotube selection procedure (*Protocol 13*), 10–100% of clones are usually positive after the second round. It is better to assess diversity earlier rather than later, as with ongoing selection, diversity becomes reduced. This provides a greater pool from which suitable antibodies can be chosen.

While binding screens are useful, they often do not take account of the final use of selected antibodies. The ability to generate large numbers of antibodies relatively quickly allows the institution of screens which go beyond the binding assay. Assays for cellular internalization, receptor activation, ligand antagonism, virus inactivation, induction or inhibition of apoptosis, can all be envisaged, and in some cases, for example, internalization (49, 68), it may prove possible to select for such functions directly.

For ELISA, phage antibodies are prepared in 96-well microtiter plates. Individual clones are picked into the wells of a microtiter plate, grown, and rescued as phagemids using helper phage. Phage production in the microtiter format is not as efficient as growth in larger volumes. This is because different clones often have very different growth rates, and as a result some are infected with helper phage at the optimum OD, whereas others are infected when the OD is either too high or too low. This can lead to vastly different phage titers in different individual wells.

## Protocol 15

## Expressing phage antibodies in microtiter plates

### Equipment and reagents

- Sterile 96-well round bottom microtiter plates made for bacterial culture, for example, Nunc 62162 (available from VWR)
- $2 \times$ TY/amp/glu media (see *Protocol 10*)
- $2 \times$ TY/amp/glu media containing 30% glycerol ($2 \times$ TY/amp/glu/gly) (see *Protocol 10*)
- $2 \times$ TY/amp/kan media (see *Protocol 11*)
- Colonies resulting from plating of selected libraries (from *Protocol 13 or 14*)

### Method

1  Use sterile toothpicks to pick individual colonies into 96-well microtiter plate wells containing 100 µl of $2 \times$ TY/amp/glu into each well of a 96-well plate.[a] Grow with shaking (300 rpm) overnight at 30 °C, preferably in a microtiter plate holder.[b, c]

2  Add 50 µl $2 \times$ TY/amp/glu/gly to each well and store the plate at −70 °C. This is the master plate and will be used for this protocol and *Protocol 17*.

3  Use a 96-well sterile transfer device or pipet to inoculate 2 µl per well from the master plate to a 96-well plate containing 150 µl $2 \times$ TY/amp/glu per well. Grow these cultures to an $A_{600\ nm}$ of approximately 0.5 (around 2.5 h), at 37 °C, shaking.

4  Add 50 µl of $2 \times$ TY/amp/glu containing $2 \times 10^9$ pfu/ml helper phage (diluted from stock) to each well. The ratio of phage to bacterium should be about 20 : 1. Incubate the plate for 30 min at 37 °C.[d]

5  Centrifuge the plate at 2700 rpm for 10 min and remove and discard the supernatant using a multichannel pipette or suction device.

6  Resuspend each bacterial pellet in 150 µl $2 \times$ TY/amp/kan. Glucose is omitted in this step, since expression of the antibody depends on leaky expression from the lacZ promoter. Incubate the plate overnight at 30 °C shaking at 300 rpm.

7  Next day, spin the plate at 2700 rpm for 10 min and use 50 µl/well of the supernatant for a phage ELISA.

[a] Controls should include wells containing (i) cells with no phage construct, (ii) cells expressing a phage construct displaying no antibody fragment, and (iii) cells expressing a phage construct displaying nonrelevant antibody fragments. If available, positive controls which bind the antigen should be included.

[b] To avoid evaporation, microtiter plates should be grown in a closed box as far as possible from the incubator fan. Place a damp paper towel in the box.

[c] The first time growth in microtiter plates is carried out, shaking conditions should be tested by placing bacteria in alternate wells, and growing overnight. Conditions are suitable if growth occurs only in inoculated wells and not in adjacent uninoculated wells.

[d] Keep the plate at 37 °C to avoid loss of pilus expression. The presence of glucose inhibits pIII production and allows pilus expression.

Phage ELISAs are quick and convenient, especially because of the signal amplification which occurs due to the multiple copies of pVIII (see Chapter 5). However, antibodies should always be subsequently tested in their soluble form as well.

## Protocol 16

## Phage antibody ELISA

### Equipment and reagents

- 96 well ELISA plates, for example, Nunc maxisorb 442404 (available from VWR)
- 2% MPBS
- PBS/Tween
- Horseradish peroxidase (HRP) conjugated mouse anti-phage monoclonal

- antibody (mAb) (Amersham Biosciences)
- 3,3′,5,5′-tetramethylbenzidine (TMB) liquid substrate system (Sigma T-8665)
- Stop solution (2 N $H_2SO_4$)
- Phage preparation from *Protocol 15*

### Method

1  Coat the wells of a microtiter plate overnight at 4 °C with 100 μl per well of protein antigen. Ten micrograms per milliliter in PBS is standard, but sometimes higher concentrations or a different binding buffer (such as carbonate buffer) are required.[a]

2  Discard the antigen solution and wash the wells twice with PBS.[b]

3  Block the wells by adding 200 μl of 2% MPBS and incubate at 37 °C for 2 h.

4  Wash the wells 3 times with PBS.

5  Add 50 μl 4% MPBS to all wells.

6  Add 50 μl of culture supernatant containing phage antibodies (see *Protocol 15*) to the wells. Mix the solution by pipetting up and down and incubate for 1 h at room temperature.

7  Discard the solution, and wash wells 3 times with PBS/Tween and 3 times with PBS.

8  Add 100 ul/well HRP-conjugated mouse anti-phage mAb diluted 1 : 5000 in 2% MPBS and incubate for 1 h at room temperature.

9  Discard the secondary antibody and wash wells 3 times with PBS/Tween and 3 times with PBS.

**10** Add 100 µl TMB system solution to each well and incubate at room temperature for 10–30 min in the dark. The blue color should appear within a few minutes.

**11** Quench the reaction by adding 50–100 µl stop solution to each well.

**12** Read the plate at 450 nm.

[a] If the antigen is precious it can be recovered and reused after coating. In this case overnight incubation at 4 °C is recommended.

[b] Wash by submerging the plate into buffer, removing the air bubbles in the wells by agitation and shaking the buffer out into the sink. To remove the last drops, bang the microtiter plate upside-down on to a pile of tissues.

An alternative to using phage(mid)s for ELISAs is to use soluble fragments. The expression of antibody (or other polypeptide) fragments from phagemid display vectors is controlled by the lacZ promoter. Glucose free conditions stop catabolic repression of the lac promoter, which can be induced by IPTG. Expressed antibody fragments leak into the supernatant (in part due to their toxicity in *E. coli*) in a passive process. It should be noted that for some antibody fragments, for example, those that are not very toxic, more protein will remain inside the periplasm (even after overnight culture) than will be found in the supernatant.

The method described here relies on the amount of glucose (0.1%) present in the starting medium being low enough to be metabolized by the time the inducer (IPTG) is added. This avoids any centrifugation steps and reduces the risk of contamination. Controls to be included are similar to those described in *Protocol 16*.

## Protocol 17

## Expressing soluble scFv in microtiter plates

### Equipment and reagents

- Sterile 96-well round bottom microtiter plates made for bacterial culture (Nunc, catalog number 62162)
- $2 \times$ TY media containing 100 µg/ml ampicillin and 0.1% glucose ($2 \times$ TY/amp/0.1%glu) (see *Protocol 10*)
- $2 \times$ TY media containing 100 µg/ml ampicillin and 3 mM IPTG ($2 \times$ TY/amp/IPTG) (see *Protocol 10*)
- Phage preparations from *Protocol 15*

### Method

**1** Use a 96-well sterile transfer device or pipet to transfer 2 µl per well from the master plate (see *Protocol 15*) to a sterile 96-well microtiter plate containing 100 µl/well $2 \times$ TY/amp/0.1% glu.

Protocol 17 continued

2 Incubate the plate at 37 °C, shaking, until the $A_{600}$ is approximately 0.9 (usually 2–3 h).

3 Add 50 µl 2 × TY/amp/IPTG to each well (final IPTG concentration of 1 mM). Continue shaking the plate at 30 °C for a further 16 h.

4 Centrifuge the plates at 2700 rpm for 15 min and use 50 µl of the supernatant for ELISA (*Protocol 18*).

ELISA using soluble scFv is similar in principle to that using phage antibodies, with the exception that detection of the soluble scFv involves the use of an anti-tag antibody with a secondary detection antibody (e.g. anti-mouse if the tag antibody is a mouse monoclonal). The anti-tag antibody described here is 9E10 which recognizes the myc tag and is incorporated into the pHEN1 vector. The anti-tag antibody 9E10 is available commercially or the hybridoma cell line can be obtained from ATCC.

# Protocol 18

## Soluble scFv ELISA in microtiter plates

### Equipment and reagents

- 9E10 mAb (Santa Cruz Biotech)
- PBS
- PBS/0.1% Tween 20 (PBS/Tween)
- 4% MPBS
- anti-mouse HRP labeled antibody (e.g. Sigma catalog number A2554)
- Antibody fragment-containing supernatants from *Protocol 17*

### Method

1 Coat the wells of a microtiter plate overnight at 4 °C with 100 µl per well of protein antigen. Ten microgram per milliliter in PBS is standard, but sometimes higher concentrations or a different binding buffer (e.g. carbonate buffer) are required.

2 Discard the antigen solution and wash the wells twice with PBS.

3 Block each well by adding 200 µl of 2% MPBS and incubate at 37 °C for 2 h.

4 Wash the wells 3 times with PBS.

5 Add 50 µl 4% MPBS to all wells.

6 Add 50 µl culture supernatant containing soluble scFv (from *Protocol 17*) to the wells. Mix the solutions by pipetting up and down and incubate the plate for 1 h at room temperature.

7 Discard the solution, and wash the wells 3 times with PBS/Tween and 3 times with PBS.

8 Add 100 µl 9E10 mAb (diluted 1 : 1000 in 2% MPBS) to each well. Incubate the plate for 1 h at room temperature.

9   Discard the antibody, and wash wells 3 times with PBS/Tween and 3 times with PBS.

10  Add 100 µl of anti-mouse HRP (diluted 1:2000 in MPBS) to each well. Incubate the plate for 1 h at room temperature.

11  Discard the secondary antibody, and wash the wells 3 times with PBS/Tween and 3 times with PBS.

12  Add 100 µl TMB system solution and incubate the plate at room temperature for 10–30 min in the dark. The blue color should appear within a few minutes

13  Quench the reaction by adding 50–100 µl stop solution.

14  Read the plate at 450 nm.

When screening indicates that a number of antigen binding antibodies have been selected, it is important to determine whether these represent different or identical clones. This is best done by sequencing. However, it is possible to first screen using PCR to estimate the number of different clones present. The scFv gene is amplified directly from bacterial colonies harboring the phagemid using primers which flank the scFv gene. The PCR product is then digested with a restriction enzyme that cuts frequently within V-genes (e.g. *Bst*NI). While this is very effective for scFv from nonimmunized libraries, it is less effective for libraries with more restricted diversity (immune libraries, affinity maturation libraries etc.), since the diversity in these libraries may not translate into gain or loss of restriction sites. Generally speaking, scFv with different restriction patterns represent unique clones. However, scFv with the same pattern may also represent different clones (approximately 10% of the time). This results from the fact that *Bst*N1 (and other enzymes) generally cut V-genes in the framework regions, and thus tend to indicate which $V_H$ gene family or member is paired with which $V_L$ gene family or member. Careful examination of fingerprint patterns may also reveal clones which are partially digested. This can be manifested by a large number of common bands with a few extra bands, the intensity of which sometimes does not correspond to that of the other bands; or band sizes whose sum does not correspond to the size of the original PCR product. It is also useful to run an aliquot of the PCR product on a gel prior to digestion. This will verify that the binding antibodies are full length. Examples exist where binding single domain ($V_H$) antibodies have been selected from phage libraries.

# Protocol 19

# Identification of different phage antibody clones by PCR

## Equipment and reagents

- *Bst*NI restriction enzyme and buffer (New England Biolabs)

- Primers (see *Table 1*)

- PCR reagents and equipment as described in *Protocol 3*
- Mineral oil (Sigma)

- Nusieve agarose
- 10 × TBE buffer (To prepare 10 × TBE, dissolve 108 g Tris base and 55 g boric acid in 900 ml water, add 40 ml of 0.5 M EDTA, pH 8.0, and bring volume to 1 L with water)

## Method

1 Make up a PCR "master mix," with 20 µl PCR mix per clone containing:

| | |
|---|---|
| Water | 15.0 µl |
| 10 × PCR buffer | 2.0 µl |
| 5 mM dNTPs | 1.0 µl |
| 5′ primer (10 µM) | 1.0 µl |
| 3′ primer (10 µM) | 1.0 µl |
| *Taq* polymerase | 0.2 µl |

If the PCR buffer does not contain $Mg^{2+}$, this should be added to a final concentration of 1.5 mM. Alternatively use the PCR reagents provided with the DNA polymerase.

2 Aliquot 20 µl of the master mix into 0.5 ml tubes, or into the wells of a 96-well PCR microplate.

3 Using a sterile toothpick gently touch a single colony and swirl in the PCR reaction mix, taking care not to transfer too much material. Excess bacteria in the PCR reaction can inhibit the PCR reaction. Throw away the toothpick, or, if required, use it to rescue the clone into medium in a separate 96-well plate, or onto an agar plate by touching the surface. One to two microliters of a glycerol stock, or the master plate culture, can be used instead of a colony.

4 Overlay the reactions with mineral oil.

5 Heat the tubes or plate to 94 °C for 10 min using the PCR block. This is needed to break open the bacteria and release the template DNA. Then cycle 30 times, incubating at 94 °C for 1 min, 55–60 °C for 1 min, and 72 °C for 1 min (2 min for screening of an Fab fragment library).

6 After the PCR it is advisable to check 5 µl of the PCR reaction on a 2% agarose gel. This will indicate how many clones contain full length insert.

7 After the PCR, add 20 µl of a master mix made as shown below, beneath the mineral oil:

| | |
|---|---|
| Acetylated BSA (10 mg/ml) | 0.2 µl |
| *Bst*NI buffer (10 ×) | 4 µl |
| Water | 15.3 µl |
| *Bst*NI (10 U/µl) | 0.5 µl |

8 Incubate the samples at 60 °C for 2–3 h.

9   Mix 2 µl sample dye with 8 µl reaction mix (taken from under the oil) and electrophorese on a 4% Nusieve agarose gel cast in 0.5 times TBE buffer containing 0.05 µg/ml ethidium bromide, or use a 2.5% agarose gel (which provides poorer resolution).

10  Run the gel at 100 V (10 V/cm) and compare the banding patterns of individual clones on a UV transilluminator.

## 6   Purification of soluble scFv fragments

The easiest way to purify scFv fragments is to append hexahistidine tags at the C-terminus and purify the proteins using immobilized metal ion affinity chromatography (IMAC). Most reagents for IMAC purification are now sold as kits with full instructions. It should be noted, however, that the protocols provided in the kits may need to be modified for different scFv vector constructs. We have found, for example, that scFv bearing the myc tag before the His tag can be washed at higher imidazole concentrations (20 mM) than those bearing other tags (5 mM).

## 7   Troubleshooting

*Table 3* summarizes problems which might occur during the procedures described in this chapter, and possible solutions.

**Table 3** Troubleshooting

| Protocol numbers | Problem | Possible cause | Solution |
|---|---|---|---|
| 1–3 | No V-region amplification | No RNA | Check the RNA on a gel, and if it is degraded or absent, prepare again |
| | | No cDNA | It can be useful to incorporate small amounts of radioactivity during the first–strand synthesis to ensure that cDNA has been made. |
| | | Annealing temperature incorrect | Should be 50–60 °C |
| | | Primers are incorrect | Check their sequence; resynthesize |
| 3 | V-region amplification in negative controls | Contamination with V-genes from laboratory environment | Replace all buffers and solutions |
| 5 | No scFv assembly | Fragment concentrations wrong | Check and correct |
| | | Annealing temperatures wrong | Should be 65 °C; also try lower temperatures (to 55 °C) |

283

**Table 3** (*Continued*)

| Protocol numbers | Problem | Possible cause | Solution |
|---|---|---|---|
| 7–10 | No clones after electroporation | Bacterial transformation efficiency is poor | Check the efficiency with a plasmid of known concentration. If efficiency is poor, make new electrocompetent cells |
| | | DNA fragments were not digested | Check the fragments on a gel—a shift should be visible; if not, repeat the digestion, checking buffer, enzyme etc. |
| | | No ligation | Check the ligation control; Check the ligation products on gel |
| 11,13 | No phage(mid) after growth of individual clones or library | Helper phage is not functional | Use a new batch of helper phage |
| | | OD of bacteria at time of helper phage infection was too high | Check carefully that the $A_{550}$ does not exceed 0.5 |
| | | Bacteria are no longer expressing F-pilus | With *E. coli* TG1 the bacteria need to be grown on minimal plates to maintain pilus expression |
| | | Lytic phage present | Place a small aliquot of the clone/library on top agar to reveal the presence of lytic phage plaques; if present, prepare new reagents |
| | | Wrong antibiotic used | Check antibiotic |
| 13,14 | No output after selection | Bacterial stock contaminated | Check for contamination, by streaking out a colony, and looking for plaques in top agar, and on various antibiotic resistant plates—if contamination is detected, change the bacteria stock |
| | | Wrong plates used | Check the antibiotic resistance of the vector used |
| | | Bacteria were grown at 37 °C | Bacteria should be grown at 30 °C, although a higher temperature is unlikely to eliminate all colonies |
| | | Glucose omitted in plate | Glucose will help the growth of some colonies, but its omission is unlikely to eliminate all colonies |
| | | Poor infection of bacteria | See solutions to problem 1 |

**Table 3** (*Continued*)

| Protocol numbers | Problem | Possible cause | Solution |
|---|---|---|---|
| | | Elution conditions were too severe | Check the concentrations of the elution reagent(s). Try adding bacteria at $A_{550} = 0.5$ for 60 min to elute. Test to see if elution conditions are killing phage |
| 15–18 | No ELISA signals | Problems with substrate, secondary antibodies, or hydrogen peroxide | Always include appropriate positive controls. Change substrates, antibodies or $H_2O_2$. |
| | | Growth temperature was too high, causing proteolytic degradation of the displayed antibody | Growth should be at 30 °C, but in some cases, even this is too high, so reduce to 25 °C. |
| | | Phage not produced (for phage ELISA) | See problem 1, above. Try scFv ELISA instead |
| | | scFv not produced (for scFv ELISA) | Check IPTG; try phage ELISA instead. Some scFv work as phage but not as soluble scFv |
| 13,14 | No antigen binding clones are selected | Selection has failed | Try presenting the antigen in a different format, or use a different selection strategy (e.g. biotinylated antigen) |
| | | Antigen did not coat well on to the surface of the microtiter plate | Try coupling the antigen in an alternative buffer—e.g. carbonate buffer instead of PBS, or vice versa |
| | | The library is not functional—that is, there is no displayed protein/antibody | Check display levels by doing a Western blot using either an anti-tag or anti-pIII antibody; the displayed protein should be 5–10% of total detected protein with the anti-tag antibody and 1–10% with the anti-pIII antibody. If the display level is poor, re-make, or re-obtain, the library |
| 18 | Selected antibody is not recognized by tag antibody | Tags have been proteolytically removed | Perform a Western blot to test presence of tag; if the tag is absent prepare the scFv again. Some scFv are recognized by protein A or protein L and these can be used as secondary reagents, although sensitivity is not very high |

**Table 3** (*Continued*)

| Protocol numbers | Problem | Possible cause | Solution |
|---|---|---|---|
| | | scFv is functional on phage but not as soluble scFv | This occasionally happens for unknown reasons. Usually it is best to test another clone |
| — | Antibody is not purified well by IMAC | The buffer conditions are wrong | Check the buffer conditions. It is important not to have EDTA in the binding buffer when using His-tag purification. Check to see whether the washing procedure is also removing antibody |

# References

1. Barbas, C. F., Kang, A. S., Lerner, R. A., and Benkovic, S. J. (1991). *Proc. Natl. Acad. Sci. USA*, **88**, 7978.
2. Williamson, R. A., Burioni, R., Sanna, P. P., Partridge, L. J., Barbas, C. F., and Burton, D. R. (1993). *Proc. Natl. Acad. Sci.*, **90**, 4141.
3. Zebedee, S. L., Barbas, C. F., Hom, Y., Caothien, R. H., Graff, R., Degraw, J., Pyati, J., LaPolla, R., Burton, D. R., Lerner, R. A., and Thornton, G. B. (1992). *Proc. Natl. Acad. Sci. USA*, **89**, 3175.
4. Orum, H., Andersen, P. S., Oster, A., Johansen, L. K., Riise, E., Bjornvad, M., Svendsen, I., and Engberg, J. (1993). *Nucl. Acids Res.*, **21**, 4491.
5. Ames, R. S., Tornetta, M. A., Jones, C. S., and Tsui, P. (1994). *J. Immunol.*, **152**, 4572.
6. Ames, R. S., Tornetta, M. A., McMillan, L. J., Kaiser, K. F., Holmes, S. D., Appelbaum, E. *et al.* (1995). *J. Immunol.*, **1545**, 6355.
7. Vaughan, T. J., Williams, A. J., Pritchard, K., Osbourn, J. K., Pope, A. R., Earnshaw, J. C. *et al.* (1996). *Nat. Biotechnol.*, **14**, 309.
8. Marks, J. D., Hoogenboom, H. R., Bonnert, T. P., McCafferty, J., Griffiths, A. D., and Winter, G. (1991). *J. Mol. Biol.*, **222**, 581.
9. Sblattero, D. and Bradbury, A. (2000). *Nat. Biotech.*, **18**, 75.
10. de Haard, H. J., van Neer, N., Reurs, A., Hufton, S. E., Roovers, R. C., Henderikx, P. *et al.* (1999). *J. Biol. Chem.*, **274**, 18218.
11. Sheets, M. D., Amersdorfer, P., Finnern, R., Sargent, P., Lindqvist, E., Schier, R. *et al.* (1998). *Proc. Natl. Acad. Sci. USA*, **95**, 6157.
12. Nissim, A., Hoogenboom, H. R., Tomlinson, I. M., Flynn, G., Midgley, C., Lane, D. *et al.* (1994). *EMBO J.*, **13**, 692.
13. Griffiths, A. D., Williams, S. C., Hartley, O., Tomlinson, I. M., Waterhouse, P., Crosby, W. L. *et al.* (1994). *EMBO J.*, **13**, 3245.
14. de Kruif, J., Boel, E., and Logtenberg, T. (1995). *J. Mol. Biol.*, **248**, 97.
15. Knappik, A., Ge, L., Honegger, A., Pack, P., Fischer, M., Wellnhofer, G. *et al.* (2000). *J. Mol. Biol.*, **296**, 57.
16. Perelson, A. S. and Oster, G. F. (1979). *J. Theor. Biol.*, **81**, 645.
17. Virnekas, B., Ge, L., Plückthun, A., Schneider, K. C., Wellnhofer, G., and Moroney, S. E. (1994). *Nucl. Acids Res.*, **22**, 5600.
18. Bird, R. E., Hardman, K. D., Jacobson, J. W., Johnson, S., Kaufman, B. M., Lee, S. M. *et al.* (1988). *Science*, **242**, 423.
19. Bird, R. E. and Walker, B. W. (1991). *Trends Biotech.*, **9**, 132.

20. Holliger, P., Prospero, T., and Winter, G. (1993). *Proc. Natl. Acad. Sci. USA*, **90**, 6444.

21. McGuinness, B. T., Walter, G., FitzGerald, K., Schuler, P., Mahoney, W., Duncan, A. R. *et al.* (1996). *Nat. Biotech.*, **14**, 1149.

22. Orlandi, R., Gussow, D. H., Jones, P. T., and Winter, G. (1989). *Proc. Natl. Acad. Sci. USA*, **86**, 3833.

23. Ruberti, F., Cattaneo, A., and Bradbury, A. (1994). *J. Immunol. Methods*, **173**, 33.

24. Hoogenboom, H. R. and Winter, G. (1992). *J. Mol. Biol.*, **227**, 381.

25. Marks, J. D., Tristem, M., Karpas, A., and Winter, G. (1991). *Eur. J. Immunol.*, **21**, 985.

26. Larrick, J. W., Danielsson, L., Brenner, C. A., Wallace, E. F., Abrahamson, M., Fry, K. E. *et al.* (1989). *BioTechnology*, **7**, 934.

27. Watkins, B. A., Davis, A. E., Fiorentini, S., and Reitz, M. S., Jr. (1995). *Scand. J. Immunol.*, **42**, 442.

28. De Boer, M., Chang, S.-Y., Eichinger, G., and Wong, H. C. (1994). *Hum. Antibodies. Hybridodomas*, **5**, 57.

29. Welschof, M., Terness, P., Kolbinger, F., Zewe, M., Dubel, S., Dorsam, H. *et al.* (1995). *J. Immunol. Methods*, **179**, 203.

30. Sblattero, D. and Bradbury, A. (1998). *Immunotechnology*, **3**, 271.

31. Dattamajumdar, A. K., Jacobson, D. P., Hood, L. E., and Osman, G. E. (1996). *Immunogenetics*, **43**, 141.

32. Krebber, A., Bornhauser, S., Burmester, J., Honeggar, A., Willuda, J., H. R., B. *et al.* (1997). *J. Immunol. Methods*, **201**, 35.

33. Cathala, G., Savouret, J. F., Mendez, B., West, B. L., Karin, M., Martial, J. A. *et al.* (1983). *DNA*, **2**, 329.

34. Sambrook, J., Fritsch, E. F., and Maniatis, T. (1989). *Molecular cloning: A laboratory manual.* 2nd edn., Cold Spring Harbour Laboratory Press, New York.

35. Kay, B., Winter, J., and McCafferty, J. (1996). *Phage display of peptides and proteins.* New York, Academic Press.

36. Malmborg, A. C., Duenas, M., Ohlin, M., Soderlind, E., and Borrebaeck, C. (1996). *J. Immunol. Methods*, **198**, 51.

37. Hawkins, R. E., Russell, S. J., and Winter, G. (1992). *J. Mol. Biol.*, **226**, 889.

38. Persic, L., Horn, I. R., Rybak, S., Cattaneo, A., Hoogenboom, H. R., and Bradbury, A. (1999). *FEBS Lett.*, **443**, 112.

39. Bradbury, A., Persic, L., Werge, T., and Cattaneo, A. (1993). *BioTechnology*, **11**, 1565.

40. Mutuberria, R., Hoogenboom, H. R., van der Linden, E., de Bruine, A. P., and Roovers, R. C. (1999). *J. Immunol. Methods*, **231**, 65.

41. Cai, X. and Garen, A. (1995). *Proc. Natl. Acad. Sci. USA*, **92**, 6537.

42. Siegel, D. L., Chang, T. Y., Russell, S. L., and Bunya, V. Y. (1997). *J. Immunol. Methods*, **206**, 73.

43. Hoogenboom, H. R., Lutgerink, J. T., Pelsers, M. M., Rousch, M. J., Coote, J., Van Neer, N. *et al.* (1999). *Eur. J. Biochem.*, **260**, 774.

44. Palmer, D. B., George, A. J., and Ritter, M. A. (1997). *Immunology*, **91**, 473.

45. Figini, M., Obici, L., Mezzanzanica, D., Griffiths, A., Colnaghi, M. I., Winter, G. *et al.* (1998). *Cancer Res.*, **58**, 991.

46. de Kruif, J., Terstappen, L., Boel, E., and Logtenberg, T. (1995). *Proc. Natl. Acad. Sci. USA*, **92**, 3938.

47. Siegel, D. L., Chang, T. Y., Russell, S. L., and Bunya, V. Y. (1997). *J. Immunol. Methods*, **206**, 73.

48. Van Ewijk, W., de Kruif, J., Germeraad, W. T. V., Berendes, P., Ropke, C., Platenburg, P. P., *et al.* (1997). *Proc. Natl. Acad. Sci. USA*, **94**, 3903.

49. Becerril, B., Poul, M. A., and Marks, J. D. (1999). *Biochem. Biophys. Res. Commun.*, **255**, 386.

50. Pasqualini, R. and Ruoslahti, E. (1996). *Nature*, **380**, 364.

51. Osbourn, J. K., Derbyshire, E. J., Vaughan, T. J., Field, A. W., and Johnson, K. S. (1998). *Immunotechnology*, **3**, 293.

52. Spada, S. and Plückthun, A. (1997). *Nat. Med.*, **6**, 694.

53. Malmborg, A. -C., Söderlind, E., Frost, L., and Borrebaeck, C. A. K. (1997). *J. Mol. Biol.*, **273**, 544.

54. de Bruin, R., Spelt, K., Mol, J., Koes, R., and Quattrocchio, F. (1999). *Nat. Biotechnol.*, **17**, 397.

55. Clackson, T. and Wells, J. A. (1994). *Trends Biotechnol.*, **12**, 173.

56. Roberts, B. L., Markland, W., Ley, A. C., Kent, R. B., White, D. W., Guterman, S. K. *et al.* (1992). *Proc. Natl. Acad. Sci. USA*, **89**, 2429.

57. Kang, A. S., Barbas, C. F., Janda, K. D., Benkovic, S. J., and Lerner, R. A. (1991). *Proc. Natl. Acad. Sci. USA*, **88**, 4363.

58. Griffiths, A. D., Malmqvist, M., Marks, J. D., Bye, J. M., Embleton, M. J., McCafferty, J. *et al.* (1993). *EMBO J.*, **12**, 725.

59. Ward, R. L., Clark, M. A., Lees, J., and Hawkins, N. J. (1996). *J. Immunol. Methods*, **189**, 73.

60. Clackson, T., Hoogenboom, H. R., Griffiths, A. D., and Winter, G. (1991). *Nature*, **352**, 624.

61. Meulemans, E. V., Slobbe, R., Wasterval, P., Ramaekers, F. C., and van Eys, G. J. (1994). *J. Mol. Biol.*, **244**, 353.

62. Schier, R. and Marks, J. D. (1996). *Hum. Antibodies Hybridomas*, **7**, 97.

63. Evan, G. I., Lewis, G. K., Ramsay, G., and Bishop, J. M. (1985). *Mol. Cell Biol.*, **5**, 3610.

64. Hanke, T., Szawlowski, P., and Randall, R. E. (1992). *J. Gen. Virol.*, **73**, 653.

65. Lindner, P., Bauer, K., Krebber, A., Nieba, L., Kremmer, E., Krebber, C. *et al.* (1997). *Biotechniques*, **22**, 140.

66. Carcamo, J., Ravera, M. W., Brissette, R., Dedova, O., Beasley, J. R., Alam-Moghe, A. *et al.* (1998). *Proc. Natl. Acad. Sci. USA*, **95**, 11,146.

67. Hochuli, E., Bannwarth, W., Döbeli, H., Gentz, R., and Stüber, D. (1988). *BioTechnology*, **6**, 1321.

68. Poul, M. A. and Marks, J. D. (1999). *J. Mol. Biol.*, **288**, 203.

# Chapter 14

# Affinity maturation of phage antibodies

Ulrik B. Nielsen and James D. Marks
Departments of Anesthesiology and Pharmaceutical Chemistry,
University of California, San Francisco, Rm 3C-38, San Francisco
General Hospital, 1001 Potrero, San Francisco, CA 94110, USA.

## 1 Introduction

Phage display can be used to increase the affinity of antibodies more than a 1000-fold (1, 2). The starting point is typically a specific antibody isolated from a phage antibody library (see Chapter 13). To accomplish affinity maturation (increased affinity) the sequence of the antibody is diversified, the mutant gene repertoire is displayed on filamentous phage, and higher affinity binders are selected on antigen. While the process is reasonably straightforward, the investigator should be reasonably certain that increasing the affinity of their particular antibody will lead to the desired biologic effect prior to undertaking *in vitro* affinity maturation. Higher affinity is especially important when trying to neutralize a circulating toxin or growth factor. In this case, the antibody distributes in the same compartment as the antigen, allowing the antigen-antibody interaction to proceed to equilibrium. The higher the affinity, the greater the amount of antigen bound. In contrast, affinity is only one factor determining the amount of antibody fragment that will accumulate in a tumor *in vivo* (3). Factors such as antibody fragment size, pharmacokinetics, and valency may have a more important impact (4).

There are four main issues to consider when performing *in vitro* antibody affinity maturation:

(a) where within the antibody gene to introduce mutations;

(b) how to introduce the mutations;

(c) how to select rare higher affinity antibodies from lower affinity antibodies; and

(d) how to identify the higher affinity antibodies.

We consider these four issues in the sections below and provide relevant protocols.

## 2 Where and how to diversify the antibody gene sequence

Mutations can be introduced into the antibody gene either randomly or at specific sites. Random introduction of mutations is the simplest approach and makes no *a priori* assumptions as to which sites are the best to mutate to increase affinity. Random mutagenesis more closely mimics the *in vivo* process of somatic hypermutation. One commonly used method for introducing random mutations has been "chain shuffling" (5), in which one of the two chains ($V_H$ and $V_L$) is fixed and combined with a repertoire of partner chains to yield a secondary library that can be searched for superior pairings (see Section 2.2 and *Figure 6*). Alternative approaches for achieving random mutagenesis of antibody genes include error-prone polymerase chain reaction (PCR) (6), DNA shuffling (7) (see Chapter 3) or propagation of phage in mutator strains of *Escherichia coli* (8). Whereas these approaches have yielded large increases in affinity for hapten antigens (up to 300 fold) (5, 8), results with protein antigens have been more modest ($< 10$-fold) (6, 9). One limitation to random mutagenesis is that little useful information is generated with respect to the location of mutations that modulate affinity. Many mutations will be present in the sequence of higher affinity binders and it will not be clear which of these increase affinity and which have no effect or actually decrease affinity. Such information could be used to guide subsequent mutagenesis efforts if the increase in affinity achieved is inadequate for the desired application. In addition, because of the random introduction of mutations into the framework regions, this approach may create problems of immunogenicity when the antibody is intended for therapeutic use. This problem could be minimized using DNA shuffling again to remove deleterious or unnecessary mutations (10).

Alternatively, mutations can be introduced into specific regions of the antibody gene using synthetic oligonucleotides. Decisions need to be made as to which residues to mutate, since randomization of only five amino acids would require $3.4 \times 10^7$ clones to completely cover the sequence space ($32^5$). [A general discussion of library size and diversity issues can be found in Chapter 2.] Several groups have shown that targeting mutations to the complementarity determining regions (CDRs) is an effective technique for increasing antibody affinity. This makes sense since during *in vivo* somatic hypermutation, mutations accumulate preferentially in the CDRs compared to framework residues (11). The location of these CDR mutations complements the locations where diversity is generated in the primary antibody repertoire (12). However many CDR residues, especially in $V_H$ CDR3 and $V_L$ CDR3, are responsible for high-energy interactions with antigen. Mutating residues in this region can in many cases abolish antigen binding. To more efficiently use sequence space, we prefer to minimize the number of nonviable structures by using nucleotide mixtures which bias for the wild-type residue. Mutant residues in the CDRs may increase affinity by introducing new contacts or by replacing low affinity or "repulsive" contacts with more favorable ones (13). However, it appears that many mutations introduced either by somatic hypermutation *in vivo*, or by random mutagenesis *in vitro*, instead exert their affect on affinity indirectly by repositioning the CDRs or the side chains of contact residues for optimal interaction with the antigen (14).

Antibodies may contain more than 50 CDR residues, raising the question of which CDRs and CDR residues to mutate. Our results and those of others indicate that targeting mutations to $V_H$ CDR3 and $V_L$ CDR3 is the most efficient means to improve affinity by phage display. For example, we increased the affinity of an anti-ErbB2 antibody more than 1200-fold by sequentially targeting mutagenesis to $V_H$ and $V_L$ CDR3. Similarly, Yang *et al.* (2) increased the affinity of an anti-gp120 Fab 420-fold by mutating four CDRs ($V_H$ CDR1, $V_L$ CDR1, $V_H$ CDR3, and $V_L$ CDR3) in five libraries and combining independently selected mutations. However, they observed the largest affinity increases when optimizing the CDR3 regions. Thus we initially select the $V_H$ and $V_L$ CDR3 for mutagenesis.

When randomizing CDR residues, sequence space can more efficiently be used by using molecular modeling of the most homologous V-domain in the protein data bank (PDB) database to distinguish between residues with solvent accessible side chains from those with buried side chains. Our previous results indicate that randomization of residues with buried side chains usually results simply in re-selection of the wild-type sequence (1, 15). We also avoid mutating glycines and tryptophans as they also invariably return as the wild-type residue after selection (1, 15). Glycines are frequently critical residues in CDR turns, and tryptophans frequently are essential structural or contact residues. Alternatively, the choice of which CDR residues to mutate may be based on studies of the sites mutated during *in vivo* affinity maturation. During the secondary immune response, mutations are frequently concentrated at hot spots in the CDRs. For example, serine residues encoded by AGY are more frequently mutated *in vivo* than are serine residues encoded by TCN (16). We have observed a similar pattern of mutation in our affinity maturation by site-directed mutagenesis and suggested targeting mutations to these hot spot residues (1). One report of affinity maturation successfully targeted only such predicted hot spots in the $V_L$ CDR3 and reported improved affinities of antibody fragments selected from very small libraries (17).

Optimization of CDRs can either be done parallely or sequentially. In the case of parallel CDR3 randomization, the two CDR3s are randomized independently and beneficial mutations in single clones are combined. Although this strategy has been successfully employed for optimization of several antibodies, not all mutations have been additive (1, 2). This is not surprising since many CDR residues pack against each other. Thus, we prefer sequential targeting of mutations to the CDR regions of the antibody.

Our lab has used both site-directed mutagenesis and chain shuffling (random mutagenesis) to increase antibody affinity and has compared these two techniques using an anti-ErbB2 single chain Fv ($V_H + V_L$) antibody fragment (scFv) (1, 9, 15). Overall, we find that site-directed mutagenesis of the CDR3s of an antibody is the more efficient means to improve affinity by phage display. However, below we describe methods for both site-directed mutagenesis (Section 2.1) and chain shuffling (Section 2.2). Our lab works exclusively with scFv antibody fragments, and so the protocols below use this format, although the general principles can also be applied to maturation of antigen-binding antibody fragment (Fab) fragments.

## 2.1 Construction of scFv libraries with diversified CDR3s by site-directed mutagenesis

The protocols below describe the methods to diversify the $V_L$ and $V_H$ CDR3 of an scFv antibody fragment and construct mutated phage libraries. The particular scFv fragment in these examples is the anti-ErbB2 scFv F5 (18). The $V_H$ and $V_L$ libraries are constructed and selected sequentially (see *Figure 1* for overview).

### 2.1.1 Construction of randomized $V_L$ CDR3 phage antibody libraries

We begin randomization with $V_L$ CDR3 as it is typically shorter than $V_H$ CDR3, and more importantly can be modeled based on homologous structures. Randomization of the $V_L$ CDR3 is also technically simpler than $V_H$ CDR3 since it is located at the 3' end of the scFv gene. The randomization can therefore be carried out in just two PCR steps. In the first PCR, a randomized primer ($V_L$ FOR) and the primer LMB3 amplify most of the scFv gene and introduce mutations into the $V_L$ CDR3. The second PCR amplifies the remainder of the scFv and appends a restriction site for cloning of the fragment (see *Figure 2*).

1. $V_L$ CDR3 of the scFv is randomized by PCR and cloned into a phagemid vector.

2. Improved scFv are selected from the phage library and the best clone identified by surface plasmon resonance.

3. The $V_H$ CDR3 of the best $V_L$ clone is randomized by PCR and cloned into a phagemid vector.

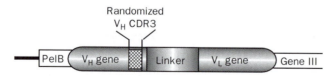

4. Improved scFv are selected from the phage library and the best clones identified by surface plasmon resonance.

**Figure 1** Overview of strategy for affinity maturation by sequentially mutating $V_L$ and $V_H$ CDR3.

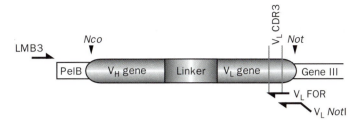

**Figure 2** Strategy for construction of a scFv gene repertoire with a diversified $V_L$ CDR3. Randomization of the $V_L$CDR3 is carried out in two PCRs. In the first PCR, a randomized primer ($V_L$ FOR) and the primer LMB3 amplify most of the scFv gene and introduce mutations into the $V_L$ CDR3. The second PCR amplifies the remainder of the scFv and appends a *Not*I restriction site for cloning of the fragment.

## Protocol 1

# Construction of an scFv gene repertoire with mutated $V_L$ CDR3

### Equipment and reagents

- Distilled (Millipore) water
- Vent DNA Polymerase and buffer (New England Biolabs (NEB))
- 20 × dNTPs (5 mM each) (NEB)
- Agarose gel box and power pack
- Geneclean kit (Qbiogene)
- PCR thermocycler

- Plasmid DNA containing starting scFv gene (single or double stranded DNA for all protocols)
- LMB3 primer (5′- CAG GAA ACA GCT ATG AC -3′), 10 pmol/μl
- $V_L$FOR random primer (see below), 10 pmol/μl

### Method

1  Design and synthesize the $V_L$ FOR primer containing CDR3 randomizing sequence. The primer must of course be designed based on the $V_L$-gene sequence of the antibody being affinity matured. In the example given here, seven amino acids of the $V_L$ CDR3 are randomized (*Figure 3*, underlined region) conserving approximately 50% wild-type amino acid at each position randomized. The resulting primer design is as follows:

$V_L$ FOR random primer: 5′- CCC TCC GCC GAA CAC CCA ACC <u>524 513 524 524 542 541 511</u> CTG GCA GTA ATA ATC AGC CTC-3′

Where the numbers indicate the molar compositions shown below:

1: A(70%), C(10%), G(10%), T(10%);
2: A(10%), C(70%), G(10%), T(10%);
3: A(10%), C(10%), G(70%), T(10%);
4: A(10%), C(10%), G(10%), T(70%);
5: G(50%), C(50%).

**Protocol 1** continued

The randomized residues were selected based on the use of molecular modeling to identify which CDR3 residues had solvent accessible side chains.

2  Set up four 50 μl PCR reaction mixes containing:

| | |
|---|---|
| Distilled water | 35.5 μl |
| 20 × dNTPs (5 mM each) | 2.5 μl |
| 10 × Vent polymerase buffer | 5.0 μl |
| LMB3 primer (10 pmol/μl) | 2.5 μl |
| V$_L$ FOR primer (10 pmol/μl) | 2.5 μl |
| Wild-type scFv gene template DNA (100 ng) | 1.0 μl |
| Vent DNA polymerase (2 units) | 1.0 μl |

3  Heat the mixture to 94 °C for 5 min in a PCR thermo-cycler with a heated lid.

4  Cycle 30 times to amplify the V$_L$ gene at 94 °C for 1 min, 42 °C for 1 min, 72 °C for 2 min.

5  Gel-purify the V$_L$ gene repertoires (app. 800 bp) on a 1% agarose gel and extract the DNA using the Geneclean kit. Resuspend each product in 20 μL of water. Determine the concentration of the DNA by analysis on a 1% agarose gel with markers of known size and concentration.

The scFv V$_L$ gene repertoire is now re-amplified with a primer which appends the restriction site *Not*I (see *Figure 2*). The re-amplified product is digested with *Nco*I and *Not*I, cloned into the phagemid display vector pHEN1 (see Chapter 13, *Figure 3*), and the products introduced into *E. coli* by electroporation.

After library construction, proceed to Section 4, selection and screening for higher affinity antibodies. We typically use the highest affinity mutant obtained from V$_L$ CDR mutagenesis as the starting antibody for V$_H$ CDR mutagenesis.

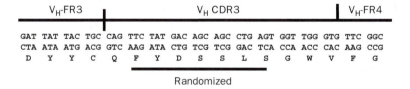

**Figure 3** Location and design of primer for diversifying V$_L$ CDR3. The primers used for the randomization of the CDR3 are based on the *V*-gene sequence of the antibody being affinity matured. In this example, seven amino acids of the V$_L$CDR3 (underlined region) were randomized. The end of framework 3 (FR3) and the beginning of framework 4 (FR4) are indicated by the solid lines.

## Protocol 2

# Reamplification of the scFv V$_L$ CDR3 gene repertoire to append the 3′ restriction site

### Equipment and reagents

- PCR reagents and equipment as described in *Protocol 1*
- Wizard PCR purification kit (Promega)
- Custom DNA primers (Genset)
- V$_L$-*Not*I primer: 5′- GAG TCA TTC TCG ACT TGC GGC CGC ACC TAG CAC GGT CAG CTT GGT CCC TCC GCC GAA CAC CCA ACC-3′

- LMB3 primer (see *Protocol 1*)
- Reagents and equipment for restriction digestion of scFv gene repertoires (Chapter 13, *Protocol 7*)
- Amplified scFv gene repertoire DNA (from *Protocol 1*)

### Method

1  Set up four 50 μl PCR reaction mixes containing:

| | |
|---|---|
| Water | 35.5 μl |
| 20 × dNTPs (5 mM each) | 2.5 μl |
| 10 × Vent polymerase buffer | 5.0 μl |
| LMB3 primer (10 pmol/μl) | 2.5 μl |
| V$_L$-*Not*I primer (10 pmol/μl) | 2.5 μl |
| scFv gene template DNA (100 ng) | 1.0 μl |
| Vent DNA polymerase (2 units) | 1.0 μl |

2  Heat the reaction mixes to 94 °C for 5 min in a PCR thermo-cycler with a heated lid.

3  Cycle 30 times to amplify the V$_L$ gene at 94 °C for 1 min, 42 °C for 1 min, 72 °C for 2 min.

4  Purify the PCR product using the Wizard PCR purification kit.

5  Digest the PCR product with *Nco*I and *Not*I as described in Chapter 13, *Protocol 7*.

## Protocol 3

# Ligation of scFv and vector DNA and transformation of *E. coli* to generate a mutant scFv library

### Equipment and reagents

- Distilled (millipore) water
- T4 DNA ligase and 10X ligase buffer (NEB)
- 16 °C water bath

- Digested scFv repertoire DNA (from *Protocol 2*)
- pHEN1 DNA, digested with *Nco*I and *Not*I, as described in Chapter 13, *Protocol 7*, steps 4 and 5 (100 ng/μl)

**Protocol 3** continued

- Electrocompetent *E. coli* strain TG1 cells (Stratagene, or prepared using the procedure described in Chapter 13, *Protocol 9*)
- Reagents and equipment for electropora-

tion of scFv libraries into *E. coli* strain TG1 (Chapter 13, *Protocol 10*)

## Method

1  Make up a 50 µl ligation mixture.[a]

| | |
|---|---|
| 10× ligation buffer | 5 µl |
| Water | 16 µl |
| *Nco*I-*Not*I-digested pHEN 1 (100 g/µl) | 20 µl |
| *Nco*I-*Not*I-digested scFv gene repertoire (100 g/µl) | 7 µl |
| T4 DNA ligase (400 U/µl) | 2 µl |

2  Incubate the reaction overnight at 16 °C.

3  Ethanol precipitate the DNA and wash the pellet twice in 70% ethanol.

4  Resuspend the DNA in 10 µl of water.

5  Electroporate the DNA as four 2.5 µl aliquots into electrocompetent *E. coli* TG1 cells, using the procedure described in Chapter 13, *Protocol 10*.[b]

6  Determine the library size,[c] and store the bacterial library stock at −70 °C, as described in Chapter 13, *Protocol 10*.

[a] In ligation experiments, the molar ratio of insert to vector should be 2 : 1. Given that the ratio of sizes of scFv (800 base pairs (bp)) to vector (4500 bp) is approximately 6 : 1, this translates in this case into a ratio of insert to vector of 1 : 3 in weight terms.

[b] An alternative and optimized bacterial strain and set of protocols for high-efficiency electroporation of bacterial cells is described in Chapter 2 (*Protocols 3* and *5*).

[c] Using this procedure we routinely obtain libraries of $>10^7$ clones. Ideally, the library size should be at least 2 times larger than the potential diversity encoded by the randomized nucleotides to ensure complete coverage of the diversity. This will not be possible if many residues are randomized, however. If the library is smaller than $10^6$, the cloning should be repeated. It should be possible to routinely achieve a transformation efficiency for *E. coli* TG1 cells of greater than $5 \times 10^9$/µg of supercoiled plasmid DNA. Refer to Chapter 13, *Table 3* for hints on how to troubleshoot inefficient ligations or transformations.

## 2.1.2 Construction of randomized $V_H$ CDR3 phage antibody libraries

Randomization of the $V_H$ CDR3 is more complicated because this sequence is located in the middle of the scFv gene. The PCR to create the repertoire is done in two steps (*Figure 4*). Initially, the $V_H$ gene is amplified with the randomized primer ($V_H$ FOR) and a primer based in the vector backbone, upstream of the scFv gene (LMB3). In a separate PCR, the $V_L$ is amplified with a primer based in gene III of the vector (FdSEQ) and a primer ($V_H$ BACK), which is complementary to a portion of the $V_H$ FOR primer. The full length $V_H$ CDR3 scFv repertoire is then constructed by PCR splicing of the amplified $V_H$ and $V_L$ gene fragments.

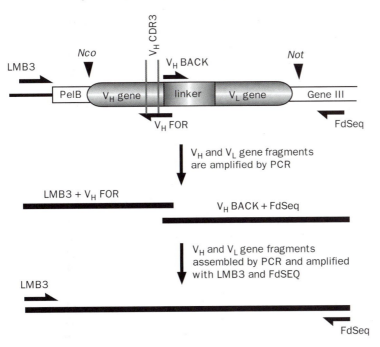

**Figure 4** Strategy for construction of a scFv gene repertoire with a diversified $V_H$ CDR3. The PCR for the $V_H$ CDR3 repertoire is done in two steps. Initially, the $V_H$ is amplified with the randomized primer ($V_H$FOR) and a primer based in the vector backbone (LMB3). In a separate PCR, the $V_L$ is amplified with a primer based in gene III of the vector (FdSEQ) plus a primer ($V_H$BACK) which is complementary to a portion of the $V_H$FOR primer. The full length $V_H$CDR3 scFv repertoire is then constructed by PCR splicing of the amplified $V_H$ and $V_L$ gene fragments and reamplified with the primers LMB3 and FdSEQ.

## Protocol 4

## Amplification of $V_H$ CDR3 gene repertoire and $V_L$ gene

### Equipment and reagents

- PCR reagents and equipment as described in *Protocol 1*
- Geneclean kit (Qbiogene)
- $V_H$FOR random primer (see below), 10 pmol/μl
- LMB3 primer (see *Protocol 1*)
- $V_H$BACK primer: 5'-TTT GAC TAC TGG GGC CAG GG -3', 10 pmol/μl
- FdSEQ primer: 5'-GAA TTT TCT GTA TGA GG -3', 10 pmol/μl
- Plasmid DNA containing starting scFv gene

297

**Protocol 4** continued

## Method

1 Design and synthesize the $V_H$ FOR primer containing CDR3 randomizing sequence. The primer must of course be designed based on the $V_H$ -gene sequence of the antibody being affinity matured. In the example given here, seven amino acids of the $V_H$ CDR3 are randomized (*Figure 5*, underlined region) conserving approximately 50% wild-type amino acid at each position randomized. The resulting primer design is as follows (the numerical codes for molar compositions are as described in *Protocol 1*):

$V_H$ FOR random primer 5'-CC CTG GCC CCA GTA GTC AAA <u>532 511 532 544 524 534 514</u> TTT CGC ACA GTA ATA AAC GGC -3'

In this example, seven consecutive CDR residues were randomized. Based on experimental results, we concluded that sequence space was more efficiently utilized when only solvent accessible residues were mutated (1,15). These residues can be identified by molecular modeling. The randomized residues in this example were selected based on the use of molecular modeling to predict which CDR3 residues had solvent accessible side chains.

2 Make up two 50 µl PCR reaction mixes, one for $V_H$ gene amplification and one for $V_L$ gene amplification, containing:

| | |
|---|---|
| Water | 35.5 µl |
| 20× dNTPs (5 mM each) | 2.5 µl |
| 10× Vent polymerase buffer | 5.0 µl |
| Back primer[a] (10 pmol/µl) | 2.5 µl |
| Forward primer[b] (10 pmol/µl) | 2.5 µl |
| scFv gene template DNA (100 ng) | 1.0 µl |
| Vent DNA polymerase (2 units) | 1.0 µl |

3 Heat the tube to 94 °C for 5 min in a PCR thermo-cycler with a heated lid. Cycle 30 times to amplify the $V_H$ and $V_L$ genes at 94 °C for 1 min, 42 °C for 1 min, 72 °C for 2 min.

4 Gel-purify the $V_H$ gene repertoire (350 base pair (bp)) and the $V_L$ gene (320 bp) on a 1.5% agarose gel and extract the DNA using the Geneclean kit. Resuspend each product in 30 µl of water. Determine the DNA concentration by analysis on a 1.5% agarose gel with markers of known size and concentration.

[a] The back primer for the $V_H$ PCR is LMB3; for the $V_L$ PCR the back primer is $V_H$ BACK.

[b] The forward primer for the $V_H$ PCR is the $V_H$ FOR random primer designed in step 1; for the $V_L$ PCR the forward primer is FdSEQ.

## Protocol 5

# PCR splicing of $V_H$ CDR3 gene repertoire and $V_L$ gene to create an scFv gene repertoire[a]

### Equipment and reagents

- Wizard PCR purification kit (Promega)
- PCR reagents and equipment as described in *Protocol 1*
- Reagents and equipment for ligation of scFv and pHEN1 DNA and electroporation into *E. coli*, as described in *Protocol 3*.
- FDSEQ primer (see *Protocol 4*)
- Amplified $V_H$CDR3 gene repertoire and $V_L$ gene from *Protocol 4*

### Method

1  Make up four 25 µL PCR reaction containing:

| | |
|---|---|
| Water | 10.25 µl |
| 10× Vent buffer | 2.5 µl |
| 20× dNTPs (5 mM each) | 1.25 µl |
| $V_H$ repertoire DNA (200 ng) | 5.0 µl |
| $V_L$ gene DNA (200 ng) | 5.0 µl |
| Vent DNA polymerase (2 units) | 1.0 µl |

Cycle 7 times without amplification in a PCR machine with a heated lid at 94 °C for 1.5 min, 65 °C for 1.5 min, and 72 °C for 1.5 min to join the fragments.

2  After 7 cycles, hold at 94 °C while adding the 25 µl mixture (below) containing the flanking primers:

| | |
|---|---|
| Water | 15.2 µl |
| 10 × Vent buffer | 2.5 µl |
| 20 × dNTP's (5 mM each) | 1.3 µl |
| FdSEQ primer (10 pmol/µl) | 2.5 µl |
| LMB3 primer (10 pmol/µl) | 2.5 µl |
| Vent DNA polymerase (2 units) | 1.0 µl |

3  Cycle 30 times to amplify the spliced fragments at 94 °C for 1 min, 42 °C for 1 min and 72 °C for 2 min. Purify the PCR product using the Wizard PCR purification kit.

4  Digest the PCR products with *Nco*I and *Not*I as described in Chapter 13, *Protocol 7*.

5  Ligate the digested PCR products into *Nco*I/*Not*I-digested pHEN1, and electroporate into *E. coli* TG1 cells, as described in *Protocol 3*.

[a] See Chapter 13, Section 2.2 for a discussion of PCR splicing procedures, and alternative sequential cloning approaches for assembling scFv libraries.

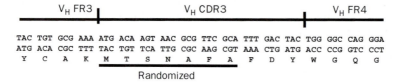

| V_H FR3 | V_H CDR3 | V_H FR4 |
|---|---|---|

```
TAC TGT GCG AAA ATG ACA AGT AAC GCG TTC GCA TTT GAC TAC TGG GGC CAG GGA
ATG ACA CGC TTT TAC TGT TCA TTG CGC AAG CGT AAA CTG ATG ACC CCG GTC CCT
 Y   C   A   K   M   T   S   N   A   F   A   F   D   Y   W   G   Q   G
```

Randomized

**Figure 5** Location and design of primer for diversifying V_H CDR3. The primers used for the randomization of the V_H CDR3 are based on the V-gene sequence of the antibody being affinity matured. In this example seven amino acids of the V_HCDR3 were randomized (underlined region). The end of framework 3 (FR3) and the beginning of framework 4 (FR4) are indicated by the solid lines.

After library construction, proceed to Section 4, selection and screening for higher affinity antibodies.

## 2.2 Construction of chain shuffled libraries for affinity maturation

For chain shuffling of scFv from phage libraries (*Figure 6*), V_H and V_L repertoires are first amplified from either preexisting naïve scFv libraries or from complementary DNA (cDNA) prepared from total RNA from volunteer human donors. Initially, we used splicing by overlap extension to splice either the wild-type V_H or V_L to a gene repertoire of the complementary chain (5, 19). However this can artifactually generate a shortened linker sequence (between V_H and V_L) leading to scFv dimers (diabodies) (20) which may be preferentially selected on the basis of avidity (9). Thus we now prefer to clone the wild-type V_H or V_L gene into a phage display vector which already contains a repertoire of the complementary chain (9). Such repertoires have been described (9) and are obtainable from the authors under the terms of a Materials Transfer Agreement (MTA), saving considerable time. Alternatively, the V_H and V_L gene repertoires can be obtained by amplifying from a preexisting naïve scFv library, or from human PBL RNA using the procedures described in Chapter 13, *Protocols 1–3*.

Below we provide protocols for chain shuffling based on the use of cloned V_H and V_L gene repertoires. The overall procedure is outlined in *Figure 6*. First, a V_H gene repertoire is generated in the vector pHEN1 by amplifying from a preexisting naïve scFv library (*Protocols 6 and 7*). Next, the wild-type V_L gene is cloned into this V_H gene repertoire-containing vector to create a heavy chain shuffled library (*Protocol 8*), that can be subjected to selection to identify tight binding clones (see Sections 3 and 4). Once such a clone is identified, a similar process can be used to identify improved light chains: a V_L gene repertoire is generated (*Protocol 9*) and then combined with the new optimal V_H gene to create a light chain shuffled library (*Protocol 10*) for further selections.

### 2.2.1 Construction of heavy chain shuffled libraries

To facilitate heavy chain shuffling, libraries are constructed in pHEN1 (21) containing human V_H gene segment repertoires (FR1 to FR3) and a cloning site at the end of V_H FR3 for inserting the V_H CDR3, V_H FR4, linker DNA and light chain from a binding scFv as a *BssHII-NotI* fragment (*Figure 7*).

1. V$_H$ gene segments are cloned by PCR into the
phagemid vector along with the V$_L$ of a binding scFv.

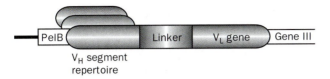

V$_H$ segment
repertoire

2. Improved scFv are selected from phage library and
the best clone identified by surface plasmon resonance.

3. The rearranged V$_H$ gene from the highest affinity
clone is cloned into a library of V$_L$ genes.

V$_L$ gene

4. Improved scFv are selected from the phage library and
the best clones identified by surface plasmon resonance.

**Figure 6** Overview of strategy
for affinity maturation by
sequential chain shuffling.

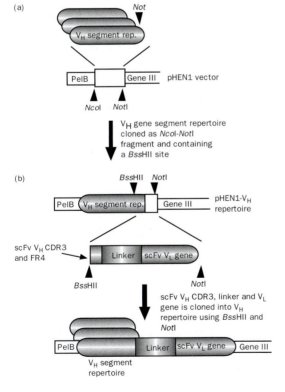

**Figure 7** Strategy for
construction of a scFv gene
repertoire with shuffled
V$_H$ chains. (a) For the heavy chain
shuffling, libraries are
constructed in pHEN1 containing
human *VH* gene segment
repertoires (FR1 to FR3). (b) The
*Bss*HII cloning site at the end of
V$_H$ FR3 allows cloning of the V$_H$
CDR3, V$_H$ FR4, linker DNA, and
light chain from the original scFv
as a *Bss*HII-*Not*I fragment to
create the V$_H$ shuffled repertoire.

301

To create the libraries, three $V_H$ gene segment repertoires enriched for human $V_H1$, $V_H3$, and $V_H5$ gene segments are amplified by PCR using as a template single stranded DNA prepared from a $1.0 \times 10^8$ member scFv phage antibody library in pHEN1 (22). This library may be obtained from our laboratory under an MTA. Alternatively, a library may be prepared as described in Chapter 13, *Protocols 1–10*. We use primers that generate repertoires derived primarily from the $V_H$ 1, 3, and 5 gene families since we have not observed any scFv phage antibodies derived from the $V_H$ 2, 4, or 6 families. Primers PVH1FOR1, PVH3FOR1, and PVH5FOR1 are designed to anneal to the consensus $V_H1$, $V_H3$, or, $V_H5$ 3′ FR3 sequence respectively (23).

## Protocol 6

## Construction of a human $V_H$ gene segment repertoire

### Equipment and reagents

- PCR reagents and equipment as described in *Protocol 1*
- scFv gene repertoire template DNA (single or double stranded DNA) prepared from a naïve scFv library in pHEN1 (see discussion above) (10 ng/µl)
- Geneclean kit (Qbiogene)

- PVH1FOR1 primer: 5′-TCG CGC GCA GTA ATA CAC GGC CGT GTC-3′
- PVH3FOR1 primer: 5′-TCG CGC GCA GTA ATA CAC AGC CGT GTC CTC-3′
- PVH5FOR1 primer: 5′-TCG CGC GCA GTA ATA CAT GGC GGT GTC CGA-3′
- LMB3 primer (see *Protocol 1*)

### Method

1  Make up three separate 50 µl PCR reaction mixes containing:

    | | |
    |---|---|
    | Water | 35.5 µl |
    | $20\times$ dNTPs (5 mM each) | 2.5 µl |
    | $10\times$ Vent polymerase buffer | 5.0 µl |
    | LMB3 primer (10 pmol/µl) | 2.5 µl |
    | FORWARD primer[a] (10 pmol/µl) | 2.5 µl |
    | scFv gene template DNA (10 ng) | 1.0 µl |
    | Vent DNA polymerase (2 units) | 1.0 µl |

2  Heat to 94 °C for 5 min in a PCR thermo-cycler with a heated lid.

3  Cycle 25 times to amplify the $V_H$ genes at 94 °C for 30s, 42 °C for 30 s, 72 °C for 1 min.

4  Gel-purify the $V_H$ gene repertoires on a 1.5% agarose gel and extract the DNA using the Geneclean kit. Resuspend each product in 20 µl of water. Determine the DNA concentration by analysis on a 1.5% agarose gel with markers of known size and concentration.

[a] Either PVH1FOR1, PVH3FOR1, or PVH5FOR1

The DNA fragments from this first PCR are then used as templates for a second PCR to introduce a *Bss*HII site at the 3'-end of FR3 followed by a *Not*I site (*Figure 7*). The *Bss*HII site corresponds to V$_H$ amino acid residues 93 and 94 (Kabat numbering (25)) and does not change the amino acid sequence (alanine--arginine).

---

## Protocol 7

## Reamplification of V$_H$ gene repertoires to append the 3' restriction site and cloning to create V$_H$ gene segment libraries

### Equipment and reagents

- PCR reagents and equipment as described in *Protocol 1*

- Wizard PCR purification kit (Promega)

- PVH1FOR2 primer: 5'-GAG TCA TTC TCG ACT TGC GGC CGC TCG CGC GCA GTA ATA CAC GGC CGT GTC-3'

- PVH3FOR2 primer: 5'-GAG TCA TTC TCG ACT TGC GGC CGC TCG CGC GCA GTA ATA CAC AGC CGT GTC CTC-3'

- PVH5FOR2 primer: 5'-GAG TCA TTC TCG ACT TGC GGC CGC TCG CGC GCA GTA ATA CAT GGC GGT GTC CGA-3'

- Gel-purified V$_H$ gene repertoires (from *Protocol 6*) (25 ng/µl)

- Reagents and equipment for restriction digestion of scFv gene repertoires (Chapter 13, *Protocol 7*)

- Reagents and equipment for ligation of scFv and pHEN1 DNA and electroporation into *E. coli*, as described in *Protocol 3*.

### Method

1 Make up three 50 µL PCR reaction mixes, one for each V$_H$ gene segment repertoire, containing:

| | |
|---|---|
| Water | 34.5 µl |
| 20 × dNTPs (5 mM each) | 2.5 µl |
| 10 × Vent polymerase buffer | 5.0 µl |
| LMB3 primer (10 pmol/µl) | 2.5 µl |
| FORWARD primer[a] (10 pmol/µl) | 2.5 µl |
| V$_H$ gene segment repertoires from *Protocol 6* (25 ng/µl) | 2.0 µl |
| Vent DNA polymerase (2 units) | 1.0 µl |

2 Heat the reactions to 94 °C for 5 min in a PCR thermo-cycler with a heated lid.

3 Cycle 25 times to re-amplify the V$_H$ gene repertoires at 94 °C for 30 s, 42 °C for 30 s, 72 °C for 1 min.

4 Purify the PCR products using the Wizard PCR purification kit.

**5** Digest the PCR products with *Nco*I and *Not*I as described in Chapter 13, *Protocol 7*.

**6** Ligate the digested PCR products into *Nco*I/*Not*I-digested pHEN1, and electroporate into *E.coli* TG1 cells, as described in *Protocol 3*, to create three V$_H$ gene segment phage libraries. Determine the library size, and store the bacterial library stock at $-70\,°C$, as described in Chapter 13, *Protocol 10*.

**7** Prepare DNA from each of the V$_H$ gene repertoire libraries by inoculating 100 ml of $2\times$ TY media containing 100 µg/ml of ampicillin and 1% glucose with bacteria from the library glycerol stock. The inoculum should be large enough so that the number of bacteria innoculated is at least 5 times larger than the library size. After overnight growth, a standard DNA plasmid preparation is performed. For subsequent digestion of the libraries, the DNA from the different repertoires can be combined.

$^a$ Either PVH1FOR2, PVH3FOR2, or PVH5FOR2

The light chain gene, linker DNA, and V$_H$ CDR3 and FR4 from the starting scFv are now amplified using PCR and cloned into the V$_H$ gene segment phage libraries generated from *Protocol 7*. For a starting scFv cloned into pHEN1, this is achieved using a specific BACK primer designed for the particular scFv, and primer FdSEQ1 which anneals within the phage gene III in pHEN1. The BACK primer used in the example here, scFvVL1BACK, anneals to the last 24 nucleotides of V$_H$ framework 3 and the first 18 nucleotides of the CDR3. The primer also contains a *Bss*HII restriction site incorporated into the six nucleotides encoding the last two amino acid residues of framework 3 of the V$_L$ gene. For any given scFv, it is likely that a unique scFvVL1BACK primer will need to be designed, incorporating the features described above. In addition, a different 3′ primer may be required, depending on the vector backbone in which the starting scFv is cloned.

## Protocol 8

## Generation of a heavy chain shuffled phage library

### Equipment and reagents

- PCR reagents and equipment as described in *Protocol 1*
- Plasmid or single-stranded DNA (ssDNA) containing starting scFv gene in the vector pHEN1
- Wizard PCR purification kit (Promega)
- FDSEQ primer (see *Protocol 4*)
- Custom BACK primer (see above). In this example protocol, this is the scFvVL1BACK primer: 5′-AGC GCC GTG TAT TTT TGC GCG CGA CAT GAC GTG GGA TAT TGC-3′

- Reagents and equipment for restriction digestion of scFv gene repertoires (Chapter 13, *Protocol 7*)
- *Bss*HII restriction enzyme (NEB)
- Reagents and equipment for ligation of scFv and pHEN1 DNA and electroporation into *E. coli*, as described in *Protocol 3*.
- Plasmid DNA from the three V$_H$ gene segment phage libraries (from *Protocol 7*)

## Method

1 Make up 50 µl PCR reaction mixes containing:

| | |
|---|---|
| Water | 34.5 µl |
| 20 × dNTPs (5 mM each) | 2.5 µl |
| 10 × Vent polymerase buffer | 5.0 µl |
| FdSEQ1 primer (10 pm/µl) | 2.5 µl |
| scFvVL1BACK (10 pm/µl) | 2.5 µl |
| scFv template DNA (50 ng/µl) | 2.0 µl |
| Vent DNA polymerase (2 units) | 1.0 µl |

2 Heat the reaction mix to 94 °C for 5 min in a PCR thermo-cycler with a heated lid.

3 Cycle 25 times to amplify the $V_H$ genes at 94 °C for 30 s, 42 °C for 30 s, 72 °C for 1 min.

4 Purify the PCR product using the Wizard PCR purification kit.

5 Digest the PCR product with *Bss*HII and *Not*I as described in Chapter 13, *Protocol 7*. The protocol will need to be modified for use of *Bss*HII instead of *Nco*I: first digest with *Not*I at 37 °C, then add the *Bss*HII enzyme and transfer to 50 °C.

6 Prepare the recipient $V_H$ gene segment repertoire vector (generated from *Protocol 8*) as described in Chapter 13, *Protocol 7* but digesting the vector with *Bss*HII/*Not*I, using the modifications described in step 5 above.

7 Ligate the digested the $V_L$ gene segment PCR product (from step 5) with the digested $V_H$ gene segment phage library vectors (from step 6) and transform *E. coli* TG1 as described in *Protocol 3*, to create heavy chain shuffled libraries.

After library construction, proceed to Section 3, selection and screening for higher affinity antibodies.

## 2.3 Construction of light chain shuffled libraries

To facilitate light chain shuffling, a library of light chain genes is constructed in the vector pHEN1-$V_\lambda$3. This vector is based on pHEN1 and contains the pelB leader, an *Nco*I cloning site, a polylinker, an *Xho*I cloning site encoding the two C-terminal amino acids of the VH framework 4, linker DNA, and a single Vλ light chain (24). The *Xho*I site can be encoded at the end of the VH FR4 without changing the amino acid sequence of residues 102 and 103 (serine-serine) (25). To create the libraries, $V_\kappa$ and $V_\lambda$ gene repertoires and linker DNA are amplified by PCR using as a template ssDNA prepared from the same $1.8 \times 10^8$ member scFv phage antibody library described in Section 2.2.1 (22). Four BACK primers are used that anneal to the first six nucleotides of the $(Gly_4Ser)_3$ linker and either the $J_H1$, 2, $J_H3$, $J_H$ 4,5, or $J_H$ 6 segments and contain the *Xho*I cloning site. The pHEN1-$V_\lambda$3 vector is digested with *Xho*I and *Not*I, removing the single Vλ3 gene, and the amplified repertoires are cloned into the digested vector.

The resulting library contains rearranged human $V_\kappa$ and $V_\lambda$ gene repertoires, linker DNA, and cloning sites for inserting a rearranged $V_H$ gene as an *NcoI-XhoI* fragment (*Figure 8*).

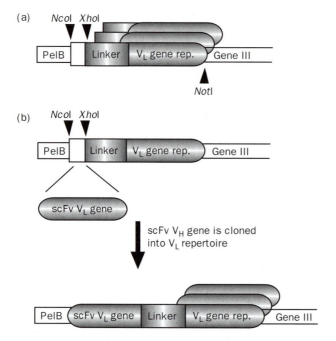

**Figure 8** Strategy for construction of a scFv gene repertoire with shuffled $V_L$ chains. (a) A light chain repertoire is constructed by PCR in the vector pHEN1. (b) The rearranged $V_H$ gene is then inserted as an *NcoI-XhoI* fragment to create the $V_L$ shuffled repertoire.

## Protocol 9

# Construction of human $V_L$ gene repertoire libraries

### Equipment and reagents

- PCR reagents and equipment as described in *Protocol 1*

- scFv gene repertoire single stranded template DNA prepared from a naïve scFv library in pHEN1 (see discussion in Section 2.2.1) (10 ng/µl)

- Geneclean kit (Qbiogene)

- Wizard PCR purification kit (Promega)

- RJH1/2Xho primer: 5′-GGC ACC CTG GTC ACC GTC TCG AGT GGT GGA-3′

- RJH3Xho primer: 5′-GGG ACA ATG GTC ACC GTC TCG AGT GGT GGA-3′

- RJH4/5Xho primer: 5′-GGA ACC CTG GTC ACC GTC TCG AGT GGT GGA-3′

- RJH6Xho primer: 5′-GGG ACC ACG GTC ACC GTC TCG AGT GGT GGA-3′

- FdSEQ primer (see *Protocol 4*)

- pHEN1-$V_\lambda$3 vector DNA (obtainable from the authors under an MTA)

306

## Method

1   Make up four separate 50 μl PCR reaction mixes containing:

| | |
|---|---|
| Water | 35.5 μl |
| 20 × dNTPs (5 mM each) | 2.5 μl |
| 10 × Vent polymerase buffer | 5.0 μl |
| FdSEQ primer (10 pmol/μl) | 2.5 μl |
| BACK primer[a] (10 pmol/μl) | 2.5 μl |
| Single stranded scFv template DNA (10 ng) | 1.0 μl |
| Vent DNA polymerase (2 units) | 1.0 μl |

2   Heat the reaction mix to 94 °C for 5 min in a PCR thermo-cycler with a heated lid.

3   Cycle 25 times to amplify the $V_L$ genes at 94 °C for 30 s, 42 °C for 30 s, 72 °C for 1 min.

4   Purify the PCR product using the Wizard PCR purification kit.

5   Digest the PCR products with *Xho*I and *Not*I as described in Chapter 13, *Protocol 7*, except substituting *Xho*I for *Nco*I and using NEB2 buffer with BSA as described by the manufacturer.

6   Digest the pHEN1-$V_\lambda$3 vector DNA with *Xho*I and *Not*I as described in Chapter 13, *Protocol 7*, Step 4, except substituting *Xho*I for *Nco*I.

7   Ligate the digested PCR products (from step 5) and the digested vector (from step 6), and electroporate the DNA into *E. coli* TG1 cells, as described in *Protocol 3*, to create four $V_L$ gene repertoire phage libraries. Determine the library size, and store the bacterial library stock at − 70 °C, as described in Chapter 13, *Protocol 10*.

8   Prepare DNA from each of the $V_L$ gene repertoire libraries by inoculating 100 ml of 2× TY media containing 100 μg/ml of ampicillin and 1% glucose with bacteria from the library glycerol stock. The inoculum should be large enough so that the number of bacteria inoculated is at least 5 times larger than the library size After overnight growth, a standard DNA plasmid preparation is performed. For subsequent digestion of the libraries, the DNA from the different repertoires can be combined.

[a] Either RJH1/2Xho, RJH3Xho, RJH4/5Xho, or RJH6Xho

To create a light chain shuffled library, the rearranged $V_H$ gene from the starting scFv is amplified using PCR, using a BACK primer that anneals upstream of the $V_H$ gene, and a FORWARD primer that anneals to the $J_H$ gene of the target scFv. In *Protocol 10*, this is detailed for a scFv gene in the vector pHEN1 using the primers LMB3 and one of the scFvJHXhoFOR primers, respectively. LMB3 anneals upstream of the pelB leader in pHEN1, and the scFvJHXhoFOR primers anneal to the $J_H$ gene of the target scFv. One of the four $J_H$FOR primers listed in the reagents section below should be a perfect match for the specific scFv $V_H$ gene and this is the FOR primer that you should use. If the starting scFv is cloned into a different vector than pHEN1, then a different 5′ primer may be required.

## Protocol 10

## Generation of a light chain shuffled phage library

### Equipment and reagents

- PCR reagents and equipment as described in *Protocol 1*
- Single-stranded template DNA of vector pHEN1 containing the starting scFv gene (50 ng/µl)
- Geneclean kit (Qbiogene)
- Wizard PCR purification kit (Promega)
- scFvJH1–2XhoFOR primer: 5′-GAG TCA TTC TCG TCT CGA GAC GGT GAC CAG GGT GCC-3′
- scFvJH3XhoFOR primer: 5′-GAG TCA TTC TCG TCT CGA GAC GGT GAC CAT TGT CCC-3′
- scFvJH4–5XhoFOR primer: 5′-GAG TCA TTC TCG TCT CGA GAC GGT GAC CAG GGT TCC-3′
- scFvJH6XhoFOR primer: 5′-GAG TCA TTC TCG TCT CGA GAC GGT GAC CGT GGT CCC-3′
- *Xho*I restriction enzyme (NEB) and manufacturer's buffer number 2 (NEB2)
- $V_L$ gene repertoire vectors (from *Protocol 9*)

### Method

1  Make up 50 µl PCR reaction mixes containing:

| | |
|---|---|
| Water | 34.5 µl |
| 20× dNTPs (5 mM each) | 2.5 µl |
| 10 × Vent polymerase buffer | 5.0 µl |
| LMB3 primer (10 pmol/µl) | 2.5 µl |
| scFvJHXhoFOR (10 pmol/µl) | 2.5 µl |
| scFv template DNA (100 ng) | 2.0 µl |
| Vent DNA polymerase (2 units) | 1.0 µl |

2  Heat the reaction mix to 94 °C for 5 min in a PCR thermo-cycler with a heated lid.

3  Cycle 25 times to amplify the $V_H$ gene at 94 °C for 30 s, 42 °C for 30 s, 72 °C for 1 min.

4  Purify the PCR product using the Wizard PCR purification kit.

5  Digest the PCR product with *Xho*I and *Nco*I as described in Chapter 13, *Protocol 7*, except substituting *Xho*I for *Nco*I, and performing the digest in NEB2 buffer.

6  Prepare the $V_L$ gene repertoire vector (generated from *Protocol 9*) as described in Chapter 13, *Protocol 7*, but digesting the vector with *Xho*I and *Nco*I.

7  Ligate the digested PCR products (from step 5) and the digested vector (from step 6), and electroporate the DNA into *E. coli* TG1 cells, as described in *Protocol 3*, to create light chain shuffled libraries.

# 3. Selection and screening for higher affinity antibodies

Selections must be carefully designed in order to select antibodies that are of higher affinity rather than those clones that display better on phage, are less toxic to *E. coli*, are more stable, or multimerize leading to a higher functional affinity due to avidity. Two approaches have been used to select rare higher affinity scFv from a background of lower affinity scFv or nonbinding scFv: selections based on binding kinetics and selections based on the equilibrium constant ($K_d$). In either case it is important to use labeled antigen in solution rather than antigen adsorbed to a solid matrix. Use of soluble antigen biases towards selections based on binding affinity or binding kinetics, rather than avidity. This is especially important when selecting scFv libraries, since it is known that some scFv can spontaneously dimerize in a sequence dependent manner (9). Failure to use soluble antigen is likely to result in the selection of dimeric scFv whose monovalent binding constant is no higher than wild-type. Even with the Fab format, soluble antigen should also be used to avoid selecting for phage displaying multiple copies of Fab which will have a higher functional affinity (avidity).

In selections based on binding kinetics, also termed off-rate ($k_{off}$) selections, the phage population is allowed to saturate the labeled antigen before a large molar excess of unlabeled antigen is added to the mix for a given amount of time. The duration of the competition with unlabeled antigen is chosen to allow the majority of the bound wild-type antibodies to dissociate while the improved mutants remain bound (6). This approach effectively selects for slower off-rates. Since a reduction in $k_{off}$ is typically the major kinetic mechanism resulting in higher affinity when V-genes are mutated, both *in vivo* and *in vitro* (5), this approach should generally result in the selection of scFv with improved $K_d$'s.

We generally prefer equilibrium selections, in which phage are incubated with an antigen concentration below the equilibrium binding constant. This approach effectively selects for antibodies with improved equilibrium constants. Reduction of the antigen concentration also helps ensure that selection for higher affinity scFv occurs, rather than selection for scFv that express well on phage or are less toxic to *E. coli*. The ability to use soluble antigen for selections affords the control over antigen concentration required for equilibrium selections. Phage are allowed to bind to biotinylated antigen and then recovered with streptavidin magnetic beads (1, 26). The optimal antigen concentration used for the selection can be estimated *a priori* (27), but we prefer to use several different concentrations of antigen for the first round of selection. The choice of antigen concentration for the following rounds of selection is guided by determining the fraction of binding phage present in the output polyclonal phage. This is most accurately done by surface plasmon resonance (SPR) in a Biacore where the binding concentration is determined under mass transport limited conditions (9). Alternatively, the percentage of binding clones can be determined by enzyme-linked immunosorbant assay (ELISA). From these data the optimal antigen concentration can be determined prior to the next round of selection. We typically lower the antigen concentration 10–100 fold after each round of selection. Three to five rounds of selection are performed and then clones are screened for improved affinity.

## Protocol 11

## Equilibrium selection of higher affinity phage antibodies

### Equipment and reagents

- Streptavidin magnetic beads (Dynabeads M280, Dynal)
- Magnetic 1.5 ml tube holder (Dynal)
- Phosphate buffered saline (PBS)
- 2% skimmed milk powder in PBS (2% MPBS)
- Biotinylation kit (e.g. Pierce catalog number 21336)
- PBS containing 0.1% Tween-20 (PBS/Tween)
- 100 mM HCl

- 1 M Tris, pH 7.4
- Exponentially growing *E. coli* TG1 cells
- Equipment and reagents for preparing phage antibodies from library glycerol stocks (see Chapter 13, *Protocol 11*)
- Library of mutated scFv derivatives generated by site-directed mutagenesis (see *Protocols 1–5*) or chain shuffling (see *Protocols 6–10*)

### Method

1  Prepare phage antibodies from the diversified phage antibody library as described in Chapter 13, *Protocol 11*.

2  Biotinylate the antigen according to the manufacturer's instructions.

3  Prior to selection, block 150 μl streptavidin-magnetic beads with 1 ml 2% MPBS for 1 h at room temperature in a 1.5 ml microcentrifuge tube. Pull the beads to one side with the magnetic tube holder. Discard the buffer.

4  Block three 1.5 ml microcentrifuge tubes with 2% MPBS for 1 h at room temperature and then discard the blocking buffer. Incubate the biotinylated antigen at three different concentrations (three tubes total) with 1 ml of phage preparation (approximately $10^{12}$ cfu/ml) in 2% MPBS (final concentration) by rocking at room temperature for 1 h. Use a separate tube for each antigen concentration. For the first round of selection, antigen concentrations should be: $K_d/1$, $K_d/10$, and $K_d/100$, where $K_d$ = the affinity of the wild-type antibody for the antigen.

5  Add 50 μl streptavidin-magnetic beads to each tube and incubate the tube on a rotator at room temperature for 15 min. Place the tube in the magnetic rack for 30 s. The beads will migrate towards the magnet.

6  Aspirate the supernatant, leaving the beads on the side of the microcentrifuge tube. This is best done with a 200 μl pipette tip on a Pasteur pipette attached to a vacuum source. Wash the beads (1 ml per wash) with PBS/Tween 7 times, followed by MPBS 2 times, then once with PBS. Transfer the beads after every second wash to a fresh 1.5 ml tube to facilitate efficient washing.

7  Elute the phage from the magnetic beads with 100 μl 100 mM HCl for 10 min at RT. Place the tube in magnetic rack for 30 s to pellet the beads. Remove the supernatant containing the eluted phage and neutralize the solution with 1 ml of 1 M Tris, pH 7.4. Save the eluate on ice.[a]

8  Add 0.75 ml of the eluted phage to 10 ml of exponentially growing *E. coli* TG1 ($A_{600}$ ~ 0.5). Store the remaining eluted phage at 4 °C.

9  Incubate the eluate–*E. coli* mixture at 37 °C for 30 min without shaking.

10  Titer the eluate by plating 1 μl and 10 μl of the infected TG1 cells onto 100 mm TYE/amp/glu plates (this is a $10^4$ and $10^3$ dilution, respectively).[b]

11  Centrifuge the remaining bacterial culture at 2700g for 10 min. Resuspend the pellet in 250 μl 2× TY, plate onto two 150 mm TYE/amp/glu plates and incubate overnight at 37 °C.

12  In the morning, add 3 ml of 2 × TY/amp/glu to each plate, then scrape the bacteria from the plate with a bent glass rod. Make glyercol stocks by mixing 1.4 ml of the bacteria with 0.6 ml of 50% glycerol (sterile filtered). Store the stock at − 70 °C. Analyze the results of the output titration (step 9) to decide which of the selections (from the antigen concentrations $= K_d/1$, $K_d/10$, and $K_d/100$) to rescue for the next round of selection. We have in the past based the decision on the percent positive binding clones as determined by phage ELISA or on the active phage concentration as determined by BIAcore (26). However, as a rule of thumb, the decision can be based on the output titers: When the output titer drops significantly more than the antigen concentration was lowered (when comparing the outputs from the concentrations $K_d/1$, $K_d/10$, and $K_d/100$) it is generally a result of loss of binding, and phage from this selection concentration should not be used for the subsequent round. Instead use one of the higher antigen concentrations. On average, we lower the antigen concentration about 10-fold per round and we often use femtomolar concentrations for the last round of the affinity maturation.

13  Prepare phage antibodies from the appropriate glycerol stock from step 11, repeating the Protocol steps from step 1.

[a] It is important not to leave the phage in HCl longer than 10 min or infectivity will be significantly reduced. The use of less stringent eluants may not remove the highest affinity phage from the antigen (26). Alternatively, NHS-SS-biotin (Pierce) can be used for biotinylation and the phage eluted with 100 mM dithiotheritol (DTT) (for details, see Chapter 13, section 4).

[b] For large, randomized libraries (complexity $>10^7$), the output number of phage from the first round of selection should be between $10^4$ and $10^6$ cfu. If the titers for all antigen concentrations are larger than $10^7$ it is likely that the antigen concentration was too high or that the washing steps were inadequate. If the titer is below $5 \times 10^4$, either too many washes were performed, or too little antigen was used.

## 4 Screening for scFv with improved off-rates

After three to five rounds of selection, random colonies should be picked into 96-well microtiter plates and soluble scFv expression induced as described in Chapter 13, *Protocol 17*. The bacterial supernatants containing soluble scFv should then be analyzed by ELISA in 96-well microtiter plates coated with the relevant antigen as described in Chapter 13, *Protocol 18*. This process will identify those scFv that bind antigen. However, even after using the stringent selections described above, only a fraction of the ELISA positive clones will have a higher affinity than wild-type. The strength of the ELISA signal is a poor indicator of which clones have a lower $K_d$, and more typically correlates with expression level of the scFv. Thus a technique is required to screen ELISA positive clones to identify those with a lower $K_d$. One such technique is a competition or inhibition ELISA (28). Alternatively, one can take advantage of the fact that a reduction in $k_{off}$ is typically the major kinetic mechanism resulting in higher affinity, and thus antibodies with improved affinity can be identified by measuring the off-rate. Since $k_{off}$ is concentration independent (unlike $k_{on}$), it is possible to measure $k_{off}$, for example, using SPR in instruments such as the BIAcore, without purifying the antibody fragment. We have found this a very useful technique for identifying higher affinity scFv and use it to rank affinity-matured clones.

First, we identify binding clones by ELISA. We then take approximately 20–50 randomly selected ELISA-positive clones and rank them by $k_{off}$ as described in *Protocol 12*. Those clones with the slowest $k_{off}$ are then subcloned, purified, and the $K_d$ is measured using SPR in a Biacore ((1) and see below). Note that in the case of scFv, the shape of the dissociation curve can indicate whether single or multiple $k_{off}$s are present. ScFv which have multiple $k_{off}$s (a curve versus a straight line when plotting ln R1/R0 versus T) are a mixture of monomer and dimer and are best avoided for subsequent characterization.

## Protocol 12

## Screening for scFv with slower off-rates by Biacore

### Equipment and reagents

- Biacore SPR instrument (Biacore Inc.)
- CM5 sensor chip (Biacore Inc.)
- Biacore EDC/NHS immobilization kit (Biacore Inc.)
- 10 mM sodium acetate, pH 4.5
- *N*-[2-hydroxyethyl] piperazine-*N'*-[2-ethanesulfonic acid] Hepes buffered saline (10 mM Hepes, 150 mM NaCl, 3.4 mM EDTA) (HBS)
- 2 × TY/amp/glu (see Chapter 13, *Protocol 10*)

- 2× TY media containing 100 µg ml ampicillin and 0.1% glucose (2× TY/amp/0.1% glu)
- 1 M Isopropyl-β-D-Thiogalactopyranoside (IPTG)
- Periplasmic lysis buffer (PPB) (20% sucrose, 1 mM EDTA, 20 mM Tris-HCl, pH = 8)
- 5 mM MgSO$_4$
- CentriSep columns (Princeton Separations)

## Method

1   For each clone, grow a 1 ml overnight culture in $2 \times$ TY/amp/glu at $30\,°$C, shaking at 250 rpm.

2   Add 50 µl of the overnight culture to 5 ml $2 \times$ TY/amp/0.1% glu in a 50 ml tube and grow at $37\,°$C shaking at 250 rpm for approximately 2 h, to absorbance at 600 nm ($A_{600}$) of about 0.9.

3   Induce scFv expression by adding 2.5 µL 1 M IPTG to each tube (final concentration of 500 µM) and grow at $30\,°$C at 250 rpm for 4 h. Collect the cells by centrifuging in a 15 ml Falcon tube at 4000g for 15 min.

4   To prepare a periplasmic extract, resuspend the bacterial pellet in 200 µl PPB buffer, transfer the resuspended pellet to 1.5 ml tubes and place them on ice for 20 min. This shrinks the cells, extracts some scFv protein, and makes the periplasm hyperosmotic, facilitating the swelling that occurs with addition of hyposomotic $MgCl_2$. Pellet the cells in a microcentrifuge at 6000 rpm for 5 min. Discard the supernatant. Use the bacterial pellet for the osmotic shock preparation.

5   To prepare an osmotic shock fraction, resuspend the pellet in 200 µl 5 mM $MgSO_4$ and incubate the resuspended pellet on ice for 20 min. Pellet the cells in a microcentrifuge at full speed (14,000 rpm) for 5 min. Save the supernatant, which is the osmotic shock fraction, in a new tube.

6   Change the buffer containing the scFv from the osmotic shock buffer to the Biacore running buffer (HBS) using the CentriSep columns, as described by the manufacturer.[a]

7   Set up the Biacore instrument according to the manufacturer's instructions. In the flow cell of the Biacore instrument, immobilize the target antigen to a CM5 sensor chip using EDC/NHS chemistry as described by the manufacturer.[b]

8   Inject the scFv preparation from step 5 over the flow cell for 1 min, followed by 2–10 min observation of the dissociation phase at a constant flow rate of 15 µl/min, with HBS as the running buffer.

9   Regenerate the chip surface and analyze the next scFv.[c] A program can be written to automate this process allowing unattended analysis of 40 to 50 scFv overnight. An apparent $k_{off}$ can be determined from the dissociation part of the sensorgram for each scFv analyzed, using the software provided by the Biacore manufacturer.

[a] The buffer needs to be changed from the osmotic shock fraction to the Biacore running buffer in order to avoid excessive refractive index change during the SPR analysis, which would obscure measurement of $k_{off}$.

[b] The appropriate antigen concentration and buffer for immobilization varies considerably among antigens. We typically start with 10 µg/ml antigen in 10 mM NaAc, pH 4.5 and adjust antigen concentration, buffer ionic strength and pH as appropriate for the antigen in question. An amount of antigen should be immobilized which results in approximately 100–300 resonance units (RU) of scFv binding.

**Protocol 12** continued

[c]The appropriate reagent for regeneration of the sensor chip between samples also has to be determined for each antigen. However, we find that 4 M MgCl$_2$ will regenerate most antigen surfaces without significant change in the sensorgram baseline after analysis of more than 100 samples.

After the above analysis, it should be possible to rank clones by $k_{off}$. A number of scFv with the slowest off rates should then be purified and the $K_d$ measured. For purification of scFv in pHEN1, we subclone the genes into a vector such as that described in reference 29, which appends a C-terminal hexahistidine tag, allowing purification of the scFv from the periplasm using immobilized metal affinity chromatography (IMAC) as described in Chapter 13, Section 6. This is not necessary for clones in phage display vectors already containing hexahistidine tags. Aggregated and dimeric scFv are then removed by gel filtration and the $K_d$ calculated from $k_{on}$ and $k_{off}$ values measured by SPR using a BIAcore instrument (1).

## References

1. Schier, R., McCall, A., Adams, G. P., Marshall, K. W., Merritt, H., Yim, M. *et al.* (1996). *J. Mol. Biol.*, **263**, 551.
2. Yang, W. P., Green, K., Pinz-Sweeney, S., Briones, A. T., Burton, D. R., and Barbas, C. F., III. (1995). *J. Mol. Biol.*, **254**, 392.
3. Adams, G. P., Schier, R., Marshall, K., Wolf, E. J., McCall, A. M., Marks, J. D. *et al.* (1998). *Cancer Res.*, **58**, 485.
4. Adams, G. P. and Schier, R. (1999). *J. Immunol. Meth.*, **231**, 249.
5. Marks, J. D., Griffiths, A. D., Malmqvist, M., Clackson, T. P., Bye, J. M., and Winter, G. (1992). *Biotechnology*, **10**, 779.
6. Hawkins, R. E., Russell, S. J., and Winter, G. (1992). *J. Mol. Biol.*, **226**, 889.
7. Crameri, A., Cwirla, S., and Stemmer, W. P. (1996). *Nat. Med.*, **2**, 100.
8. Low, N., Holliger, P., and Winter, G. (1996). *J. Mol. Biol.*, **260**, 359.
9. Schier, R., Bye, J., Apell, G., McCall, A., Adams, G. P., Malmqvist, M. *et al.* (1996). *J. Mol. Biol.*, **255**, 28.
10. Stemmer, W. P. (1994). *Nature*, **370**, 389.
11. Loh, D. Y., Bothwell, A. L. M., White-Scharf, M. E., Imanishi-Kari, T., and Baltimore, D. (1983). **33**, 85.
12. Tomlinson, I. M., Walter, G., Jones, P. T., Dear, P. H., Sonnhammer, E. L., and Winter, G. (1996). *J. Mol. Biol.*, **256**, 813.
13. Novotny, J., Bruccoleri, R. E., and Saul, F. A. (1989). *Biochemistry*, **28**, 4735.
14. Foote, J. and Winter, G. (1992). *J. Mol. Biol.*, **224**, 487.
15. Schier, R., Balint, R. F., McCall, A., Apell, G., Larrick, J. W., and Marks, J. D. (1996). *Gene*, **169**, 147.
16. Neuberger, M. S. and Milstein, C. (1995). *Curr. Opin. Immunol.*, **7**, 248.
17. Chowdhury, P. and Pastan, I. (1999). *Nat. Biotech.*, **17**, 568.
18. Poul, M. A. and Marks, J. D. (1999). *J. Mol. Biol.*, **288**, 203.
19. Clackson, T., Hoogenboom, H. R., Griffiths, A. D., and Winter, G. (1991). *Nature*, **352**, 624.
20. Holliger, P., Prospero, T., and Winter, G. (1993). *Proc. Natl. Acad. Sci. USA.*, **90**, 6444.

21. Hoogenboom, H. R., Griffiths, A. D., Johnson, K. S., Chiswell, D. J., Hudson, P., and Winter, G. (1991). *Nucl. Acids Res.*, **19**, 4133.

22. Marks, J. D., Hoogenboom, H. R., Bonnert, T. P., McCafferty, J., Griffiths, A. D., and Winter, G. (1991). *J. Mol. Biol.*, **222**, 581.

23. Tomlinson, I. M., Walter, G., Marks, J. D., Llewelyn, M. B., and Winter, G. (1992). *J. Mol. Biol.*, **227**, 776.

24. Hoogenboom, H. R. and Winter, G. (1992). *J. Mol. Biol.*, **227**, 381.

25. Kabat, E. A., Wu, T. T., Perry, H. M., Gottesman, K. S., and Foeller, C. (1991). *Sequences of proteins of immunological interest*, US Department of Health and Human Services, US Government Printing Office.

26. Schier, R. and Marks, J. D. (1996). *Hum. Antibodies Hybridomas*, **7**, 97.

27. Boder, E. and Wittrup, K. (1998). *Biotech. Pogress*, **14**, 55.

28. Friguet, B., Chaffotte, A. F., Djavadi-Ohaniance, L., and Goldberg, M. E. (1985). *J. Immunol. Meth.*, **77**, 305.

29. Schier, R., Marks, J. D., Wolf, E. J., Apell, G., Wong, C., McCartney, J. E. *et al.* (1995). *Immunotechnology*, **1**, 73.

# Non-standard suppliers

**Amersham Biosciences**
800 Centennial Avenue, Piscataway, NJ 08855, USA
Tel: 800 526 3593 or (+1) 732 457 8000
Fax: 800 526 3593 or (+1) 732 457 0557
Web site: www.amershambiosciences.com

**Biacore AB**
*Biacore, Inc.*, Suite 100, 200 Centennial Avenue, Piscataway, NJ 08854, USA
Tel: 800 242 2599 or (+1) 732 885 5618
Fax: (+1) 732 885 5669
*Biacore AB United Kingdom*, 2 Medway Court, Meadway Technology Park,
Stevenage, Herts, SG1 2EF, UK
Tel: (+44) 1438 846200
Fax: (+44) 1438 846201
Web site: www.biacore.com

**Boehringer Mannheim Biochemicals**
See *Roche Molecular Biochemicals*

**BTX**
11199 Sorrento Valley Road, San Diego, CA 92121, USA
Tel: (+1) 858 597 6006
Fax: (+1) 858 597 9594
Web site: www.btxonline.com

**Corning Inc**.
45 Nagog Park, Acton, MA 01720, USA
Tel: (+1) 978 635 2200
Fax: (+1) 978 635 2476
Web site: www.scienceproducts.corning.com

**Costar Corp**.
See *Corning Inc.*

**DAKO Corporation**
6392 Via Real, Carpinteria, CA 93013, USA
Tel: 800 235 5763 or (+1) 805 566 6655
Fax: 800 566 6688 or (+1) 805 566 6688
Web site: www.dako.com

**Difco**
Available from *Becton Dickinson and Co.*

**Dynal Biotech**
*Dynal, Inc.*, 5 Delaware Drive, Lake Success, NY 11042, USA
Tel:  (+1) 516 326 3270
Fax: (+1) 516 326 3298
*Dynal Biotech ASA*, P.O.Box 114, Smestad, N-0309 Oslo, Norway
Tel: (+47) 22 06 10 00
Fax: (+47) 22 50 70 15
Web site: www.dynal.no

**Dynex Technologies, Inc**.
14340 Sullyfield Circle, Chantilly, VA 20151—1683, USA
Tel: (+1) 703 631 7800
Fax (+1) 703 631 7816
Web site: www.dynextechnologies.com

**EM Science**
Division of *EM Industries, Inc.*, 480 South Democrat Road, Gibbstown,
NJ 08027, USA
Tel: 800 222 0342 or (+1) 856 423 6300
Fax: (+1) 856 423 4389
Web site: www.emscience.com

**Equibio Ltd**.
The Wheelwrights, Boughton Monchelsea, Kent, ME17 4LT, UK
Tel: (+44) 1622 746 300
Fax: (+44) 1622 747 060
Web site: www.equibio.com

**Falcon**
Available from  *Becton Dickinson and Co.*

**Gelman Sciences**
*Pall Gelman Laboratory*, 600 South Wagner Road, Ann Arbor, MI 48103, USA
Tel: 800 521 1520 or (+1) 734 665 0651
Fax:  (+1) 734 913 6114
Web site: www.pall.com/gelman

**Gene Codes Corporation**
PO Box 3369, Ann Arbor, MI 48106, USA
Tel: (+1) 313 769 9249
Web site: www.genecodes.com

**Genetix Ltd**.
*Genetix Limited*, Queensway, New Milton, Hampshire BH25 5NN, UK
Tel: (+44) 1425 624600
Fax: (+44) 1425 624700
*Genetix USA Inc.*, 56 Roland Street, Suite 106, Boston MA 02129, USA

Tel: 877 436 3849
Fax: 888 522 7499
Web site: www.genetix.co.uk

**Marvel, Premier Brands UK Ltd**.
PO Box 8, Moreton, Wirral, Merseyside CH46 8XF, UK
Tel: ( + 44) 800 783 2196

**MBI Fermentas Inc**.
Suite 29, 4240 Ridge Lea Rd., Amherst, NY 14226, USA
Tel: 800 340 9026
Fax: 800 472 8322
Web site: www.fermentas.com

**Miles Scientific (A Division of Bayer Corporation Diagnostics)**
511 Benedict Avenue, Tarrytown, NY 10591, USA
Tel: 800 431 1970 or (+1) 914 631 8000
Fax: 914 524 2132
Web site: www.bayerdiag.com

**Molecular Dynamics**
See *Amersham Biosciences*

**Nalge Nunc International**
*Nalge Nunc International*, 75 Panorama Creek Drive, Rochester, NY 14625, USA
Tel: (+1) 716 586 8800
*Nalge (Europe) Limited*,
Foxwood Court, Rotherwas Industrial Estate,
Hereford HR2 6JQ, UK
Tel: (+44) 1432 263933
Fax: (+44) 44 1432 351923
Web site: www.nalgenunc.com

**New Brunswick Scientific Co., Inc**
PO Box 4005, 44 Talmadge Road, Edison, NJ, 08818-4005, USA
Tel: (+1) 732 287 1200
Fax: (+1) 732 287 4222
Web site: www.nbsc.com

**Novagen, Inc**.
601 Science Drive, Madison, WI 53711, USA
Tel: (+1) 608 238 6110
Fax: (+1) 608 238 1388
Web site: www.novagen.com

**Out Patient Services**
1320 Scott Street, Petaluma, CA 94954, USA
Tel: 800 648 1666

**Qbiogene, Inc.**
2251 Rutherford Road, Carlsbad, CA 92008, USA
Tel: 800 424 6101 or (+1) 760 929 1700
Fax: (+1) 760 918 9313
Web site: www.qbiogene.com

**Pall Corporation**
2200 Northern Boulevard, East Hills, NY 11548, USA
Tel: 800 717 7255
Web site: www.pall.com

**Pierce Chemical Co**.
3747 North Meridian Road, Rockford, IL 61105, USA
Tel: 800 874 3723 or (+1) 815 968 0747
Fax: 800 842 5007 or (+1) 815 968 8148
Web site: www.piercenet.com

**Princeton Separations**
P.O. Box 300 Adelphia, NJ 07710, USA
Tel: 800 223 0902 or  or (+1) 732 431 3338
Fax: (+1) 732 431 3768
Web site: www.prinsep.com

**Roche Molecular Biochemicals (Roche Applied Science;
Roche Diagnostics GmbH)**
Sandhofer Strasse 116, D-68305 Mannheim, Germany
Tel: (+49) 621 759 8540
Fax: (+49) 621 759 4083
Web site: www.biochem.roche.com

**Sakura Finetek U.S.A., Inc**.
1750 West 214th Street, Torrance, CA 90501, USA
Tel: 800 725 8723 or (+1) 310 972 7800
Fax: (+1) 310 972 7888
Web site: www.sakuraus.com

**Santa Cruz Biotech**
2161 Delaware Avenue, Santa Cruz, CA 95060, USA
Tel: 800 457 3801 or (+1) 831 457 3800
Fax: (+1) 831 457 3801
Web site: www.scbt.com

**Sartorius AG**
*Sartorius AG*, Weender Landstrasse 94–108, D-37075 Goettingen, Germany
Tel: (+49) 551 308 0
Fax: (+49) 551 308 3289
*Sartorius North America Inc.*, 131 Heartland Blvd., Edgewood, NY 11717, USA
Tel: 800 635 2906 or (+1) 631 254 4249

Fax: (+1) 631 254 4253
Web site: www.sartorius.com

**Skatron Instruments Inc**.
*Molecular Devices Corp.*
1311 Orleans Drive, Sunnyvale
CA 94089, USA
Tel: 800 635 5577 or (+1) 408 747 1700
Fax: 408 747 3601
Web site: www. moleculardevices.com

**Sorvall Products (Kendro Laboratory Products)**
31 Pecks Lane, Newtown, CT 06470, USA
Tel: (+1) 203 270 2080
Fax: (+1) 203 270 2166
Web site: www.kendro.com

**Surgipath Medical Instruments, Inc**.
5205 Route 12, Richmond, IL 60071, USA
Tel: 800 225 8867 or (+1) 815 678 2000
Fax: (+1) 815 678 2216
Web site: www.surgipath.com

**Synergy Software**
2457 Perkiomen Ave., Reading, PA 19606, USA
Tel: (+1) 610 779 0522
Fax: (+1) 610 370 0548
Web site: www.synergy.com

**Thermo Labsystems Oy**
Ratastie 2, PO Box 100, Vantaa, 01620, Finland
Tel: (+358) 9 329100
Fax: (+358) 9 32910500
Web site: www.thermo.com

**Trevigen, Inc.**
8405 Helgerman Court, Gaithersburg, MD 20877, USA
Tel: 800 873 8443 or (+1) 301 216 2800
Fax: (+1) 301 216 2801
Web site: www.trevigen.com

**Vector Laboratories, Inc**.
30 Ingold Road, Burlingame, CA 94010, USA
Tel: 800 227 6666 or (+1) 650 697 3600
Fax: (+1) 650 697 0339
Web site: www.vectorlabs.com

**Whatman plc**
*Whatman International Ltd*., Whatman House, St. Leonard's Road, 20/20 Maidstone,
Kent ME16 0LS, UK

Tel: (+44) 1622 676670
Fax: (+44) 1622 677011
*Whatman Inc.*, 9 Bridewell Place, Clifton, NJ 07014, USA
Tel: 800 441 6555 or (+1) 973 773 5800
Fax: (+1) 973 472 6949
Web site: www.whatman.com

**Zymed Laboratories, Inc**.
458 Carlton Court, South San Francisco, CA 94080, USA
Tel: 800 874 4494 or (+1) 650 871 4494
Fax: (+1) 650 871 4499
Web site: www.zymed.com

# Index